PARTIAL DIFFERENTIAL EQUATIONS

An Introduction

David Colton

University of Delaware

DOVER PUBLICATIONS, INC.

Mineola, New York

Bibliographical Note

This Dover edition, first published in 2004, is an unabridged, slightly corrected republication of the edition originally published by Random House, Inc., New York, 1988.

Library of Congress Cataloging-in-Publication Data

Colton, David L., 1943–
Partial differential equations : an introduction / David Colton.
p. cm.
Originally published: New York : Random House, c1988, in series: Random House/Birkhäuser mathematics series.
Includes bibliographical references and index.
ISBN 0-486-43834-1 (pbk.)
1. Differential equations, Partial. I. Title.

QA374.C649 2004
515'.353—dc22

2004056229

Manufactured in the United States of America
Dover Publications, Inc., 31 East 2nd Street, Mineola, N.Y. 11501

Preface

I have designed this book as a course for senior year or first-year graduate students in mathematics and mathematically-oriented beginning graduate students in engineering and the applied sciences. Its aim is to fill the gap between elementary textbooks that deal almost exclusively with separation of variables and integral transform techniques and those more advanced books which assume a mathematical sophistication at the level of a second- or third-year graduate student. In particular, this book is accessible to students having only a knowledge of advanced calculus and ordinary differential equations and will provide a glimpse of the beauty of the theory of partial differential equations at an earlier stage in their mathematical career than is normally the case.

Having determined the level at which I wanted to write this book, the next problem was the choice of topics. Here I was guided by my own interests and the desire not to put the book beyond the reach of a well-prepared senior-year student. In the interest of honesty, let me first make some of my prejudices known. I feel that an entire course based primarily on the method of separation of variables and integral transforms does not provide a proper introduction to the area of partial differential equations. With an advanced calculus background, but without knowing any functional analysis, I think that a first course on partial differential equations can cover a wealth of interesting and useful concepts well beyond the method of separation of variables and integral transforms. At the same time, trying to do too much simply overwhelms the student. I believe a first course should concentrate on the classical linear equations of mathematical physics (the wave equation, the heat equation, Laplace's equation, and the Helmholtz equation) and provide a reasonably complete mathematical treatment of these equations; it also should present some related results on linear equations with variable coefficients. However, the student should not leave a first course in partial differential equations with the idea that nothing of interest can be said about nonlinear equations, or that only closed form or series solutions of linear equations are of interest. I therefore have included introductory material on the nonlinear equations of gas dynamics, the Stefan problem in heat conduction, the approach to steady state of nonlinear parabolic equations, and the inverse scattering problem, as well as a brief description of various numerical methods, e.g., Trefftz's method for the heat equation, finite difference methods for Laplace's equation, and the Nyström method for solving Fredholm integral equations of the second kind. Having mastered the material in this book, the student should be well motivated and prepared to take more advanced courses in partial differential equations.

In writing a book at this level, I was also faced with the problem of what to do about topics in analysis outside the usual domain of advanced calculus and ordinary differential equations as presented in the typical undergraduate curriculum but that are nevertheless essential to a systematic treatment of the classical partial differential equations of mathematical physics—for example, what about Fourier series and integrals, integral equations, special functions? I have taken the view that an introductory course in partial differential equations is an ideal place to treat such topics and indeed should be a part of such a course. Many important topics in classical analysis thus can be covered which would normally only be treated in a superficial way, if at all, in an undergraduate curriculum. In this book I have included such topics as Fourier series and Fourier integrals, the Weierstrass approximation theorem, the method of stationary phase, the Fredholm alternative, and Hilbert Schmidt theory, as well as the basic properties of Legendre polynomials, the gamma function, and Bessel and Hankel functions. In addition, without assuming any knowledge of functional analysis other than a willingness to accept the fact that $L^2[a,b]$ is complete, I have introduced and made use of such concepts as the inner product, norm, orthonormal sets, complete sets, compact sets of functions, the definition of a Sobolev space, continuous functionals, linear operators, and contraction mappings. This should provide motivation for the student to seek a course on functional analysis and the "modern" theory of partial differential equations.

The subject of analytic function theory now arises. When I teach a one-semester version of this course I have always assumed that the students had no knowledge of analytic function theory. However, my experience in teaching a two-semester course on partial differential equations at the senior level is that by the second semester all of the students have taken a course in function theory. Hence, in order to make this book as flexible as possible, I have included in Chapter 1 a brief, self-contained discussion of analytic function theory requiring only a knowledge of power series and line integrals, which the student knows from advanced calculus. The presentation is deliberately short and aims only to prove the identity theorem and the residue theorem. These are the only results from analytic function theory that are needed in this book. If the starred (*) sections of the book are omitted, no analytic function theory is needed at all. Thus, the section on analytic function theory in this book can be viewed either as supplementary material to the student's knowledge of advanced calculus or as a review of the student's course on complex function theory. Of course, the subject of analytic function theory fits naturally into a course on partial differential equations in the sense that analytic functions are the solution of a first order system of partial differential equations (the Cauchy-Riemann equations). A considerable amount of recent research has been done on generalized analytic functions and elliptic systems in the plane, and I have given lip service to this area by exhibiting in Section 1.5.4 an example of a linear partial differential equation having no solution in any neighborhood of the origin.

I have mentioned my motivations for choosing the major themes of this book. As do all human beings, I have my own special interests and I have not hesitated to include a taste of these in the text. A book on mathematics should not be devoid of the writer's personality! In particular, in Section 3.6 I have included a brief example of the use of integral operators in the theory of partial differential equations and in Section 6.6 I have given a concise treatment of the inverse scattering problem for time harmonic acoustic and electromagnetic waves. At various points in the text I have also stressed the fact that not all problems in mathematical physics are properly posed and have exhibited examples of improperly posed problems arising in applications, in particular the inverse Stefan problem and the inverse scattering problem. My hope is that this material, together with the other problems considered in the text, will impress upon the student the diversity of issues actively under consideration by mathematicians working in the area of partial differential equations.

The teacher may use the material in this book in a variety of ways. For many years, I have taught a one-semester senior-year course at the University of Delaware based on the first three sections of Chapter 1, Chapter 2, the first four sections of Chapter 3, and Chapter 4. For such a course, no analytic function theory is needed. On the other hand, at Drexel University I taught a two-semester course that covered the entire book with the exception of Sections 1.4–1.6, 2.6, 3.6–3.8, and 6.6.

ACKNOWLEDGMENTS

No book is ever written in a vacuum. I have benefitted from the comments of the many students who have sat through my course on partial differential equations both at the University of Delaware and at Drexel University, as well as from the pleasant atmosphere and working environment in the Department of Mathematical Sciences at the University of Delaware. In addition, I would like to thank the manuscript reviewers for their helpful suggestions for improvements: William Fitzgibbon (University of Houston), Allen Pipkin (Brown University), Charles Groetsch (University of Cincinnati), and Michael Williams (Virginia Polytechnic Institute and State University). The material in this book is clearly related to my research interests, particularly Section 6.6, and in this regard I would like to gratefully acknowledge financial support from the Air Force Office of Scientific Research, the National Science Foundation, and the Office of Naval Research. Finally, I would like to thank Alison Chandler for her expert typing of the manuscript.

Contents

Chapter 1

Introduction

In this chapter we prepare the groundwork for our study of partial differential equations. We show how the wave equation, the heat equation, and Laplace's equation arise in the modeling of certain basic physical phenomena and how these equations are in fact the simplest examples of the three main classes of second order partial differential equations. In addition, we introduce the elementary theory of first order linear partial differential equations as well as the tools beyond advanced calculus that are needed in our investigations—i.e., the elementary theory of Fourier series, the Fourier integral, and analytic functions. Section 1.5, Analytic Functions, is kept brief and is needed only for the starred sections of this volume. In particular, if you are willing to forgo Sections 3.5 and 3.6, and Chapter 6 on scattering theory, you need not read anything about analytic function theory at all. However, at the end of our presentation of analytic function theory, we do use the identity theorem to establish the existence of a linear partial differential equation that has no solution at all. We conclude by giving a brief history of the theory of partial differential equations, which will serve as a guide through the analysis contained in the following chapters. Since the area of partial differential equations encompasses almost all areas of analysis (and is intimately connected with the physical sciences), we can only touch upon the highlights of its history. For a more comprehensive history, we recommend the monumental and scholarly treatise by Kline, upon which we have based our own short presentation.

First, we need to introduce some notation and a few elementary definitions. We will denote n-dimensional Euclidean space by R^n. A point in R^n will be written as $\mathbf{x} = (x_1,\ldots,x_n)$, where $x_1,x_2,\ldots,x_n$ are the Cartesian coordinates of $\mathbf{x}$. For $n = 2$ or $n = 3$, we may also use different letters for coordinates, e.g., $(x,y) \in R^2$, $(x,y,z) \in R^3$, etc. The Euclidean distance between two points $\mathbf{x},\mathbf{y} \in R^n$ is denoted by $|\mathbf{x} - \mathbf{y}|$. The boundary of a set S will be denoted by ∂S and the closure of a set S by $\bar{S}$. The complement of a set S in R^n is written as $R^n \backslash S$. An open interval in R^1 will be written as (a,b) and its closure as $[a,b]$. The set of functions defined on a set S and having all its partial derivatives of order less than or equal to k continuous in S will be denoted by $C^k(S)$. If the function $f(\mathbf{x})$ is continuous in S, we write $f \in C(S)$, whereas if $f(\mathbf{x})$ has continuous derivatives of all orders in S, we write $f \in C^\infty(S)$. A surface S in R^n is said to be in class C^k if for any point $\mathbf{x}_0 \in S$ there exists an n-dimensional neighborhood $N(\mathbf{x}_0)$ of $\mathbf{x}_0$ and a function $f \in C^k(N(\mathbf{x}_0))$ such that $\nabla f(\mathbf{x}_0) \neq 0$ and the set $S \cap N(\mathbf{x}_0)$ is described by the equation $f(\mathbf{x}) = 0$. The partial derivatives of a function $f(\mathbf{x})$ with respect to x_i will be denoted by $\partial f/\partial x_i$ or f_{x_i}. If $f = f(x)$ is a function of a single variable x, we will denote the derivative of $f(x)$ by $df/dx, f'$, or $\dot{f}$. Finally, if $f(\mathbf{x})$ is a complex valued function, its complex conjugate will be written as $\overline{f(\mathbf{x})}$.

A **partial differential equation** is an equation relating an unknown function $u(x_1,\ldots,x_n)$, the independent variables $x_1,\ldots,x_n$, and a finite number of the partial derivative of u, i.e.,

$$F\left(x_1,x_2,\ldots,x_n,\frac{\partial u}{\partial x_1},\ldots,\frac{\partial u}{\partial x_n},\ldots,\frac{\partial^k u}{\partial x_1^{k_1}\ldots\partial x_n^{k_n}}\right) = 0.$$

The **order** of a partial differential equation is that of the derivative of the highest order. A partial differential equation is called **quasilinear** if it is linear in all the highest order derivatives of the unknown function, e.g.,

$$A(x,y,u,u_x,u_y)\frac{\partial^2 u}{\partial x^2} + B(x,y,u,u_x,u_y)\frac{\partial^2 u}{\partial x \partial y} + C(x,y,u,u_x,u_y)\frac{\partial^2 u}{\partial y^2} + F(x,y,u,u_x,u_y) = 0$$

is a quasilinear second order equation. A partial differential equation is called **linear** if it is linear in the unknown function and its partial derivatives, e.g.,

$$A(x,y)\frac{\partial^2 u}{\partial x^2} + B(x,y)\frac{\partial^2 u}{\partial x \partial y} + C(x,y)\frac{\partial^2 u}{\partial y^2} + D(x,y)\frac{\partial u}{\partial x} + E(x,y)\frac{\partial u}{\partial y} + G(x,y)u = F(x,y)$$

is a second order linear equation. We shall be mainly concerned with second order linear equations, especially the **wave equation**

$$\frac{\partial^2 u}{\partial x^2} + \frac{\partial^2 u}{\partial y^2} + \frac{\partial^2 u}{\partial z^2} = \frac{\partial^2 u}{\partial t^2},$$

Laplace's equation

$$\frac{\partial^2 u}{\partial x^2} + \frac{\partial^2 u}{\partial y^2} + \frac{\partial^2 u}{\partial z^2} = 0,$$

and the **heat equation**

$$\frac{\partial^2 u}{\partial x^2} + \frac{\partial^2 u}{\partial y^2} + \frac{\partial^2 u}{\partial z^2} = \frac{\partial u}{\partial t}.$$

Except for Sections 1.5.4, 4.8, and 6.3–6.6, we shall always assume that solutions of initial or boundary value problems for partial differential equations are real valued.

1.1 PHYSICAL EXAMPLES

We shall now briefly examine how the wave equation, Laplace's equation, and the heat equation apply to certain physical situations. This discussion is merely illustrative rather than representative of the variety of applications of these equations in the physical and biological sciences.

We first consider the theory of heat conduction. Let D be a body in R^3 with boundary ∂D. Let $T(\mathbf{x},t)$, $\mathbf{x} \in D$, denote the temperature in the body at the point $\mathbf{x}$ and at time t. A difference in temperature in D creates a heat flow given by

$$\mathbf{q} = -\kappa \nabla T \tag{1.1}$$

where κ is the coefficient of heat conductivity in D, which is assumed to be constant, and the gradient ∇ is taken with respect to the space variables. Then through a surface $d\sigma$ with unit normal $\boldsymbol{\nu}$ the amount of heat

$$\begin{aligned} \Delta Q &= \mathbf{q} \cdot \boldsymbol{\nu} d\sigma \Delta t \\ &= -\kappa \nabla T \cdot \boldsymbol{\nu}\, d\sigma\, \Delta t \\ &= -\kappa \frac{\partial T}{\partial \nu} d\sigma\, \Delta t \end{aligned} \tag{1.2}$$

flows in time Δt in the direction $\boldsymbol{\nu}$. Let c denote the specific heat of D, where c is assumed to be constant. If we consider a portion D_1 of D which is bounded by ∂D_1, then the heat content of D_1 is given by

$$Q_{D_1}(t) = c \iiint_{D_1} T(\mathbf{x},t)\, d\mathbf{x} \tag{1.3}$$

and from (1.2) we have

$$\Delta Q_{D_1} = -\kappa \iint\limits_{\partial D_1} \frac{\partial T}{\partial \nu}\, d\sigma\, \Delta t \tag{1.4}$$

if we assume that the heat content of D_1 can be changed only by a flow of heat through ∂D_1. From (1.3) and (1.4) we have

$$c \iiint\limits_{D_1} \frac{\partial T}{\partial t}\, d\mathbf{x} - \kappa \iint\limits_{\partial D_1} \frac{\partial T}{\partial \nu}\, d\sigma = 0 \tag{1.5}$$

where $\boldsymbol{\nu}$ is the unit outward normal to ∂D_1. Using the **divergence theorem**

$$\iiint\limits_{D_1} \operatorname{div} \mathbf{F}(\mathbf{x})\, d\mathbf{x} = \iint\limits_{\partial D_1} \mathbf{F}(\mathbf{x}) \cdot \boldsymbol{\nu}(\mathbf{x})\, d\sigma$$

we can rewrite (1.5) as

$$\iiint\limits_{D_1} \left[c\frac{\partial T}{\partial t} - \kappa \Delta_3 T \right] d\mathbf{x} = 0 \tag{1.6}$$

where we have used the identity

$$\begin{aligned} \Delta_3 T &= \frac{\partial^2 T}{\partial x^2} + \frac{\partial^2 T}{\partial y^2} + \frac{\partial^2 T}{\partial z^2} \\ &= \operatorname{div} \nabla T; \qquad \mathbf{x} = (x, y, z). \end{aligned}$$

Since D_1 is arbitrary, we have

$$\Delta_3 T = \frac{c}{\kappa} \frac{\partial T}{\partial t} \tag{1.7}$$

throughout D, so T satisfies the heat equation. We note that the change of variables $t' = (\kappa/c)\, t$ reduces (1.7) to

$$\Delta_3 T = \frac{\partial T}{\partial t'},$$

which is the form of (1.7) we shall later investigate. On ∂D we can either prescribe the temperature T or the heat flux $-\kappa\ (\partial T/\partial \nu)$, and in addition we must prescribe the initial temperature at time $t = 0$.

If T is independent of time, then T satisfies Laplace's equation

$$\Delta_3 T = 0. \tag{1.8}$$

For example, this will be true when the temperature of the body has reached a steady state. In such situations, it is no longer necessary to prescribe the initial temperature. Laplace's equation also appears in problems in fluid dynamics and electrostatics (cf. Bergman and Schiffer).

We now look at the propagation of sound waves of small amplitude viewed as a problem in fluid dynamics. Let $\mathbf{v}(\mathbf{x},t)$ be the velocity vector of a fluid particle in an inviscid fluid and let $p(\mathbf{x},t)$ and $\rho(\mathbf{x},t)$ denote the pressure and density, respectively, of the fluid. If no external forces are acting on the fluid, **Euler's equations** of motion are

$$\frac{\partial \rho}{\partial t} + \nabla \cdot (\rho \mathbf{v}) = 0 \qquad \text{(equation of continuity)}$$

$$\frac{\partial \mathbf{v}}{\partial t} + (\mathbf{v} \cdot \nabla)\mathbf{v} + \frac{1}{\rho} \nabla p = 0 \qquad \text{(equation of conservation of momentum)}$$

$$p = f(\rho) \quad \text{(equation of state)},$$

where $f(\rho)$ is a function depending on the fluid. Assuming $\mathbf{v}(\mathbf{x},t)$, $p(\mathbf{x},t)$, and $\rho(\mathbf{x},t)$ are small, we perturb these quantities around an equilibrium state $\mathbf{v} = 0$, $p = p_0$, and $\rho = \rho_0$ (with $p_0 = f(\rho_0)$) and write

$$\begin{aligned} \mathbf{v}(\mathbf{x},t) &= \epsilon \mathbf{v}_1(\mathbf{x},t) + \ldots \\ p(\mathbf{x},t) &= p_0 + \epsilon p_1(\mathbf{x},t) + \ldots \\ \rho(\mathbf{x},t) &= \rho_0 + \epsilon \rho_1(\mathbf{x},t) + \ldots, \end{aligned} \tag{1.9}$$

where $0 < \epsilon << 1$ and the dots refer to higher order terms in ϵ. Inserting (1.9) into Euler's equations and retaining only terms of order ϵ gives (assuming that ρ_0 is a constant) the linearized Euler equations

$$\frac{\partial \rho_1}{\partial t} + \rho_0 \nabla \cdot \mathbf{v}_1 = 0 \tag{1.10a}$$

$$\frac{\partial \mathbf{v}_1}{\partial t} + \frac{1}{\rho_0} \nabla p_1 = 0 \tag{1.10b}$$

$$p_1 = f'(\rho_0)\rho_1. \tag{1.10c}$$

Assume that at time $t = 0$ the fluid flow is irrotational, i.e., curl $\mathbf{v}_1(\mathbf{x},0) = 0$. Then there exists a potential $\psi(\mathbf{x})$ such that $\mathbf{v}_1(\mathbf{x},0) = -\nabla\psi(\mathbf{x})$ and hence from (1.10b) we have

$$\begin{aligned} \mathbf{v}_1(\mathbf{x},t) &= -\nabla\left[\psi(\mathbf{x}) + \frac{1}{\rho_0}\int_0^t p_1(\mathbf{x},t)dt\right] \\ &= -\nabla\phi(\mathbf{x},t) \end{aligned} \tag{1.11}$$

where $\phi(\mathbf{x},t)$ is the term in brackets in (1.11). Inserting (1.11) into (1.10b) gives

$$\nabla\left[-\frac{\partial \phi}{\partial t} + \frac{1}{\rho_0} p_1\right] = 0$$

and hence we can choose

$$p_1(\mathbf{x},t) = \rho_0 \frac{\partial \phi(\mathbf{x},t)}{\partial t} . \tag{1.12}$$

Equation (1.10c) now implies that

$$\rho_1(\mathbf{x},t) = \left[\frac{\rho_0}{f'(\rho_0)}\right] \frac{\partial \phi}{\partial t}(\mathbf{x},t),$$

and from (1.10a) we have

$$\frac{\partial^2 \phi}{\partial t^2} - [f'(\rho_0)]\Delta_3 \phi = 0.$$

Setting $c^2 = f'(\rho_0)$ and assuming that $f'(\rho_0) > 0$ gives the wave equation

$$\Delta_3 \phi = \frac{1}{c^2} \frac{\partial^2 \phi}{\partial t^2} . \tag{1.13}$$

Note that from (1.12) the pressure $p_1(\mathbf{x},t)$ also satisfies the wave equation (1.13). The change of variables $t' = ct$ reduces (1.13) to

$$\Delta_3 \phi = \frac{\partial^2 \phi}{\partial t'^2} , \tag{1.14}$$

which is the form of (1.13) we shall consider in the next chapter.

In order to determine $\phi(\mathbf{x},t)$ from (1.13) or (1.14), we must prescribe initial and boundary data. In particular, at time $t = 0$ we must be given the initial velocity and acceleration, i.e. $\phi(\mathbf{x},0)$ and $\partial\phi(\mathbf{x},0)/\partial t$. On the boundary of the region containing (or contained in) the fluid there are various possibilities. For example, if the boundary is rigid and impenetrable, then $\mathbf{v}_1 \cdot \boldsymbol{\nu} = 0$ on the boundary where $\boldsymbol{\nu}$ is the normal vector, i.e., $\partial\phi/\partial\nu = 0$. At the other extreme, if the boundary is a fixed surface which is a site of pressure release so that the pressure vanishes there, then it suffices to set $\phi = 0$ on the boundary. Note that if $\phi(\mathbf{x},t)$ is time harmonic, i.e.,

$$\phi(\mathbf{x},t) = \phi(\mathbf{x})e^{-i\omega t},$$

then $\phi(\mathbf{x})$ satisfies the **Helmholtz equation**

$$\Delta_3 \phi + k^2 \phi = 0,$$

where $k = \omega/c$. In this case it is no longer necessary to prescribe initial data.

The wave equation also appears in the theory of vibrations and electromagnetic wave motion (cf. Baldock and Bridgeman).

1.2 FIRST ORDER LINEAR EQUATIONS

The simplest type of partial differential equation in two independent variables is the first order linear equation. Our reason for studying such equations right at the beginning is for the pedagogical value in introducing characteristic curves and their importance in solving initial value problems, as well as for the fact that our results on first order equations will be needed shortly when we want to reduce second order equations in two independent variables to canonical form. Our aim is to show how the first order linear partial differential equation

$$a(x,y)\frac{\partial u}{\partial x} + b(x,y)\frac{\partial u}{\partial y} + c(x,y)u = f(x,y) \tag{1.15}$$

can be solved by reducing it to an ordinary differential equation. We make the assumption that the coefficients $a(x,y)$, $b(x,y)$, $c(x,y)$, and $f(x,y)$ are continuously differentiable functions of x and y in some domain D, and that $a(x,y)$ and $b(x,y)$ never both vanish at the same point. We further assume that the coefficients $a(x,y)$, $b(x,y)$, $c(x,y)$, and $f(x,y)$ have real values. We shall show that by a change of variables we can reduce (1.15) to an equation of the form

$$\frac{\partial w}{\partial \xi} + s(\xi,\eta)w = t(\xi,\eta), \tag{1.16}$$

which is an ordinary differential equation in ξ depending on the parameter η, where ξ and η are new independent variables. If $a(x,y)$ or $b(x,y)$ is identically zero, then (1.15) is already in the form (1.16), and hence we suppose that neither $a(x,y)$ nor $b(x,y)$ is identically zero.

Let $\phi \in C^1(D)$, $\psi \in C^1(D)$ be such that the Jacobian

$$J(\phi,\psi) = \frac{\partial \phi}{\partial x}\frac{\partial \psi}{\partial y} - \frac{\partial \phi}{\partial y}\frac{\partial \psi}{\partial x}$$

is not identically zero, and define

$$\xi = \phi(x,y)$$
$$\eta = \psi(x,y).$$

Then if $u(x,y) = w(\xi,\eta)$ we have

$$\frac{\partial u}{\partial x} = \frac{\partial w}{\partial \xi}\frac{\partial \phi}{\partial x} + \frac{\partial w}{\partial \eta}\frac{\partial \psi}{\partial x}$$
$$\frac{\partial u}{\partial y} = \frac{\partial w}{\partial \xi}\frac{\partial \phi}{\partial y} + \frac{\partial w}{\partial \eta}\frac{\partial \psi}{\partial y}.$$

Then (1.15) becomes

$$\left(a\frac{\partial \phi}{\partial x} + b\frac{\partial \phi}{\partial y}\right)\frac{\partial w}{\partial \xi} + \left(a\frac{\partial \psi}{\partial x} + b\frac{\partial \psi}{\partial y}\right)\frac{\partial w}{\partial \eta} + cw = f. \tag{1.17}$$

Hence if we choose $\psi(x,y)$ such that

$$a(x,y)\frac{\partial \psi}{\partial x} + b(x,y)\frac{\partial \psi}{\partial y} = 0, \tag{1.18}$$

then we arrive at an equation of the form (1.16). The problem now is to construct a solution $\psi(x,y)$ of (1.18) and a function $\phi(x,y)$ such that the Jacobian $J(\phi,\psi)$ is not identically zero.

Suppose for the moment that a solution $\psi(x,y)$ of (1.18) exists such that $\partial\psi/\partial y$ is not identically zero. Define the curve $y = y(x)$ implicitly by $\psi(x,y) = \gamma$ where γ is a constant. Then

$$\frac{\partial\psi}{\partial x}dx + \frac{\partial\psi}{\partial y}dy = 0,$$

which implies that

$$\begin{aligned}\frac{dy}{dx} &= -\frac{\partial\psi/\partial x}{\partial\psi/\partial y}\\ &= \frac{b(x,y)}{a(x,y)}.\end{aligned}$$

Thus $\psi(x,y) = \gamma$ implicitly defines a solution of the ordinary differential equation

$$\frac{dy}{dx} = \frac{b(x,y)}{a(x,y)}. \tag{1.19}$$

Conversely, if $\psi(x,y) = \gamma$ implicitly defines a solution of (1.19) such that $\partial\psi/\partial y$ is not identically zero, then ψ satisfies (1.18). Equation (1.19) is called the **characteristic equation** of the partial differential equation (1.15) and defines a one-parameter family of curves called the **characteristic curves** of the partial differential equation (1.15).

Having found $\psi(x,y)$ as the solution of (1.19), we can choose the function $\phi(x,y)$ arbitrarily such that the Jacobian doesn't vanish. For example, choose $\phi(x,y) = x$. The ordinary differential equation (1.17) for w is then

$$\frac{\partial w}{\partial\xi} + \frac{c(x,y)}{a(x,y)}w = \frac{f(x,y)}{a(x,y)}$$

and from the elementary theory of ordinary differential equations the solution of this equation is

$$w(\xi,\eta) = \exp\left(-\int^{\xi}\frac{c(\xi,\eta)}{a(\xi,\eta)}d\xi\right)\left[\int^{\xi}\frac{f(\xi,\eta)}{a(\xi,\eta)}\exp\left(\int^{\xi}\frac{c(\xi,\eta)}{a(\xi,\eta)}d\xi\right)d\xi + d(\eta)\right]$$

where $d(\eta)$ is an arbitrary function of η and $a(\xi,\eta) = a(\xi,y(\xi,\eta))$, $c(\xi,\eta) = c(\xi,y(\xi,\eta))$, etc., where the function $y(\xi,\eta)$ is determined by solving for y in the equations $\xi = x$, $\eta = \psi(x,y)$. Hence, the general solution of (1.15) is

$$u(x,y) = \frac{1}{v(x,\psi(x,y))}\{w(x,\psi(x,y)) + d(\psi(x,y))\} \tag{1.20}$$

where

$$v(\xi,\eta) = \exp\left(\int^{\xi} \frac{c(\xi,\eta)}{a(\xi,\eta)}\, d\xi\right)$$

$$w(\xi,\eta) = \int^{\xi} \frac{f(\xi,\eta)}{a(\xi,\eta)}\, v(\xi,\eta)d\xi.$$

☐ EXAMPLE 1

Find the general solution of $xu_x - yu_y + u = x$.

■ ***Solution.*** The characteristic equation is

$$\frac{dy}{dx} = -\frac{y}{x}.$$

This is a separable ordinary differential equation, i.e.,

$$\frac{dy}{y} = -\frac{dx}{x}.$$

Integrating both sides gives

$$\log y = -\log x + c$$

where c is an arbitrary constant, or

$$xy = \gamma$$

where γ is (another) arbitrary constant. Hence, setting

$$\xi = x$$
$$\eta = xy$$

in our first order partial differential equation yields

$$\frac{\partial w}{\partial \xi} + \frac{1}{\xi} w = 1,$$

whose solution is

$$w(\xi,\eta) = \frac{\xi}{2} + \frac{1}{\xi} d(\eta),$$

where $d(\eta)$ is an arbitrary function of η. The solution of $xu_x - yu_y + u = x$ is now given by

$$u(x,y) = \frac{x}{2} + \frac{1}{x} d(xy).$$

In order to guarantee that u is continuously differentiable we require that the arbitrary function d be continuously differentiable. ■

We now turn to the problem of solving initial value problems for first order linear partial differential equations. The **Cauchy problem** for the partial differential equation (1.15) is to find a solution of (1.15) taking on prescribed values $\phi(x)$ on a specified curve C: $y = y(x)$ in the xy plane. The curve C must not be a characteristic curve of (1.15) (or tangent to such a curve). On a characteristic curve we have $\psi(x,y) = \gamma$ where γ is a constant and hence on this curve (1.20) becomes

$$u(x,y) = \frac{w(x,\gamma)}{v(x,\gamma)} + \frac{d(\gamma)}{v(x,\gamma)},$$

i.e., $u(x,y)$ does not equal $\phi(x)$ on C: $y = y(x)$ unless $\phi(x)$ is of the special form

$$\phi(x) = \frac{w(x,\gamma)}{v(x,\gamma)} + \frac{k}{v(x,\gamma)}$$

where k is a constant. On the other hand, if $\phi(x)$ is of this form, there exist infinitely many solutions of (1.15) given by (1.20) where $d(\eta)$ is any differentiable function such that $d(\gamma) = k$.

□ EXAMPLE 2

Find the solution of $xu_x - yu_y + u = x$ such that $u(x,y) = x$ on the curve $y = x^2$.

■ ***Solution.*** As we have shown in the previous example, the general solution of $xu_x - yu_y + u = x$ is given by

$$u(x,y) = \frac{x}{2} + \frac{1}{x}\,d(xy)$$

where d is an arbitrary differentiable function. Hence, on $y = x^2$ we want

$$x = \frac{x}{2} + \frac{1}{x}\,d(x^3),$$

i.e.,

$$\frac{x^2}{2} = d(x^3),$$

or

$$d(x) = \frac{1}{2}x^{2/3}.$$

Therefore

$$u(x,y) = \frac{x}{2} + \frac{1}{2x}(xy)^{2/3}.$$

Note that we must not allow (x,y) to be on the lines $x = 0$ or $y = 0$ because there $u(x,y)$ is not continuously differentiable. In particular, no solution to our Cauchy problem exists in a neighborhood of the origin. This is a consequence of the fact that at the origin the curve $y = x^2$ is tangent to the characteristic curve $y = 0$. ■

For a more complete exposition of first order partial differential equations, including quasilinear equations, the reader is referred to Courant and Hilbert 1961, Garabedian, and John.

1.3 CLASSIFICATION OF SECOND ORDER EQUATIONS AND CANONICAL FORMS

1.3.1 Types of Second Order Equations

We now consider second order quasilinear partial differential equations in Euclidean n space R^n and show how at a given point in R^n they can be classified into four distinct types. As we shall see later, the type of a partial differential equation dictates to a large extent the behavior of solutions to these equations. In this sense our classification procedure can be compared in spirit to that of a zoologist who classifies animals into mammals, birds, fishes, and reptiles and then proceeds to study representative examples from each of these classes.

Consider the second order partial differential equation

$$\sum_{i,j=1}^{n} a_{ij} \frac{\partial^2 u}{\partial x_i \partial x_j} + f\left(x_1, \ldots, x_n, u, \frac{\partial u}{\partial x_1}, \ldots, \frac{\partial u}{\partial x_n} \right) = 0 \qquad \textbf{(1.21)}$$

where $a_{ij} = a_{ij}(x_1, \ldots, x_n)$ are given real valued functions defined in a domain $D \subset R^n$ and without loss of generality assume that $a_{ij} = a_{ji}$. Let $(x_1^0, \ldots, x_n^0)$ be a fixed point of D and consider the quadratic form

$$\sum_{i,j=1}^{n} a_{ij}(x_1^0, \ldots, x_n^0) t_i t_j. \qquad \textbf{(1.22)}$$

Then

(1) Equation (1.21) is of **elliptic** type at $(x_1^0, \ldots, x_n^0)$ if at this point the quadratic form (1.22) is non-singular and definite, i.e., it can be reduced by a real linear transformation to a sum of n squares all of the same sign.

(2) Equation (1.21) is of **hyperbolic** type at $(x_1^0, \ldots, x_n^0)$ if at this point (1.22) is non-singular, indefinite, and can be reduced by a real linear transformation to the sum of n squares, $n - 1$ of which are the same sign.

(3) Equation (1.21) is of **ultra-hyperbolic** type at $(x_1^0, \ldots, x_n^0)$ if at this point (1.22) is non-singular, indefinite, and can be reduced by a real linear transformation to the sum of n ($n \geq 4$) squares with more than one coefficient of either sign.

(4) Equation (1.21) is of **parabolic** type at $(x_1^0,\ldots,x_n^0)$ if at this point (1.22) is singular, i.e., it can be reduced by a real linear transformation to the sum of fewer than n squares, not necessarily all of the same sign.

Equation (1.21) is of one of these types in D if it is of that type at each point in D. There is no similar classification scheme for partial differential equations of order greater than two and in the remainder of this book we shall consider only second order equations. (This does not imply that higher order equations are not of interest from both a physical and mathematical viewpoint!)

☐ EXAMPLES 3–6

The wave equation

$$\frac{\partial^2 u}{\partial x^2} + \frac{\partial^2 u}{\partial y^2} + \frac{\partial^2 u}{\partial z^2} = \frac{\partial^2 u}{\partial t^2}$$

is hyperbolic in any domain, the Laplace equation

$$\frac{\partial^2 u}{\partial x^2} + \frac{\partial^2 u}{\partial y^2} + \frac{\partial^2 u}{\partial z^2} = 0$$

is of elliptic type in any domain, and the heat equation

$$\frac{\partial^2 u}{\partial x^2} + \frac{\partial^2 u}{\partial y^2} + \frac{\partial^2 u}{\partial z^2} = \frac{\partial u}{\partial t}$$

is of parabolic type in any domain. The equation

$$\frac{\partial^2 u}{\partial x^2} + \frac{\partial^2 u}{\partial y^2} = \frac{\partial^2 u}{\partial z^2} + \frac{\partial^2 u}{\partial t^2}$$

is of ultra-hyperbolic type in any domain. ☐

1.3.2 Reduction of Second Order Equations with Constant Coefficients to Canonical Form

This section shows how a partial differential equation with constant coefficients can be reduced to a particularly simple form by a linear change of variables. In particular, consider the partial differential equation with real valued constant coefficients

$$\sum_{i,j=1}^{n} a_{ij} \frac{\partial^2 u}{\partial x_i \partial x_j} + \sum_{i=1}^{n} b_i \frac{\partial u}{\partial x_i} + cu = f(x_1,\ldots,x_n). \tag{1.23}$$

Define the linear change of variables

$$\xi_k = \sum_{i=1}^{n} c_{ki} x_i \qquad (k = 1,2,\ldots,n),$$

where the real constants c_{ki} are to be chosen later such that det $|c_{ki}|$ does not vanish.

Then (1.23) becomes

$$\sum_{k,l=1}^{n} \tilde{a}_{kl} \frac{\partial^2 u}{\partial \xi_k \partial \xi_l} + \sum_{i=1}^{n} \tilde{b}_i \frac{\partial u}{\partial \xi_i} + cu = \tilde{f}(\xi_1,\ldots,\xi_n) \tag{1.24}$$

where

$$\tilde{a}_{kl} = \sum_{i,j=1}^{n} a_{ij} c_{ki} c_{lj},$$

and the constants $\tilde{b}_i$ and function $\tilde{f}(\xi_1,\ldots,\xi_n)$ can be expressed in terms of the b_i and $f(x_1,\ldots,x_n)$, respectively. From linear algebra it is known that we can always choose the c_{ij} such that

$$\tilde{a}_{kl} = \begin{cases} 0 & \text{when} \quad k \neq l \\ \lambda_k & \text{when} \quad k = l \end{cases}$$

where $\lambda_k = 0$, 1, or -1 and (1.24) becomes

$$\sum_{k=1}^{n} \lambda_k \frac{\partial^2 u}{\partial \xi_k^2} + \sum_{i=1}^{n} \tilde{b}_i \frac{\partial u}{\partial \xi_i} + cu = \tilde{f}(\xi_1,\ldots,\xi_n). \tag{1.25}$$

The signs and/or vanishing of the λ_k clearly determine the type of equation (1.25) and hence of equation (1.23), since from the definition of type a real linear change of variables cannot change the type of a partial differential equation. Equation (1.25) is called the **canonical form** of (1.23). In the case of equation (1.23) with variable coefficients, it is possible to reduce the equation to canonical form at each given point $(x_1^0,\ldots,x_n^0)$ of D, but one cannot in general find a single transformation that will work for all points in a neighborhood of $(x_1^0,\ldots,x_n^0)$. An exception is the case $n = 2$, if we allow our transformation to be nonlinear. We shall now show how this can be done.

1.3.3 Reduction of Second Order Equations in Two Independent Variables to Canonical Form

Consider the partial differential equation

$$A \frac{\partial^2 u}{\partial x^2} + 2B \frac{\partial^2 u}{\partial x \partial y} + C \frac{\partial^2 u}{\partial y^2} + F\left(x, y, u, \frac{\partial u}{\partial x}, \frac{\partial u}{\partial y}\right) = 0 \tag{1.26}$$

where $A = A(x,y)$, $B = B(x,y)$, and $C = C(x,y)$ are real-valued twice continuously differentiable functions of x and y in some domain $D \subset R^2$ and A, B, and C do not all vanish at the same point. The factor of two appearing in the coefficient of $\partial^2 u/\partial x \partial y$ is purely for notational convenience to avoid having a factor of four appearing repeatedly in our subsequent formulas. (The reader is warned to remember this factor when doing the exercises at the end of the chapter.) The quadratic form corresponding to (1.26) is

$$At_1^2 + 2Bt_1t_2 + Ct_2^2\,; \tag{1.27}$$

(1.26) is of hyperbolic type if $B^2 - AC > 0$, of parabolic type if $B^2 - AC = 0$, and of elliptic type if $B^2 - AC < 0$.

Now make the change of variables

$$\begin{aligned} \xi &= \xi(x,y) \\ \eta &= \eta(x,y) \end{aligned} \tag{1.28}$$

such that $\xi(x,y)$, $\eta(x,y) \in C^1(D)$ and the Jacobian

$$J = \begin{vmatrix} \dfrac{\partial \xi}{\partial x} & \dfrac{\partial \xi}{\partial y} \\[2ex] \dfrac{\partial \eta}{\partial x} & \dfrac{\partial \eta}{\partial y} \end{vmatrix}$$

does not vanish for $(x,y) \in D$. Then (1.26) becomes

$$\tilde{A}\frac{\partial^2 u}{\partial \xi^2} + 2\tilde{B}\frac{\partial^2 u}{\partial \xi \partial \eta} + \tilde{C}\frac{\partial^2 u}{\partial \eta^2} + \tilde{F}\left(\xi,\eta,u,\frac{\partial u}{\partial \xi},\frac{\partial u}{\partial \eta}\right) = 0 \tag{1.29}$$

where

$$\tilde{A}(\xi,\eta) = A\frac{\partial \xi}{\partial x}\frac{\partial \xi}{\partial x} + 2B\frac{\partial \xi}{\partial x}\frac{\partial \xi}{\partial y} + C\frac{\partial \xi}{\partial y}\frac{\partial \xi}{\partial y}$$

$$\tilde{C}(\xi,\eta) = A\frac{\partial \eta}{\partial x}\frac{\partial \eta}{\partial x} + 2B\frac{\partial \eta}{\partial x}\frac{\partial \eta}{\partial y} + C\frac{\partial \eta}{\partial y}\frac{\partial \eta}{\partial y}$$

$$\tilde{B}(\xi,\eta) = A\frac{\partial \xi}{\partial x}\frac{\partial \eta}{\partial x} + B\left(\frac{\partial \xi}{\partial x}\frac{\partial \eta}{\partial y} + \frac{\partial \xi}{\partial y}\frac{\partial \eta}{\partial x}\right) + C\frac{\partial \xi}{\partial y}\frac{\partial \eta}{\partial y},$$

and it can be verified directly that

$$\tilde{B}^2 - \tilde{A}\tilde{C} = (B^2 - AC)J^2;$$

thus the type remains unchanged under the change of variables (1.28)! We shall show that $\xi(x,y)$ and $\eta(x,y)$ can be chosen such that one and only one of the following conditions is satisfied:

(1) $\tilde{A} = 0, \quad \tilde{C} = 0$

(2) $\tilde{A} = 0, \quad \tilde{B} = 0$

(3) $\tilde{A} = \tilde{C}, \quad \tilde{B} = 0.$

Case 1 ($B^2 - AC > 0$). In this case (1.26) is of hyperbolic type in D. We shall show how to choose $\xi(x,y)$ and $\eta(x,y)$ such that $\tilde{A} = \tilde{C} = 0$. Assume that this is not already the case in equation (1.26) and without loss of generality assume that A does not vanish in a neighborhood of $(x_0,y_0) \in D$. Motivated by the equations for $\tilde{A}$ and

$\tilde{C}$, consider the first order partial differential equation

$$A\frac{\partial\phi}{\partial x}\frac{\partial\phi}{\partial x} + 2B\frac{\partial\phi}{\partial x}\frac{\partial\phi}{\partial y} + C\frac{\partial\phi}{\partial y}\frac{\partial\phi}{\partial y} = 0. \tag{1.30}$$

Since $B^2 - AC > 0$ we can write (1.30) as

$$\left(A\frac{\partial\phi}{\partial x} + (B + \sqrt{B^2 - AC})\frac{\partial\phi}{\partial y}\right)\left(A\frac{\partial\phi}{\partial x} + (B - \sqrt{B^2 - AC})\frac{\partial\phi}{\partial y}\right) = 0,$$

and solutions of

$$A\frac{\partial\phi}{\partial x} + (B + \sqrt{B^2 - AC})\frac{\partial\phi}{\partial y} = 0 \tag{1.31}$$

and

$$A\frac{\partial\phi}{\partial x} + (B - \sqrt{B^2 - AC})\frac{\partial\phi}{\partial y} = 0 \tag{1.32}$$

will be solutions of (1.30). These equations can be solved by the methods given in Section 1.2. Let $\phi_1(x,y)$ and $\phi_2(x,y)$ be non-constant solutions of (1.31) and (1.32), respectively. Then $\phi_1(x,y)$ = constant and $\phi_2(x,y)$ = constant define the **characteristic curves** of (1.26), and (1.30) is called the **characteristic equation** of (1.26). Since $B^2 - AC > 0$ and $\phi_1(x,y)$ and $\phi_2(x,y)$ are not identically equal to a constant, $\phi_1(x,y)$ is not equal to $\phi_2(x,y)$, and so we have two different families of real characteristics.

Now let

$$\begin{aligned} \xi &= \phi_1(x,y) \\ \eta &= \phi_2(x,y) \end{aligned} \tag{1.33}$$

such that $\partial\phi_1(x_0,y_0)/\partial y$ and $\partial\phi_2(x_0,y_0)/\partial y$ are not equal to zero (this is always possible by imposing appropriate initial conditions on $\phi_1(x,y)$ and $\phi_2(x,y)$). Then from (1.31) and (1.32) we have

$$J = \begin{vmatrix} \dfrac{\partial\phi_1}{\partial x} & \dfrac{\partial\phi_1}{\partial y} \\ \dfrac{\partial\phi_2}{\partial x} & \dfrac{\partial\phi_2}{\partial y} \end{vmatrix} = -\frac{2\sqrt{B^2 - AC}}{A}\frac{\partial\phi_1}{\partial y}\frac{\partial\phi_2}{\partial y}$$

and hence the Jacobian J does not vanish in a neighborhood of (x_0,y_0). Thus under the change of variables (1.33) we have $\tilde{A} = \tilde{C} = 0$ in (1.29), and since $\tilde{B}^2 - \tilde{A}\tilde{C} = (B^2 - AC)J^2$ does not vanish, $\tilde{B}$ does not equal zero in a neighborhood of (x_0,y_0). Dividing (1.29) by $2\tilde{B}$ now gives

$$\frac{\partial^2 u}{\partial\xi\partial\eta} = \tilde{F}\left(\xi,\eta,u,\frac{\partial u}{\partial\xi},\frac{\partial u}{\partial\eta}\right), \tag{1.34}$$

which is the canonical form for hyperbolic equations. If we now set

$$\xi = \alpha + \beta$$
$$\eta = \alpha - \beta$$

in (1.34) we arrive at an equation of the form

$$\frac{\partial^2 u}{\partial \alpha^2} - \frac{\partial^2 u}{\partial \beta^2} = \Phi\left(\alpha,\beta,u,\frac{\partial u}{\partial \alpha},\frac{\partial u}{\partial \beta}\right), \tag{1.35}$$

which is the second canonical form for hyperbolic equations.

Case 2 ($B^2 - AC = 0$). In this case (1.36) is of parabolic type. Since A, B, and C do not vanish at the same time and $B^2 - AC = 0$, it follows that A and C cannot both be zero at the same time. Without loss of generality assume that A does not vanish at $(x_0,y_0) \in D$. Then (1.31) and (1.32) are identical,

$$A\frac{\partial \phi}{\partial x} + B\frac{\partial \phi}{\partial y} = 0, \tag{1.36}$$

and since $B^2 - AC = 0$ any solution of (1.36) also satisfies

$$B\frac{\partial \phi}{\partial x} + C\frac{\partial \phi}{\partial y} = 0. \tag{1.37}$$

Then we can find a solution $\phi(x,y)$ of (1.36) such that $\partial\phi(x_0,y_0)/\partial y$ does not equal zero (and hence $\partial\phi(x_0,y_0)/\partial x$ also does not equal zero) and the characteristic curves of (1.26) are given by $\phi(x,y) = \text{constant}$.

Now let

$$\xi = \phi(x,y)$$

and let $\eta = \eta(x,y)$ be any twice continuously differentiable function such that the Jacobian J of ξ and η does not vanish at (x_0,y_0). We then have $\tilde{A} = 0$ in a neighborhood of (x_0,y_0) and

$$\tilde{B} = \left(A\frac{\partial \phi}{\partial x} + B\frac{\partial \phi}{\partial y}\right)\frac{\partial \eta}{\partial x} + \left(B\frac{\partial \phi}{\partial x} + C\frac{\partial \phi}{\partial y}\right)\frac{\partial \eta}{\partial y} = 0$$

in a neighborhood of (x_0,y_0). Furthermore,

$$\tilde{C} = \frac{1}{A}\left(A\frac{\partial \eta}{\partial x} + B\frac{\partial \eta}{\partial y}\right)^2$$

is not equal to zero in a neighborhood of (x_0,y_0) since otherwise from (1.36) we would have $J = 0$. Dividing by $\tilde{C}$ we have that (1.29) can be written in the form

$$\frac{\partial^2 u}{\partial \eta^2} = \tilde{F}\left(\xi,\eta,u,\frac{\partial u}{\partial \xi},\frac{\partial u}{\partial \eta}\right), \tag{1.38}$$

which is the canonical form for parabolic equations.

Case 3 ($B^2 - AC < 0$). In this case (1.26) is of elliptic type. Assume that A, B, and C are analytic functions of x and y in a neighborhood of $(x_0,y_0) \in D$, i.e., A, B, and C have Taylor series expansions in x and y which converge in some neighborhood of (x_0,y_0). Then the coefficients of (1.31) and (1.32) are also analytic and it can be shown that (1.31) (which is a first order system of partial differential equations written in complex form) has an analytic solution (in the sense defined above)

$$\phi(x,y) = \phi_1(x,y) + i\phi_2(x,y)$$

defined in a neighborhood of (x_0,y_0) such that $|\partial\phi/\partial x| + |\partial\phi/\partial y|$ does not vanish in this neighborhood. Let

$$\xi = \phi_1(x,y) = \text{Re}\,\phi(x,y)$$

$$\eta = \phi_2(x,y) = \text{Im}\,\phi(x,y).$$

Then it can be shown that the Jacobian of $\phi_1(x,y)$ and $\phi_2(x,y)$ does not vanish in a neighborhood of (x_0,y_0), and if we separate the real and imaginary parts in (1.30) we obtain

$$A\left(\frac{\partial\xi}{\partial x}\right)^2 + 2B\frac{\partial\xi}{\partial x}\frac{\partial\xi}{\partial y} + C\left(\frac{\partial\xi}{\partial y}\right)^2 = A\left(\frac{\partial\eta}{\partial x}\right)^2 + 2B\frac{\partial\eta}{\partial x}\frac{\partial\eta}{\partial y} + C\left(\frac{\partial\eta}{\partial y}\right)^2$$

and

$$A\frac{\partial\xi}{\partial x}\frac{\partial\eta}{\partial x} + B\left(\frac{\partial\xi}{\partial x}\frac{\partial\eta}{\partial y} + \frac{\partial\xi}{\partial y}\frac{\partial\eta}{\partial x}\right) + C\frac{\partial\xi}{\partial y}\frac{\partial\eta}{\partial y} = 0.$$

Hence it follows that $\tilde{A} = \tilde{C}$ and $\tilde{B} = 0$. But since $\tilde{B}^2 - \tilde{A}\tilde{C} < 0$, we have that $\tilde{A}$ and $\tilde{C}$ do not vanish in a neighborhood of (x_0,y_0), and if we divide (1.29) by $\tilde{A}$ we have

$$\frac{\partial^2 u}{\partial\xi^2} + \frac{\partial^2 u}{\partial\eta^2} = \tilde{F}\left(\xi,\eta,u,\frac{\partial u}{\partial\xi},\frac{\partial u}{\partial\eta}\right) \qquad \textbf{(1.39)}$$

which is the canonical form for elliptic equations.

Note that it may turn out that (1.26) is of different type in different parts of the domain D. Equation (1.26) is then said to be of **mixed type.** For example, the Tricomi equation

$$y\frac{\partial^2 u}{\partial x^2} + \frac{\partial^2 u}{\partial y^2} = 0$$

is elliptic for $y > 0$ and hyperbolic for $y < 0$.

☐ EXAMPLE 7

Reduce the equation

$$\frac{\partial^2 u}{\partial x^2} - 2\sin x\frac{\partial^2 u}{\partial x\partial y} - \cos^2 x\frac{\partial^2 u}{\partial y^2} - \cos x\frac{\partial u}{\partial y} = 0 \qquad \textbf{(1.40)}$$

to canonical form.

■ ***Solution.*** $B^2 - AC = \sin^2 x + \cos^2 x = 1$ and hence (1.40) is of hyperbolic type. The equations for $\phi_1(x,y)$ and $\phi_2(x,y)$ are

$$\frac{\partial\phi}{\partial x} - (\sin x - 1)\frac{\partial\phi}{\partial y} = 0$$

$$\frac{\partial\phi}{\partial x} - (\sin x + 1)\frac{\partial\phi}{\partial y} = 0.$$

To solve these equations consider the characteristic equations

$$\frac{dy}{dx} = -(\sin x - 1)$$

$$\frac{dy}{dx} = -(\sin x + 1)$$

and solve them to obtain $y = x + \cos x + c$ and $y = -x + \cos x + c$, respectively, where c is an arbitrary constant. If we now make the change of variables

$$\xi = \phi_1(x,y) = x - y + \cos x$$

$$\eta = \phi_2(x,y) = x + y - \cos x$$

in (1.40) we arrive at the equation

$$\frac{\partial^2 u}{\partial\xi\partial\eta} = 0. \tag{1.41}$$

We can now actually solve this equation since (1.41) implies that

$$\frac{\partial u}{\partial\eta} = f(\eta) \tag{1.42}$$

where $f(\eta)$ is an arbitrary function of η, and (1.42) implies that

$$u(\xi,\eta) = \phi(\eta) + \psi(\xi),$$

where $\phi(\eta)$ and $\psi(\xi)$ are arbitrary functions of their indicated variables. Hence the general solution of (1.40) is

$$u(x,y) = \phi(x + y - \cos x) + \psi(x - y + \cos x)$$

where we now require that the arbitrary functions $\phi(\eta)$ and $\psi(\xi)$ be twice continuously differentiable so that $u(x,y)$ also has this property. ■

1.4 FOURIER SERIES AND INTEGRALS

For future use, we want to know when a function $f(x)$ of period 2π can be expanded in a **Fourier series,** which has the form

$$f(x) = \sum_{n=-\infty}^{\infty} a_n e^{inx}. \tag{1.43}$$

If we can integrate termwise, it follows from the orthogonality of the set $\{e^{inx}\}$ that

$$a_n = \frac{1}{2\pi}\int_{-\pi}^{\pi} f(x)e^{-inx}dx \tag{1.44}$$

and henceforth we shall assume that the coefficients $\{a_n\}$ in (1.43) are given by (1.44). Our problem is to determine under what conditions the equality (1.43) is valid. More precisely, if we define the partial sums of (1.43) by

$$S_N(x) = \sum_{n=-N}^{N} a_n e^{inx}, \tag{1.45}$$

when does

$$\lim_{N\to\infty} S_N(x) = f(x)?$$

We begin by noting that from (1.44) and (1.45) we have

$$S_N(x) = \frac{1}{2\pi}\int_{-\pi}^{\pi} f(t) \sum_{n=-N}^{N} e^{in(x-t)}dt. \tag{1.46}$$

The function

$$D_N(x) = \frac{1}{2\pi}\sum_{n=-N}^{N} e^{inx}$$

is known as **Dirichlet's kernel.** It is easily seen that

$$D_N(-x) = D_N(x) \tag{1.47}$$

and

$$\int_{-\pi}^{\pi} D_N(x)dx = 1. \tag{1.48}$$

Furthermore, since

$$(1 - e^{ix})\sum_{n=-N}^{N} e^{inx} = \sum_{n=-N}^{N} (e^{iNx} - e^{i(N+1)x}) = e^{-iNx} - e^{i(N+1)x}$$

we have

$$\begin{aligned} D_N(x) &= \frac{1}{2\pi}\,\frac{e^{-iNx} - e^{i(N+1)x}}{1 - e^{ix}} = \frac{1}{2\pi}\,\frac{e^{-i(N+(1/2))x} - e^{i(N+(1/2))x}}{e^{-i(x/2)} - e^{i(x/2)}} \\ &= \frac{1}{2\pi}\,\frac{\sin\left(N + \dfrac{1}{2}\right)x}{\sin\dfrac{x}{2}}. \end{aligned} \tag{1.49}$$

DEFINITION 1

A function $f(x)$ is **piecewise continuous** on $[a,b]$ if there exist finitely many points a_j, $a = a_0 < a_1 < \ldots < a_n = b$ such that $f(x)$ is continuous on the open intervals (a_j,a_{j+1}) and the one-sided limits $f(a_j+)$, $0 \leq j < n$, $f(a_j-)$, $0 < j \leq n$, exist. If in addition $f'(x) = df(x)/dx$ is piecewise continuous, we say that $f(x)$ is **piecewise differentiable.**

We shall show that if $f(x)$ is piecewise differentiable, then at points of continuity the Fourier series of $f(x)$ converges to $f(x)$, and at points of discontinuity the Fourier series converges to a value equal to the average of the left- and right-hand limits. In order to prove this theorem we need the following lemma known as the Riemann-Lebesgue lemma.

Lemma (Riemann-Lebesgue): If $f(x)$ is piecewise differentiable on $a \leq x \leq b$, then

$$\lim_{A\to\infty} \int_a^b f(x) \sin Ax\, dx = 0.$$

■ ***Proof.*** Let (a_j,a_{j+1}) $(j = 0,1,\ldots,n)$ be the intervals in which $f(x)$ has a continuous derivative. Then

$$\begin{aligned}\int_a^b f(x) \sin Ax\, dx &= \sum_{j=0}^{n} \int_{a_j}^{a_{j+1}} f(x) \sin Ax\, dx \\ &= \frac{1}{A} \sum_{j=0}^{n} [f(a_j+) \cos Aa_j - f(a_{j+1}-) \cos Aa_{j+1}] \\ &\quad + \frac{1}{A} \sum_{j=0}^{n} \int_{a_j}^{a_{j+1}} f'(x) \cos Ax\, dx\end{aligned}$$

which tends to zero as A tends to infinity. ■

We can now prove Fourier's theorem.

THEOREM 1 Fourier's Theorem

Let $f(x)$ be periodic of period 2π and piecewise differentiable. Then at each point x,

$$\frac{f(x+) + f(x-)}{2} = \sum_{n=-\infty}^{\infty} a_n e^{inx}$$

where the coefficients $\{a_n\}$ are the Fourier coefficients given by (1.44).

■ ***Proof.*** From (1.47) and (1.48) we have

$$f(x\pm) = 2 \int_0^{\pi} D_N(t)\, f(x\pm)\, dt$$

so from (1.46) we see that

$$S_N(x) - \frac{1}{2}[f(x+) + f(x-)] = \int_0^{\pi} D_N(t)[f(x+t) - f(x+)]\,dt + \int_0^{\pi} D_N(t)[f(x-t) - f(x-)]\,dt, \quad \textbf{(1.50)}$$

where we have made a change of variables in (1.46) and used the periodicity of $f(x)$. We shall only show that the first integral on the right side of (1.50) goes to zero; the same proof will also apply to the second integral.

Let δ_0 be such that $\delta_0 < \pi$ and $f'(t)$ is continuous for $x < t \leq x + \delta_0$. Then for $0 < \delta \leq \delta_0$ we write

$$\int_0^{\pi} D_N(t)[f(x+t) - f(x+)]\,dt = I_1(\delta,N) + I_2(\delta,N), \quad \textbf{(1.51)}$$

where

$$I_1(\delta,N) = \int_0^{\delta} D_N(t)[f(x+t) - f(x+)]\,dt$$

$$I_2(\delta,N) = \int_{\delta}^{\pi} D_N(t)[f(x+t) - f(x+)]\,dt.$$

Let $M > 0$ be such that $|f'(t)| \leq M$ for $x < t \leq x + \delta_0$. Then by the mean value theorem

$$|f(x+t) - f(x+)| \leq Mt$$

for $0 < t \leq \delta_0$, and since

$$t \leq \pi \sin\frac{t}{2}$$

for $0 \leq t \leq \pi$, we have from (1.49) that

$$|I_1(\delta,N)| \leq \int_0^{\delta} Mt|D_N(t)|\,dt \leq \frac{M}{2\pi}\int_0^{\delta} \frac{t}{\sin\frac{t}{2}}\,dt \leq \frac{1}{2}M\delta.$$

Hence, for every $\epsilon > 0$ and $\delta \leq 2\epsilon/M$ we have that $|I_1(\delta,N)| < \epsilon$. Since ϵ is arbitrary, by (1.51) the proof will be complete if we can show that $I_2(\delta,N)$ tends to zero as N tends to infinity. But this follows from (1.49), the definition of $I_2(\delta,N)$, and the Riemann-Lebesgue lemma. The theorem is now proved. ■

Let us briefly discuss some trivial modifications of the Fourier series. For a function $f(x)$ defined on $(0, L)$, the **Fourier cosine series** on $(0, L)$ is

$$\begin{aligned} &\frac{1}{2}A_0 + \sum_{n=1}^{\infty} A_n \cos\frac{n\pi x}{L} \\ &A_n = \frac{2}{L}\int_0^L f(x)\cos\frac{n\pi x}{L}\,dx, \end{aligned} \tag{1.52}$$

and the **Fourier sine series** on $(0, L)$ is

$$\begin{aligned} &\sum_{n=1}^{\infty} B_n \sin\frac{n\pi x}{L} \\ &B_n = \frac{2}{L}\int_0^L f(x)\sin\frac{n\pi x}{L}\,dx. \end{aligned} \tag{1.53}$$

Furthermore, if $f(x)$ has period $2L$, then

$$g(x) = f\left(\frac{xL}{\pi}\right)$$

has period 2π and hence the Fourier series for $g(x)$ on $(-\pi,\pi)$ yields the Fourier series for $f(x)$ on $(-L, L)$ defined by

$$\begin{aligned} &\sum_{n=-\infty}^{\infty} a_n e^{in(\pi(x/L))} \\ &a_n = \frac{1}{2L}\int_{-L}^{L} f(x)e^{-in(\pi(x/L))}\,dx. \end{aligned} \tag{1.54}$$

By Fourier's theorem, (1.54) converges to $1/2[f(x+) + f(x-)]$ on $(-L, L)$ if f is piecewise differentiable on $(-L, L)$. If f is an even function on $(-L, L)$, i.e., $f(x) = f(-x)$, then from (1.54) we can easily deduce that $a_n = a_{-n}$ and hence

$$\sum_{n=-N}^{N} a_n e^{in(\pi(x/L))} = a_0 + \sum_{n=0}^{N} 2a_n \cos\frac{n\pi x}{L}$$

where we have used the identity $e^{ix} = \cos x + i \sin x$. Since the partial sums on the left side are convergent, so are the partial sums on the right side, i.e., the Fourier cosine series (where $A_n = 2a_n$) converges to $1/2[f(x+) + f(x-)]$. Similarly, if $f(x)$ is an odd function, i.e., $f(x) = -f(-x)$, then the Fourier sine series (where $B_n = 2ia_n$) converges to $1/2[f(x+) + f(x-)]$.

Finally, if $f(x)$ is defined only on the interval $(0, L)$, we can extend $f(x)$ to $(-L, L)$ by making $f(x)$ even and then over the whole line by making it periodic. For such $f(x)$, if piecewise differentiable, the Fourier cosine series converges to $(1/2)[f(x+) + f(x-)]$ on $(0, L)$. Similarly, if $f(x)$ is extended to $(-L, L)$ by making $f(x)$ odd and then over the whole line by making it periodic, then if $f(x)$ is piecewise differentiable the Fourier sine series converges to $(1/2)[f(x+) + f(x-)]$ on $(0, L)$.

In summary, the Fourier cosine or sine series of a piecewise differentiable function $f(x)$ converges to $(1/2)[f(x+) + f(x-)]$ on $(0, L)$ (where at $x = 0$ and $x = L$, $f(x-)$ and $f(x+)$ are defined, respectively, by the periodic extension of $f(x)$).

In many cases, the pointwise convergence result of Fourier's theorem is more than is needed. The following theorem deals with **mean-square convergence** of the Fourier series (1.43) to the function $f(x)$ under the assumption that $f(x)$ is merely continuous. (In fact, it suffices to assume only that $f(x)$ is square integrable, but we shall not prove this here.)

THEOREM 2

Let $f(x)$ be a continuous function defined on $[-\pi,\pi]$, $\{a_n\}$ be the Fourier coefficients of $f(x)$, and $S_N(x)$ be defined by (1.45). Then

(1) $\displaystyle \lim_{N\to\infty} \int_{-\pi}^{\pi} |S_N(x) - f(x)|^2\, dx = 0.$

(2) If $T_N(x) = \sum_{n=-N}^{N} b_n e^{inx}$ with arbitrary coefficients $\{b_n\}$, then

$$\int_{-\pi}^{\pi} |T_N(x) - f(x)|^2\, dx > \int_{-\pi}^{\pi} |S_N(x) - f(x)|^2\, dx$$

unless $b_n = a_n$ for $-N \le n \le N$.

(3) $\displaystyle \sum_{n=-\infty}^{\infty} |a_n|^2 = \frac{1}{2\pi}\int_{-\pi}^{\pi} |f(x)|^2\, dx$ **(Parseval's equality).**

■ ***Proof.*** We first prove part 2. We have

$$\int_{-\pi}^{\pi} |T_N(x) - f(x)|^2\, dx = 2\pi \sum_{n=-N}^{N} (|b_n|^2 - b_n\bar{a}_n - \bar{b}_n a_n) + \int_{-\pi}^{\pi} |f(x)|^2\, dx.$$

Setting $b_n = a_n$ in the above gives

$$\int_{-\pi}^{\pi} |S_N(x) - f(x)|^2\, dx = \int_{-\pi}^{\pi} |f(x)|^2\, dx - 2\pi \sum_{n=-N}^{N} |a_n|^2 \tag{1.55}$$

and hence

$$\int_{-\pi}^{\pi} |T_N(x) - f(x)|^2\, dx - \int_{-\pi}^{\pi} |S_N(x) - f(x)|^2\, dx = 2\pi \sum_{n=-N}^{N} |b_n - a_n|^2,$$

which is positive if $b_n \ne a_n$ for some n.

To prove part 1 we first use the Weierstrass approximation theorem (whose proof we defer until Chapter 3) to approximate $f(x)$ by a polynomial on $[-\pi,\pi]$. After possibly redefining this polynomial at an endpoint of $[-\pi,\pi]$, we can extend it to a piecewise differentiable function of period 2π on the whole x-axis. By Fourier's theorem we can approximate this function on $(-\pi,\pi)$ by a trigonometric polynomial

$$P(x) = \sum_{n=-M}^{M} b_n e^{inx}.$$

Then from part 2 of the theorem (proved above) we see that for every $\epsilon > 0$ there exists an integer N $(= M)$ such that

$$\int_{-\pi}^{\pi} |S_N(x) - f(x)|^2 dx \leq \int_{-\pi}^{\pi} |P(x) - f(x)|^2 dx \leq \epsilon.$$

Since ϵ is arbitrary, this proves part 1.

To prove part 3, we simply note that from (1.55) and part 1 we have

$$\frac{1}{2\pi}\int_{-\pi}^{\pi} |f(x)|^2 dx = \sum_{n=-\infty}^{\infty} |a_n|^2 \tag{1.56}$$

by letting N tend to infinity in (1.55). ■

We conclude our brief study of Fourier series by introducing a few simple concepts and results on inner products, which will be used in Chapter 5 for more general orthogonal series than these Fourier series.

DEFINITION 2

Let $f(x)$ and $g(x)$ be square integrable complex valued functions defined on $[a,b]$. Then we define the **inner product** by

$$(f,g) = \int_a^b f(x)\overline{g(x)}\,dx$$

and the **norm** by

$$\|f\| = \sqrt{(f,f)}$$

where $\overline{g(x)}$ denotes the complex conjugate of $g(x)$. The functions $f(x)$ and $g(x)$ are said to be **orthogonal** if $(f,g) = 0$.

It is easily verified that $(f,g) = \overline{(g,f)}$, and if c_1 and c_2 are constants, then

$$(c_1 f_1 + c_2 f_2, g) = c_1(f_1,g) + c_2(f_2,g).$$

THEOREM 3 Schwarz's Inequality

$$|(f,g)| \leq \|f\|\,\|g\|$$

■ ***Proof.*** Without loss of generality assume g is not zero. Then the inequality is equivalent to $|(f,g/\|g\|)| \leq \|f\|$. Since $\|g/\|g\|\| = 1$ we can restrict our attention to showing that $|(f,h)| \leq \|f\|$ for $\|h\| = 1$. But this follows from the inequalities

$$\begin{aligned} 0 \leq \|f - (f,h)h\|^2 &= (f - (f,h)h, f - (f,h)h) \\ &= (f,f) - 2(f,h)\overline{(f,h)} + (f,h)\overline{(f,h)}(h,h) \\ &= \|f\|^2 - |(f,h)|^2. \end{aligned}$$

■

THEOREM 4 Triangle Inequality

$$\|f + g\| \leq \|f\| + \|g\|$$

■ ***Proof.*** By Schwarz's inequality,

$$\begin{aligned} \|f + g\|^2 = (f + g, f + g) &= \|f\|^2 + (f,g) + (g,f) + \|g\|^2 \\ &\leq \|f\|^2 + 2\|f\|\,\|g\| + \|g\|^2 \\ &= (\|f\| + \|g\|)^2 \end{aligned}$$

and the theorem follows. ■

Recalling the definition of the inner product and norm, we see that Schwarz's inequality and the triangle inequality can be explicitly written as

$$\left|\int_a^b f(x)\overline{g(x)}\,dx\right|^2 \leq \int_a^b |f(x)|^2\,dx \int_a^b |g(x)|^2\,dx$$

and

$$\int_a^b |f(x) + g(x)|^2\,dx \leq \left(\left[\int_a^b |f(x)|^2\,dx\right]^{1/2} + \left[\int_a^b |g(x)|^2\,dx\right]^{1/2}\right)^2,$$

respectively. In particular, let $a = -\pi$, $b = \pi$, and let $\{a_n\}$, $\{b_n\}$ be square summable sequences. Then if we define

$$f(\theta) = \sum_{n=-N}^{N} a_n e^{in\theta}$$

$$g(\theta) = \sum_{n=-N}^{N} b_n e^{in\theta}$$

and let N tend to infinity, the above inequalities take the form

$$\left|\sum_{n=-\infty}^{\infty} a_n\overline{b}_n\right|^2 \leq \sum_{n=-\infty}^{\infty} |a_n|^2 \sum_{n=-\infty}^{\infty} |b_n|^2$$

and

$$\sum_{n=-\infty}^{\infty} |a_n + b_n|^2 \leq \left(\left[\sum_{n=-\infty}^{\infty} |a_n|^2\right]^{1/2} + \left[\sum_{n=-\infty}^{\infty} |b_n|^2\right]^{1/2}\right)^2,$$

respectively.

We conclude this section by proving the Fourier integral theorem, which can be viewed as a continuous analogue of the Fourier series representation of a function—the infinite series representation is replaced by an integral representation. In order to do this we need two lemmas.

Lemma 1: $\displaystyle\int_0^\infty \frac{\sin x}{x}\,dx = \frac{\pi}{2}.$

■ ***Proof.*** We delay the proof of this lemma until the following section on analytic function theory. ■

Lemma 2: Let $f(x)$ be piecewise differentiable for all x and let the integral

$$\int_{-\infty}^{\infty} |f(x)|\,dx$$

exist. Then for ξ such that $f'(\xi) = df(\xi)/d\xi$ exists we have

$$\lim_{A\to\infty} \int_{-\infty}^{\infty} \frac{f(x) - f(\xi)}{x - \xi} \sin A(x - \xi)\,dx = 0.$$

■ ***Proof.*** We first show that the integral exists. Assume x_1 and x_2 such that $-\infty < x_1 < \xi < x_2 < \infty$. Since $f'(\xi)$ exists, the limit comparison test implies the existence of the integral

$$\int_{x_1}^{x_2} \frac{f(x) - f(\xi)}{x - \xi} \sin A(x - \xi)\,dx.$$

We know that the integral

$$\int_{x_2}^{\infty} \frac{f(x)}{x - \xi} \sin A(x - \xi)\,dx$$

exists by the comparison test, because the integrand is dominated by $|f(x)|/(x_2 - \xi)$. Finally, we have

$$\int_{x_2}^{\infty} \frac{f(\xi)}{x - \xi} \sin A(x - \xi)\,dx = f(\xi) \int_{A(x_2-\xi)}^{\infty} \frac{\sin t}{t}\,dt$$

which exists by Lemma 1. Hence the integral

$$\int_{x_2}^{\infty} \frac{f(x) - f(\xi)}{x - \xi} \sin A(x - \xi)\,dx$$

exists, and the same analysis shows that

$$\int_{-\infty}^{x_1} \frac{f(x) - f(\xi)}{x - \xi} \sin A(x - \xi)\,dx$$

also exists. Thus the integral appearing in the lemma is convergent.

We now consider the limit of this integral as A tends to infinity. Let $\epsilon > 0$ and restrict x_1 and x_2 such that $x_1 < \xi - 1$, $x_2 > \xi + 1$, and

$$\int_{-\infty}^{x_1} |f(x)|\, dx < \epsilon$$

$$\int_{x_2}^{\infty} |f(x)|\, dx < \epsilon.$$

We write

$$\int_{-\infty}^{\infty} \frac{f(x) - f(\xi)}{x - \xi} \sin A(x - \xi)\, dx = I_1 + I_2 + I_3 + I_4 + I_5$$

where

$$I_1 = \int_{-\infty}^{x_1} \frac{f(x)}{x - \xi} \sin A(x - \xi)\, dx$$

$$I_2 = -\int_{-\infty}^{x_1} \frac{f(\xi)}{x - \xi} \sin A(x - \xi)\, dx$$

$$I_3 = \int_{x_1}^{x_2} \frac{f(x) - f(\xi)}{x - \xi} \sin A(x - \xi)\, dx$$

$$I_4 = \int_{x_2}^{\infty} \frac{f(x)}{x - \xi} \sin A(x - \xi)\, dx$$

$$I_5 = -\int_{x_2}^{\infty} \frac{f(\xi)}{x - \xi} \sin A(x - \xi)\, dx.$$

Then $|I_1| < \epsilon$ and $|I_4| < \epsilon$ regardless of the value of A. By making the substitution $t = A(x - \xi)$ we see that for A sufficiently large $|I_2| < \epsilon$ and $|I_5| < \epsilon$. Now consider the integral I_3. Choose δ such that $x_1 < \xi - \delta < \xi + \delta < x_2$ and

$$\int_{\xi-\delta}^{\xi+\delta} \frac{f(x) - f(\xi)}{x - \xi} \sin A(x - \xi)\, dx < \frac{\epsilon}{2}$$

which can be done independently of A since $|(f(x) - f(\xi))/(x - \xi)|$ is absolutely integrable on $[x_1,x_2]$. Applying the Riemann-Lebesgue lemma to

$$\int_{x_1}^{\xi-\delta} \frac{f(x) - f(\xi)}{x - \xi} \sin A(x - \xi)\, dx$$

and

$$\int_{\xi+\delta}^{x_2} \frac{f(x) - f(\xi)}{x - \xi} \sin A(x - \xi)\, dx,$$

we find that for A sufficiently large, $|I_3| < \epsilon$. Hence for A sufficiently large,

$$|I_1 + I_2 + I_3 + I_4 + I_5| \leq |I_1| + |I_2| + |I_3| + |I_4| + |I_5| < 5\epsilon$$

and since ϵ can be arbitrarily small, this establishes the lemma. ■

We can now prove Fourier's integral theorem.

THEOREM 5 Fourier's Integral Theorem

Let $f(x)$ and ξ be the same as in Lemma 2. Then the integral

$$g(\lambda) = \int_{-\infty}^{\infty} f(x)e^{-i\lambda x}dx$$

exists, is a continuous function of λ, and

$$f(\xi) = \frac{1}{2\pi} \lim_{A\to\infty} \int_{-A}^{A} g(\lambda)e^{i\lambda\xi}\, d\lambda.$$

■ ***Proof.*** The hypotheses regarding $f(x)$ imply that the integral that defines $g(\lambda)$ is uniformly convergent, and hence $g(\lambda)$ is continuous for $-\infty < \lambda < \infty$. We now have

$$\begin{aligned}\frac{1}{2\pi}\int_{-A}^{A} g(\lambda)e^{i\lambda\xi}\, d\lambda &= \frac{1}{2\pi}\int_{-A}^{A}\left[\int_{-\infty}^{\infty} f(x)e^{-i\lambda x}dx\right] e^{i\lambda\xi}\, d\lambda \\ &= \frac{1}{2\pi}\int_{-\infty}^{\infty} f(x)\left[\int_{-A}^{A} e^{-i\lambda(x-\xi)}\, d\lambda\right] dx \qquad \textbf{(1.57)} \\ &= \frac{1}{\pi}\int_{-\infty}^{\infty} f(x)\,\frac{\sin A(x-\xi)}{x-\xi}\, dx\end{aligned}$$

where the interchange of orders of integration is justified by the integrability of $|f(x)|$. From Lemma 1 we have

$$f(\xi) = \frac{1}{\pi}\int_{-\infty}^{\infty} f(\xi)\,\frac{\sin A(x-\xi)}{x-\xi}\, d\xi\ ;$$

subtracting this from (1.57) gives

$$\frac{1}{2\pi}\int_{-A}^{A} g(\lambda)e^{i\lambda\xi}\, d\lambda - f(\xi) = \frac{1}{\pi}\int_{-\infty}^{\infty} \frac{f(x) - f(\xi)}{x-\xi}\sin A(x-\xi)\, dx. \qquad \textbf{(1.58)}$$

Applying Lemma 2 to (1.58) now proves the theorem. ■

For more information on Fourier series and Fourier integrals we refer the reader to Tolstov and Sneddon. In this text, we shall make use of Fourier series and Fourier integrals in a variety of places, particularly in Chapters 2 and 3.

* 1.5 ANALYTIC FUNCTIONS

We now present the few elements of analytic function theory that are needed for the starred sections of this book and give a simple but important application to partial differential equations. Our presentation is based on a knowledge of power series and line integrals from advanced calculus; loosely speaking, our aim is to present a clever method for integrating power series along closed curves in the complex plane. We make no attempt to give a complete treatment of analytic function theory, which would be the topic of a course in itself. Instead, we shall take the shortest possible route to the two key theorems we shall need in this book: the **identity theorem** and the **residue theorem.** Our aim is not to establish expertise in analytic function theory, but only to provide enough information to understand the proofs in the starred sections of this book.

* 1.51 Power Series and Analytic Functions

We consider power series (or Taylor series) of the form

$$\sum_{n=0}^{\infty} a_n(z - z_0)^n \tag{1.59}$$

where $\{a_n\}$ and z_0 are real or complex numbers and z is a complex variable. From advanced calculus, we know that there exists a number R (which may be infinite) such that the series (1.59) converges for $|z - z_0| < R$ and diverges for $|z - z_0| > R$ where

$$\frac{1}{R} = \overline{\lim}\, |a_n|^{1/n}.$$

The number R is called the **radius of convergence** and the circle $|z - z_0| = R$ is called the **circle of convergence.** Assuming that $R > 0$, it is an exercise in calculus to establish these facts:

(1) The series (1.59) converges absolutely on $|z - z_0| < R$ and uniformly on $|z - z_0| \leq r$ where $r < R$.

(2) The series (1.59) can be differentiated term by term with the resulting series having radius of convergence R. In particular it follows that if $f(z)$ is defined by

$$f(z) = \sum_{n=0}^{\infty} a_n(z - z_0)^n \qquad (|z - z_0| < R)$$

then

$$a_n = \frac{1}{n!} f^{(n)}(z_0) \qquad (n = 0,1,2,\ldots)$$

where

$$f^{(n)}(z) = \frac{d^n f(z)}{dz^n}.$$

In what follows Ω will always denote an open set in the complex z plane.

DEFINITION 3

A function $f(z)$ defined on Ω is called **analytic** (or **holomorphic**, or **regular**) if for every point z_0 in Ω there is a neighborhood of z_0 on which $f(z)$ can be expanded in a power series:

$$f(z) = \sum_{n=0}^{\infty} \frac{1}{n!} f^{(n)}(z_0)(z - z_0)^n \qquad (|z - z_0| < R).$$

A function that is analytic in the entire complex plane is called an **entire function.**

□ EXAMPLES 8–9

(8) Polynomials, i.e. $f(z) = a_n z^n + a_{n-1} z^{n-1} + \ldots + a_0$ are entire functions.

(9) $f(z) = 1/(\zeta - z)$, with ζ fixed, is analytic on the set Ω obtained by deleting the point ζ from the complex plane. To see this, let z_0 not be equal to ζ. Then for z not equal to ζ we have

$$\begin{aligned}\frac{1}{\zeta - z} = \frac{1}{\zeta - z_0 - (z - z_0)} &= \frac{1}{\zeta - z_0} \frac{1}{1 - \dfrac{z - z_0}{\zeta - z_0}} \\ &= \frac{1}{\zeta - z_0} \sum_{n=0}^{\infty} \left(\frac{z - z_0}{\zeta - z_0}\right)^n \\ &= \sum_{n=0}^{\infty} \frac{1}{(\zeta - z_0)^{n+1}} (z - z_0)^n \qquad (|z - z_0| < |\zeta - z_0|).\end{aligned}$$

Note that analytic functions are infinitely differentiable and the derivatives are also analytic functions. □

DEFINITION 4

An open connected set is called a **domain.**

THEOREM 6 The Identity Theorem

Let $f_1(z)$ and $f_2(z)$ be analytic functions in a domain Ω such that $f_1(z) = f_2(z)$ on a set of points with an accumulation point in Ω. Then $f_1(z) = f_2(z)$ throughout Ω.

■ ***Proof.*** Set $f(z) = f_1(z) - f_2(z)$. Our problem is to show that if $f(z)$ is analytic in Ω and vanishes on a sequence of points $\{z_k\}_{k=1}^{\infty}$ such that as k tends to infinity z_k

tends to $z_0 \in \Omega$, then $f(z)$ is identically zero in Ω. Since $f(z)$ is analytic, we have

$$f(z) = \sum_{n=0}^{\infty} c_n(z - z_0)^n \qquad (|z - z_0| < r)$$

for r sufficiently small. We shall show that all of the coefficients c_n are zero. Indeed, if this were not true there would be a first nonzero coefficient c_p. Then for $|z - z_0| < r$, define $g(z)$ by

$$g(z) = \sum_{n=p}^{\infty} c_n(z - z_0)^{n-p}$$
$$= \begin{cases} \dfrac{f(z)}{(z - z_0)^p} & \text{when} \quad z \neq z_0 \\ c_p & \text{when} \quad z = z_0. \end{cases}$$

Then $g(z_k) = 0$ $(k = 1,2,3,\ldots)$, where z_k tends to z_0 as k tends to infinity, so

$$c_p = g(z_0) = \lim_{k\to\infty} g(z_k) = 0,$$

which is a contradiction. Hence all of the c_n are zero. We now know that $f(z) = 0$ for $|z - z_0| < r$.

Let Ω_1 be the subset of Ω consisting of all points ζ such that $f(z) = 0$ on some disk with center at ζ. Then Ω_1 is clearly nonempty and open. However, the set $\Omega' = \Omega \backslash \Omega_1$ is also open since if $z' \in \Omega'$ then z' cannot be an accumulation point for Ω_1 (if it were, then z' would be the limit of a sequence of points on which $f(z)$ vanishes and hence $f(z)$ would vanish on a disk centered at z'). But the complement of an open subset of a connected set cannot be open unless it is the empty set, so Ω' is the empty set. Hence $\Omega_1 = \Omega$, and the theorem is proved. ■

The identity theorem enables us analytically to continue functions defined by power series for real values of x in a unique way into the complex plane. For example, for real x we have the well-known expansion

$$e^x = \sum_{n=0}^{\infty} \frac{1}{n!} x^n \qquad (|x| < \infty),$$

and hence we can uniquely define the analytic function e^z for complex z by

$$e^z = \sum_{n=0}^{\infty} \frac{1}{n!} z^n \qquad (|z| < \infty), \tag{1.60}$$

i.e., e^z, as defined by (1.60), is the unique analytic function that agrees with e^x for real z. Similarly,

$$e^z = e^{z_0} e^{z - z_0} \tag{1.61}$$

for all complex z and z_0, because the identity is valid for real z and z_0, and both sides of (1.61) are analytic functions of z and z_0 in the entire complex plane. In the same fashion we can define the analytic functions cos z and sin z by

$$\cos z = \sum_{n=0}^{\infty} \frac{(-1)^n}{(2n)!} z^{2n} \qquad (|z| < \infty)$$
$$\sin z = \sum_{n=0}^{\infty} \frac{(-1)^n}{(2n+1)!} z^{2n+1} \qquad (|z| < \infty), \tag{1.62}$$

and from (1.60) and (1.62) we see that

$$e^{iz} = \cos z + i \sin z.$$

In particular, if (r,θ) are polar coordinates then the polar form of the complex number $z = x + iy$ is

$$z = r\cos\theta + ir\sin\theta = re^{i\theta}.$$

An analytic function is often defined simply as a function which is differentiable everywhere in Ω, i.e., for $z_0 \in \Omega$

$$\lim_{z\to z_0} \frac{f(z) - f(z_0)}{z - z_0} = f'(z_0) \tag{1.63}$$

exists *independently of how* z *approaches* z_0. We shall soon see that every differentiable function on an open set in the complex plane can be represented locally by a power series, so this definition is consistent with our previous definition. In (1.63) let $z = z_0 + h$ where h tends to zero through real values, i.e., we approach z_0 along the line y = constant. Then setting $z_0 = x_0 + iy_0$ we have

$$f'(z_0) = \lim_{h\to 0} \frac{f(x_0 + h + iy_0) - f(x_0 + iy_0)}{h} = \frac{\partial f}{\partial x}. \tag{1.64}$$

Similarly, setting $z = z_0 + ik$ with k real gives

$$f'(z_0) = \lim_{k\to 0} \frac{f(x_0 + i(y_0 + k)) - f(x_0 + iy_0)}{ik} = \frac{1}{i}\frac{\partial f}{\partial y}. \tag{1.65}$$

Writing $f(z) = u(x,y) + iv(x,y)$ (i.e., $u(x,y) = \text{Re } f(z)$, $v(x,y) = \text{Im } f(z)$), we see that equations (1.64) and (1.65) imply that the real and imaginary parts of a differentiable (i.e., analytic) function of a complex variable z must satisfy the **Cauchy-Riemann equations**

$$\frac{\partial u}{\partial x} = \frac{\partial v}{\partial y}$$
$$\frac{\partial u}{\partial y} = -\frac{\partial v}{\partial x}. \tag{1.66}$$

Conversely, we can easily establish that if $u(x,y)$ and $v(x,y)$ are continuously differentiable (as functions of the real variables x and y) and satisfy the Cauchy-

Riemann equations, then $f(z) = u(x,y) + iv(x,y)$ is differentiable (as a function of the complex variable z) and hence analytic. By the chain rule we can also write the Cauchy-Riemann equations in the complex form

$$\frac{\partial f}{\partial \bar{z}} = 0,$$

that is,

$$\frac{\partial f}{\partial x} + i\frac{\partial f}{\partial y} = 0 \qquad \textbf{(1.67)}$$

where $z = x + iy$, $\bar{z} = x - iy$. Hence, from the point of view of partial differential equations, an analytic function is simply a complex valued solution of the first order linear partial differential equation with complex coefficients (1.67), or, equivalently, a real valued solution of the first order system (1.66).

Since analytic functions are infinitely differentiable, (1.66) implies that the real and imaginary parts of an analytic function are **harmonic,** i.e., if $f(z) = u(x,y) + iv(x,y)$, $z = x + iy$, then

$$\Delta_2 u = \Delta_2 v = 0$$

where $\Delta_2 u = \partial^2 u/\partial x^2 + \partial^2 u/\partial y^2$. From the definition (1.63) it follows that all the usual rules for differentiable functions on the real axis are also valid for differentiable functions in the complex domain, so $(f(z) + g(z))' = f'(z) + g'(z)$, $(f(z)g(z))' = f'(z)g(z) + g'(z)f(z)$, $(d/dz)(f(g(z)) = f'(g(z))g'(z)$, etc., where the prime denotes differentiation with respect to z.

* 1.5.2 Integration of Analytic Functions

Let C be an oriented piecewise smooth curve in the complex plane and $f(z)$ a function defined and continuous on C. Then if we parameterize C by

$$z = g(t) = g_1(t) + ig_2(t) \qquad (a \leq t \leq b)$$

we can define the integral of $f(z)$ along C by

$$\int_C f(z)dz = \int_a^b f[g(t)]g'(t)dt.$$

It can be shown that the value of the integral is independent of the particular parameterization used and that

$$\int_{-C} f(z)dz = -\int_C f(z)dz$$

where $-C$ denotes the curve C with the orientation reversed,

$$\int_{C_1+C_2} f(z)dz = \int_{C_1} f(z)dz + \int_{C_2} f(z)dz$$

and

$$\left|\int_C f(z)dz\right| \leq \max_{z\in C} |f(z)|\, L(C)$$

where $L(C)$ denotes the length of C.

☐ EXAMPLE 10

Let C be the positively oriented circle with center z_0 and radius r. Then C can be parameterized by

$$z = z_0 + re^{it} \qquad (0 \leq t \leq 2\pi)$$

and for every integer n not equal to -1 we have

$$\begin{aligned}\int_C (z - z_0)^n dz &= \int_0^{2\pi} r^n e^{\text{int}} ire^{it} dt \\ &= \left[r^{n+1} \frac{e^{(n+1)it}}{n+1} \right]_0^{2\pi} \\ &= 0.\end{aligned}$$

On the other hand, if $n = -1$ we have

$$\int_C \frac{1}{z - z_0} dz = \int_0^{2\pi} i\, dt = 2\pi i.$$

☐

The following theorem is the most important theorem in analytic function theory.

THEOREM 7 Cauchy's Theorem

Let Ω be an open set in the complex plane and let $f(z)$ be a continuously differentiable function of the complex variable z for z in Ω. Let C be any simple, closed, piecewise smooth curve in Ω whose interior belongs entirely to Ω. Then

$$\int_C f(z)dz = 0.$$

■ ***Proof.*** The Cauchy-Riemann equations show that

$$\frac{\partial f}{\partial x} = -i \frac{\partial f}{\partial y}$$

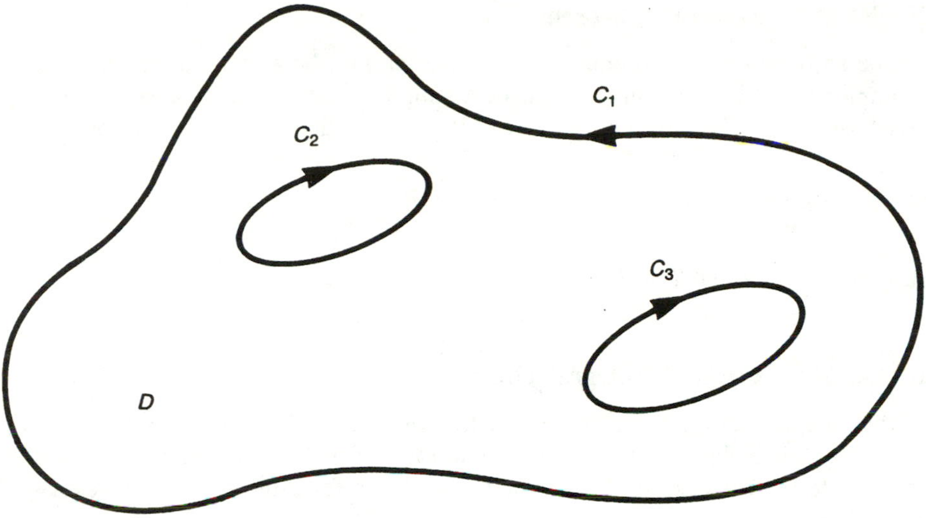

Figure 1.1

and hence by Green's theorem

$$\int_C f(z)dz = \int_C f(z)dx + if(z)dy$$

$$= \iint_D \left(i\frac{\partial f}{\partial x} - \frac{\partial f}{\partial y} \right) dxdy = 0$$

where D denotes the interior of C. ■

A more careful analysis (cf. Titchmarsh) shows that Cauchy's theorem is valid if $f(z)$ is assumed only to be differentiable in Ω; we don't need to assume that $f(z)$ is *continuously* differentiable in Ω. Henceforth we shall always assume this weaker hypothesis when we refer to Cauchy's theorem.

The above proof of Cauchy's theorem can be used to show that if a domain D is bounded by a finite number of nonintersecting, simple, closed, piecewise smooth curves $C_1,\ldots,C_k$ of finite length oriented such that D is to the left of all of them and $f(z)$ is continuous in $\overline{D}$ and differentiable in D, then

$$\int_{C_1} f(z)dz + \ldots + \int_{C_k} f(z)dz = 0.$$

The following theorem, which we state without proof (see Exercise 31), can be viewed as a converse to Cauchy's theorem.

THEOREM 8 Morera's Theorem

Let Ω be an open set in the complex plane and let $f(z)$ be a continuous function of the complex variable z for z in Ω. Let C be a simple, closed, piecewise smooth curve in Ω whose interior belongs entirely to Ω. If for any such curve C we have

$$\int_C f(z)dz = 0$$

then $f(z)$ is differentiable in Ω.

THEOREM 9 Cauchy's Integral Theorem

Let D be the interior of a positively oriented, simple, closed, piecewise smooth curve C in the complex plane. Let $f(z)$ be a function defined and continuous in $\overline{D}$ and differentiable with respect to the complex variable z for z in D. Then for every $z_0 \in D$ we have

$$f(z_0) = \frac{1}{2\pi i}\int_C \frac{f(z)}{z - z_0}\,dz.$$

■ ***Proof.*** Let $D_{z_0} = \{z: |z - z_0| < r\}$ be contained in D and positively orient $\gamma = \partial D_{z_0}$. Then by Cauchy's theorem

$$\int_C \frac{f(z)}{z - z_0}\,dz = \int_\gamma \frac{f(z)}{z - z_0}\,dz,$$

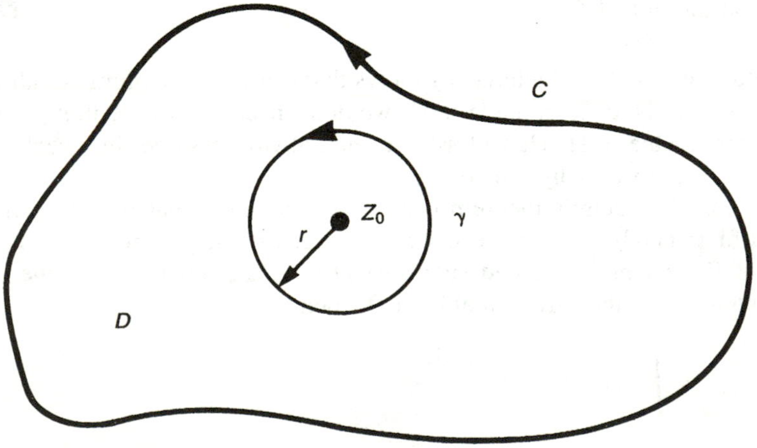

Figure 1.2

i.e., we can deform the contour C onto γ. But

$$\int_\gamma \frac{f(z)}{z - z_0}\, dz = \int_\gamma \frac{f(z_0)}{z - z_0}\, dz + \int_\gamma \frac{f(z) - f(z_0)}{z - z_0}\, dz = 2\pi i f(z_0) + E_\gamma \tag{1.68}$$

by our previous example where

$$E_\gamma = \int_\gamma \frac{f(z) - f(z_0)}{z - z_0}\, dz. \tag{1.69}$$

Since $f(z)$ is differentiable at $z = z_0$, the integrand in (1.69) is bounded by a constant M for r sufficiently small, and hence for such values of r

$$|E_\gamma| \leq M L(\gamma) = 2\pi M r.$$

Letting r tend to zero in (1.68) now establishes the theorem. ■

COROLLARY

The uniform limit of analytic functions on an open set Ω is analytic in Ω.

■ ***Proof.*** Let $z_0 \in \Omega$ and $D_{z_0} \subset \Omega$ as in the above proof. Suppose $f_n(z)$ is uniformly convergent to $f(z)$ in Ω. Let $\gamma = \partial D_{z_0}$, where γ is positively oriented. Then

$$f_n(z_0) = \frac{1}{2\pi i} \int_\gamma \frac{f_n(z)}{z - z_0}\, dz,$$

since $f_n(z)$ is analytic and hence differentiable and letting n tend to infinity gives

$$f(z_0) = \frac{1}{2\pi i} \int_\gamma \frac{f(z)}{z - z_0}\, dz, \tag{1.70}$$

where $f(z)$ is continuous by the uniform convergence of the sequence $\{f_n(z)\}$. Expanding $1/(z - z_0)$ in a power series (see the proof of the holomorphy theorem below) now shows that $f(z)$ is analytic in D_{z_0}, and since z_0 was arbitrary, $f(z)$ is analytic in Ω. ■

Note that if $C_1,\ldots,C_k$ are defined as in the comments following the proof of Cauchy's theorem (see pp. 34–35), for $z_0 \in D$,

$$f(z_0) = \frac{1}{2\pi i} \int_{C_1} \frac{f(z)}{z - z_0}\, dz + \ldots + \frac{1}{2\pi i} \int_{C_k} \frac{f(z)}{z - z_0}\, dz$$

for $f(z)$ continuous in $\overline{D}$ and differentiable in D.

THEOREM 10 Holomorphy Theorem

Let Ω be an open set in the complex plane, and let $f(z)$ be differentiable with respect to the complex variable z for z in Ω. Then $f(z)$ is analytic in Ω.

■ ***Proof.*** Let $z_0 \in \Omega$ and $D_{z_0} = \{z: |z - z_0| < r\}$ be contained in Ω. Then if $\gamma = \partial D_{z_0}$ is positively oriented, we have

$$f(z) = \frac{1}{2\pi i}\int_\gamma \frac{f(\zeta)}{\zeta - z}\,d\zeta \qquad (z \in D_{z_0}). \tag{1.71}$$

From part (a) of this section we have

$$\frac{1}{\zeta - z} = \sum_{n=0}^{\infty} \frac{1}{(\zeta - z_0)^{n+1}}(z - z_0)^n \qquad (|z - z_0| < r), \tag{1.72}$$

where $r < |\zeta - z_0|$. The series (1.72) is uniformly convergent with respect to ζ, and substituting it into (1.71) and interchanging orders of summation and integration now gives

$$f(z) = \sum_{n=0}^{\infty} c_n(z - z_0)^n \qquad (|z - z_0| < r),$$

where

$$c_n = \frac{1}{2\pi i}\int_\gamma \frac{f(\zeta)}{(\zeta - z_0)^{n+1}}\,d\zeta. \tag{1.73}$$

Hence $f(z)$ is analytic in Ω. ■

We note that by Cauchy's theorem we can deform the curve γ in (1.73) to another curve C provided the deformation stays in Ω. Furthermore, from part (a) of this section,

$$c_n = \frac{1}{n!} f^{(n)}(z_0).$$

Hence we derive **Cauchy's formula**

$$f^{(n)}(z_0) = \frac{n!}{2\pi i}\int_C \frac{f(z)}{(z - z_0)^{n+1}}\,dz.$$

This formula can also be obtained by differentiation under the integral sign in Cauchy's integral formula.

THEOREM 11 Maximum Modulus Theorem

Let $f(z)$ be analytic on a bounded, open set Ω and continuous on $\overline{\Omega}$. Then the maximum value of $|f(z)|$ on $\overline{\Omega}$ is not assumed in Ω unless $f(z)$ is identically equal to a constant.

■ ***Proof.*** Since $|f(z)|$ is continuous on $\overline{\Omega}$, it achieves its maximum on this set. Suppose $|f(z)|$ achieves its maximum at an interior point $z_0 \in \Omega$ and let $D_{z_0} = \{z: |z - z_0| < r\}$ be contained in Ω with $\gamma = \partial D_{z_0}$. Then by Cauchy's integral formula,

$$f(z_0) = \frac{1}{2\pi i}\int_\gamma \frac{f(z)}{z - z_0}\,dz. \tag{1.74}$$

Setting $z = z_0 + re^{i\theta}$ in (1.74) gives

$$f(z_0) = \frac{1}{2\pi} \int_0^{2\pi} f(z_0 + re^{i\theta})\, d\theta$$

or

$$|f(z_0)| \leq \frac{1}{2\pi} \int_0^{2\pi} |f(z_0 + re^{i\theta})|\, d\theta.$$

But by assumption $|f(z_0 + re^{i\theta})| \leq |f(z_0)|$, and it follows that

$$|f(z_0)| = \frac{1}{2\pi} \int_0^{2\pi} |f(z_0 + re^{i\theta})|\, d\theta$$

or $|f(z_0)| = |f(z_0 + re^{i\theta})|$ for $0 \leq \theta \leq 2\pi$. Since r can be arbitrarily small we have $|f(z)| = |f(z_0)|$ for all z in D_{z_0}. From the Cauchy-Riemann equations we can now conclude that $f(z)$ is equal to a constant in D_{z_0} and hence, by the identity theorem, in all of Ω. The theorem is now proved. ■

* 1.5.3 Singularities of Analytic Functions and the Residue Theorem

We now consider two types of singularities for analytic functions: branch points and isolated singularities. We are concerned mainly with isolated singularities, but before defining and discussing them, we want to give some brief examples of analytic functions with branch points. A branch point occurs in a multi-valued function. For example, consider the equation

$$e^w = z$$

where w and z ($z \neq 0$) are complex numbers. Any solution w of this equation is denoted by Log z. Writing $w = u + iv$ shows that

$$e^u e^{iv} = Re^{i\theta} \qquad (R \neq 0), \tag{1.75}$$

where $z = Re^{i\theta} = R\cos\theta + iR\sin\theta = x + iy$. Then since $|e^{iv}| = |e^{i\theta}| = 1$ we have $e^u = R$, so since $R = |z|$,

$$u = \log |z|.$$

Equation (1.75) will be satisfied if and only if $v = \theta \pm 2\pi n$, where n is an integer, i.e., v is equal to the argument of z ($v = \arg z$). Hence

$$\operatorname{Log} z = \log |z| + i \arg z. \tag{1.76}$$

It is easily seen that Log z is differentiable for $z \neq 0$ and hence is analytic for z not equal to zero. However, Log z is multi-valued! To make Log z single valued we restrict arg z to satisfy

$$-\pi < \arg z \leq \pi. \tag{1.77}$$

We call the negative real axis a **branch cut** and the endpoints $z = 0$, $z = \infty$ of this branch cut are called **branch points.** The choice of the negative real axis is arbitrary and any other ray emanating from the origin could also be used as a branch cut. Equations (1.76) and (1.77) define a single-valued analytic function which is called the **principal value** of Log z and agrees with log z for z real and positive.

As a second example, consider the function

$$f(z) = z^\alpha$$

where z and α are complex numbers. What does this mean if α is not an integer? To answer this question we define z^α by the formula

$$z^\alpha = e^{\alpha \operatorname{Log} z},$$

where again $-\pi < \arg z \leq \pi$. Then $f(z) = z^\alpha$ is a single-valued analytic function in the complex plane cut along the negative real axis and which agrees with the positive root of z^α for z real and positive.

We have no need to pursue further study of multi-valued analytic functions, but now turn to our primary aim, the analysis of isolated singularities of analytic functions. Let Ω be an open set in the complex plane and Ω' the set obtained by deleting the point $z_0 \in \Omega$. Let $f(z)$ be a function defined (in particular single-valued) and analytic in Ω'. We say that such a function $f(z)$ has an **isolated singularity** at $z = z_0$. In order to understand the behavior of an analytic function near an isolated singularity we need to derive the Laurent series for $f(z)$. To this end we note that for ρ sufficiently small $f(z)$ is analytic for $0 < |z - z_0| < \rho$. Let γ and γ' be positively oriented circles with centers at $z = z_0$ and radii r and r', where $0 < r' < r < \rho$. Then for z in the annulus $r' < |z| < r$, Cauchy's integral formula implies that

$$f(z) = \frac{1}{2\pi i}\int_\gamma \frac{f(\zeta)}{\zeta - z}\, d\zeta - \frac{1}{2\pi i}\int_{\gamma'} \frac{f(\zeta)}{\zeta - z}\, d\zeta. \tag{1.78}$$

In the first integral we write

$$\frac{1}{\zeta - z} = \sum_{n=0}^{\infty} \frac{1}{(\zeta - z_0)^{n+1}}(z - z_0)^n \qquad (|z - z_0| < r)\,, \tag{1.79}$$

and in the second integral

$$\begin{aligned} \frac{1}{\zeta - z} &= -\frac{1}{z - z_0 - (\zeta - z_0)} \\ &= -\sum_{n=1}^{\infty} (\zeta - z_0)^{n-1}(z - z_0)^{-n} \qquad (|z - z_0| > r'). \end{aligned} \tag{1.80}$$

Substituting (1.79) and (1.80) into (1.78) and interchanging the orders of integration and summation gives

$$f(z) = \sum_{n=0}^{\infty} c_n(z - z_0)^n + \sum_{n=-\infty}^{-1} c_n(z - z_0)^{-n}$$
$$= \sum_{n=-\infty}^{\infty} c_n(z - z_0)^n \qquad (0 < |z - z_0| < \rho), \tag{1.81}$$

where (after deforming γ' onto γ and changing notation slightly)

$$c_n = \frac{1}{2\pi i} \int_{\gamma} \frac{f(z)}{(z - z_0)^{n+1}} \, dz. \tag{1.82}$$

The series (1.81) with coefficients given by (1.82) is called the **Laurent series** for $f(z)$.

DEFINITION 5

If in the Laurent series for $f(z)$ we have $c_n = 0$ for $n < 0$, then $f(z)$ is analytic at $z = z_0$, and $z = z_0$ is said to be a **removable singularity**. If there exists a positive integer p such that in the Laurent series for $f(z)$ we have $c_{-p} \neq 0$ but $c_n = 0$ for $n < -p$, then $f(z)$ is said to have a **pole of order *p*** at $z = z_0$. If $p = 1$, $f(z)$ is said to have a **simple pole** at $z = z_0$. If no such integer p exists, then $f(z)$ is said to have an **essential singularity** at $z = z_0$.

□ **EXAMPLES 11–14**

(11) $f(z) = (\sin z)/z$ has a removable singularity at $z = 0$.

(12) $f(z) = 1/z$ has a simple pole at $z = 0$.

(13) $f(z) = 1/(z - 1)^2$ has a pole of order two at $z = 1$.

(14) $f(z) = e^{1/z}$ has an essential singularity at $z = 0$. □

DEFINITION 6

Assume that $f(z)$ has an isolated singularity at $z = z_0$. Then the **residue** of $f(z)$ at $z = z_0$, denoted by res (f,z_0), is defined by

$$\text{res}\,(f,z_0) = c_{-1} = \frac{1}{2\pi i} \int_{\gamma} f(z)\, dz.$$

Note that in the special case where $f(z)$ has a pole of order one at $z = z_0$,

$$\text{res}\,(f,z_0) = \lim_{z \to z_0} (z - z_0) f(z).$$

Suppose now that $f(z)$ is analytic in a simple connected region D with piecewise smooth, positively oriented boundary C except for isolated singularities at $z = z_1, z_2, \ldots, z_k, z_j \in D$ for $j = 1, \ldots, k$. By applying Cauchy's theorem to the multiply connected region bounded externally by C and internally by small, positively oriented circles γ_j with centers at $z = z_j$, we have

$$\int_C f(z)\,dz = \sum_{j=1}^{k} \int_{\gamma_j} f(z)\,dz = 2\pi i \sum_{j=1}^{k} \operatorname{res}\,(f,z_j). \tag{1.83}$$

This equation is known as the **residue theorem** and is a fundamental tool in the evaluation of a wide class of definite real integrals as the following examples illustrate.

☐ EXAMPLE 15

Consider the real integral

$$\begin{aligned} I &= \int_{-\infty}^{\infty} \frac{dx}{1+x^2} \\ &= \lim_{R\to\infty} I_R, \end{aligned}$$

where

$$I_R = \int_{-R}^{R} \frac{dx}{1+x^2}.$$

In order to use the residue theorem we consider

$$I_1 = \int_C \frac{dz}{1+z^2},$$

and C is as in Figure 1.3. Then $I_1 = I_R + I_2$ where I_2 is given by

$$I_2 = \int_{C_R} \frac{dz}{1+z^2},$$

where C_R is the semicircular part of the contour C. By the residue theorem

$$I_1 = 2\pi i \operatorname{res}\left(\frac{1}{1+z^2}, i\right),$$

since the only singularity within C is a pole of order one at $z = i$ (noting that $(z - i)/(1 + z^2)$ has a removable singularity at $z = i$). To compute I_1 we simply note that

$$\operatorname{res}\left(\frac{1}{1+z^2}, i\right) = \lim_{z\to i} (z-i)\frac{1}{1+z^2} = \lim_{z\to i} \frac{1}{z+i} = \frac{1}{2i},$$

and therefore

$$I_1 = \pi.$$

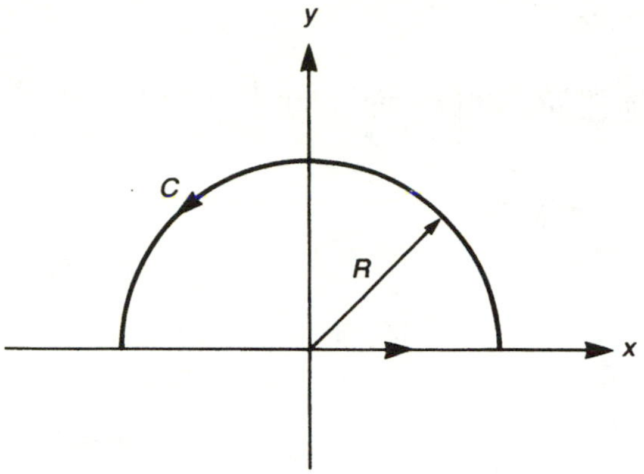

Figure 1.3

Returning now to I_2, we note that

$$|I_2| \leq \max_{|z|=R} \frac{1}{1+z^2} \pi R \leq \frac{\pi R}{R^2-1}$$

for $R > 1$, and this tends to zero as R tends to infinity. Hence

$$\pi = \lim_{R\to\infty} I_R = \int_{-\infty}^{\infty} \frac{dx}{1+x^2} .$$ □

The concept of "completing the contour" (here by the semicircle C_R) is basic in applying the residue theorem to the evaluation of real integrals. Often, more refined methods are needed to show that the integral along a circular arc of radius R tends to zero as R tends to infinity. A useful result in this regard is the following theorem.

THEOREM 12 Jordan's Lemma

Let C_R be the contour $z = Re^{i\theta}$, $0 \leq \theta \leq \pi$, and suppose that $f(z)$ is analytic for $\text{Im } z \geq 0$, $|z| \geq R_0$, such that

$$\lim_{R\to\infty} \sup_{|z|=R} |f(z)| = 0,$$

where Im z denotes the imaginary part of z and R_0 is a positive number. Then if $\sigma > 0$,

$$\lim_{R\to\infty} \int_{C_R} f(z)e^{i\sigma z}\, dz = 0.$$

■ ***Proof.*** We first note that for $R \geq R_0$,

$$\left| \int_{C_R} f(z) e^{i\sigma z}\, dz \right| = \left| \int_0^{\pi} f(Re^{i\theta}) e^{i\sigma Re^{i\theta}} iRe^{i\theta} d\theta \right| \leq \sup_{|z|=R} |f(z)| \int_0^{\pi} e^{-\sigma R \sin\theta} R\, d\theta$$

$$= 2 \sup_{|z|=R} \int_0^{\pi/2} e^{-\sigma R \sin\theta} R\, d\theta,$$

since $\sin(\pi - \theta) = \sin\theta$. But

$$\sin\theta \geq \frac{2}{\pi}\theta \qquad \left(0 \leq \theta \leq \frac{\pi}{2}\right),$$

and hence

$$\left| \int_{C_R} f(z) e^{i\sigma z}\, dz \right| \leq 2R \sup_{|z|=R} |f(z)| \int_0^{\pi/2} e^{-2\sigma\theta R/\pi}\, d\theta$$

$$= -\frac{\pi}{\sigma} \sup_{|z|=R} |f(z)| e^{-2\sigma\theta R/\pi} \Bigg|_{\theta=0}^{\theta=\pi/2}$$

$$< \frac{\pi}{\sigma} \sup_{|z|=R} |f(z)|,$$

which by hypothesis tends to zero as R tends to infinity, so the theorem is proved. ■

As an application of Jordan's lemma and the residue theorem, we now prove Lemma 1 of the previous section, i.e.,

$$\int_0^{\infty} \frac{\sin x}{x}\, dx = \frac{\pi}{2}.$$

Because the integrand is even, it suffices to show that

$$\int_{-\infty}^{\infty} \frac{\sin x}{x}\, dx = \pi.$$

□ EXAMPLE 16

Consider the integral

$$I = \int_{-\infty}^{\infty} \frac{e^{ix} - 1}{x}\, dx,$$

noting that the integrand is continuous for $-\infty < x < \infty$ and

$$\operatorname{Im} I = \int_{-\infty}^{\infty} \frac{\sin x}{x}\, dx.$$

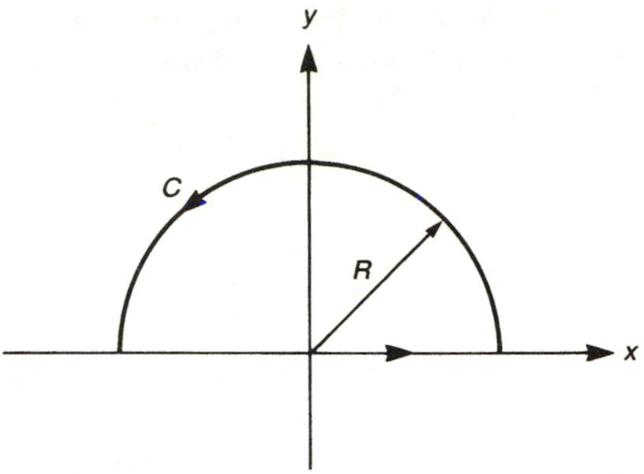

Figure 1.4

Consider the contour integral

$$I_1 = \int_C \frac{e^{iz} - 1}{z}\, dz,$$

where C is as in Figure 1.4. Then by the residue theorem (or Cauchy's theorem) $I_1 = 0$, and

$$0 = \int_C \frac{e^{iz} - 1}{z}\, dz = \int_{-R}^{R} \frac{e^{ix} - 1}{x}\, dx + \int_{C_R} \frac{e^{iz} - 1}{z}\, dz,$$

where C_R is the semicircular part of the contour C. Hence

$$\int_{-R}^{R} \frac{e^{ix} - 1}{x}\, dx = \int_{C_R} \frac{1 - e^{iz}}{z}\, dz = \int_{C_R} \frac{dz}{z} - \int_{C_R} \frac{e^{iz}}{z}\, dz. \qquad \textbf{(1.84)}$$

We now have

$$\int_{C_R} \frac{dz}{z} = i \int_0^{\pi} d\theta = \pi i,$$

and by Jordan's lemma the second integral on the right hand side of (1.84) tends to zero as R tends to infinity, thus showing that $I = \pi i$ and

$$\int_{-\infty}^{\infty} \frac{\sin x}{x}\, dx = \pi.$$

□

□ EXAMPLE 17

Consider

$$I = \int_{-\pi}^{\pi} \frac{d\theta}{1 - 2r\cos\theta + r^2} \qquad (0 \le r < 1).$$

Set $z = e^{i\theta}$ and note that $z = e^{i\theta}$, $-\pi \leq \theta \leq \pi$ describes the positively oriented circle $C = \{z: |z| = 1\}$. Then $dz = ie^{i\theta}\, d\theta$, i.e., $d\theta = dz/iz$, and $\cos\theta = (1/2)(z + 1/z)$. Hence

$$I = \int_C \frac{dz}{iz\left(1 - r\left(z + \frac{1}{z}\right) + r^2\right)}$$

$$= \frac{1}{i}\int_C \frac{dz}{(z - r)(1 - rz)} .$$

The function

$$f(z) = \frac{1}{(z - r)(1 - rz)}$$

is analytic in the closed disk $|z| \leq 1$ except for a pole of order one at $z = r$, and

$$\text{res}\,(f,r) = \lim_{z \to r} (z - r) f(z) = \frac{1}{1 - r^2} .$$

We can now conclude by the residue theorem that

$$I = \frac{1}{i}\, 2\pi i\left(\frac{1}{1 - r^2}\right) = \frac{2\pi}{1 - r^2} .$$

This integral can also be evaluated by other methods (cf. Section 4.4). □

For other examples of the uses of the residue theorem, see the exercises for this chapter and the text of Chapter 6.

We close this section by noting that of the numerous textbooks on analytic function theory, two "classics" are Ahlfors and Titchmarsh, to which the reader is referred for a more detailed presentation of this elegant and beautiful theory.

* 1.5.4 A Linear Partial Differential Equation with No Solution

It is natural to assume that any linear partial differential equation has a solution, in fact enough solutions so that it is necessary to prescribe additional data in order to pick out a unique one. After a little thought the reader will realize that this assumption must be modified to exclude partial differential equations with singular points, i.e., points at which the coefficients of the equation simultaneously vanish. For example, the equation

$$x\frac{\partial u}{\partial x} + y\frac{\partial u}{\partial y} = 1$$

has no solution in a neighborhood of the origin (cf. Exercise 3). It may come as a surprise (as it did to many mathematicians!) that there exist linear partial differential equations without singular points that have no solution. The first example of such an equation was given by Lewy in 1957 and has inspired much work in the theory of partial differential equations (Hormander). Due to its surprise value, as well as its influence on some of the modern developments in the general theory of partial differential equations, we shall now use the identity theorem for analytic functions to show that for certain infinitely differentiable functions $f(x,y)$ the equation

$$\frac{\partial w}{\partial x} + ix\frac{\partial w}{\partial y} = f(x,y) \tag{1.85}$$

has no solution in a neighborhood of the origin (Garabedian 1970). Note that this single equation with complex valued coefficients for a complex valued solution $w(x,y)$ is equivalent to a system of two equations for the real and imaginary parts of $w(x,y)$.

We first define a function $f(x,y)$ for which we shall subsequently show that no solution of (1.85) exists. Let D_n, $n = 1,2,\ldots$ be a sequence of closed, nonintersecting disks in the half plane $x > 0$ with centers at $(x_n,0)$ such that x_n tends to zero as n tends to infinity. Let $f(x,y)$ be an infinitely differentiable function of compact support (i.e., $f(x,y)$ vanishes for (x,y) outside of some disk centered at the origin) such that

(1) $f(x,y)$ is an even function of x.

(2) $f(x,y) = 0$ for $x \geq 0$ and (x,y) not in $\bigcup_{n=1}^{\infty} D_n$.

(3) $\iint_{D_n} f(x,y)\, dxdy$ does not equal zero for any n.

We than have the following theorem:

Theorem 13

For $f(x,y)$ satisfying conditions 1, 2, and 3 above there does not exist a continuously differentiable solution $w(x,y)$ of (1.85) in any neighborhood of the origin.

■ ***Proof.*** Suppose there exists a continuously differentiable solution $w(x,y)$ of (1.85) defined in a neighborhood of the origin. Let $u(x,y)$ and $v(x,y)$ be the odd and even parts, respectively, of $w(x,y)$ with respect to x. Then, taking the even part of (1.85), we have

$$\frac{\partial u}{\partial x} + ix\frac{\partial u}{\partial y} = f(x,y). \tag{1.86}$$

Now consider (1.86) in the half plane $x \geq 0$ and note that since $u(x,y)$ is an odd function of x,

$$u(0,y) = 0. \tag{1.87}$$

Making the change of variables $s = (1/2)x^2$ we know from (1.86) and (1.87) that $U(s,y) = u(\sqrt{2s},y)$ satisfies

$$\frac{\partial U}{\partial s} + i\frac{\partial U}{\partial y} = \frac{1}{\sqrt{2s}}f(\sqrt{2s},y) \tag{1.88}$$

$$U(0,y) = 0 \tag{1.89}$$

and hence, in the right half plane outside the images of the disks D_n under the mapping $x \to \sqrt{2s}$, $U(s,y)$ satisfies the Cauchy-Riemann equations. Thus $U(s,y)$ is an analytic function of $s + iy$ in this region.

We now want to show that $U(s,y)$ is in fact analytic in a domain containing a portion of the axis $s = 0$. To this end we apply Cauchy's integral theorem (deforming a smooth contour in the right half plane onto the piecewise smooth closed curve C shown in Figure 1.5) to obtain

$$U(s,y) = \frac{1}{2\pi i}\int_C \frac{U(\zeta,\eta)}{\xi - z}\,d\xi, \tag{1.90}$$

where $\xi = \zeta + i\eta$, $z = s + iy$ is in the right half plane, and C is the half circle with diameter lying on $s = 0$ and lying outside the images of the disks D_n. But from (1.89) and (1.90) we see that

$$U(s,y) = \frac{1}{2\pi i}\int_{C^+} \frac{U(\zeta,\eta)}{\xi - z}\,d\xi, \tag{1.91}$$

where C^+ is the circular part of C. This now shows that $U(s,y)$ is analytic in a neighborhood of the axis $s = 0$, i.e., in a domain containing a portion of the axis $s = 0$.

The complement of $\bigcup_{n=1}^{\infty} D_n$ is connected and so is its image under the mapping $x \to \sqrt{2s}$. We can now conclude from (1.89) and the identity theorem for analytic functions that $U(s,y)$ is identically zero in its domain of definition. Hence $u(x,y)$

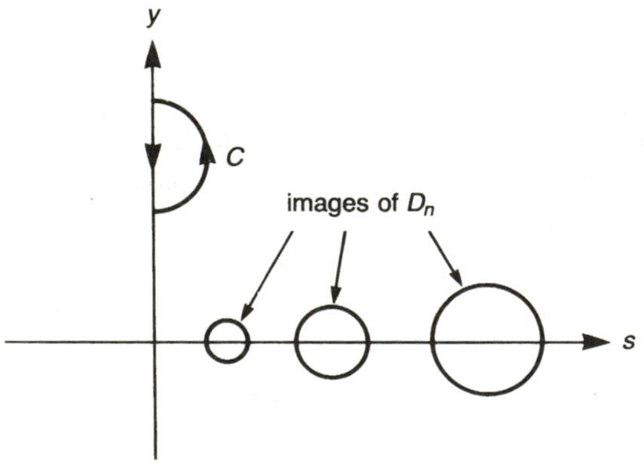

Figure 1.5

vanishes on ∂D_n for $n = 1,2,\ldots$. But from (1.86) and Green's theorem we have

$$\iint_{D_n} f(x,y)dxdy = \iint_{D_n} \left(\frac{\partial u}{\partial x} + ix\frac{\partial u}{\partial y}\right)dxdy$$

$$= \int_{\partial D_n} u(x,y)dy - ixu(x,y)dx = 0,$$

which contradicts condition 3 satisfied by $f(x,y)$. Thus no continuously differentiable solution $w(x,y)$ of (1.85) exists in any neighborhood of the origin. ■

1.6 A BRIEF HISTORY OF THE THEORY OF PARTIAL DIFFERENTIAL EQUATIONS

Mathematicians did not spontaneously decide to create the theory of partial differential equations, but rather were initially led to study certain particular equations arising in the mathematical formulation of specific physical phenomena. The first significant progress in solving partial differential equations occurred in the middle of the eighteenth century when Euler (1707–1783) and d'Alembert (1717–1783) investigated the wave equation

$$\frac{\partial^2 u}{\partial x^2} = \frac{1}{c^2}\frac{\partial^2 u}{\partial t^2}.$$

Both were led to the solution

$$u(x,t) = f(x + ct) + g(x - ct)$$

where f and g are "arbitrary" functions, and a debate continued to rage until the 1770s on what "arbitrary" meant. Euler also took up the problem of the vibrations of a rectangular and circular drum governed by the two-dimensional wave equation

$$\frac{\partial^2 u}{\partial x^2} + \frac{\partial^2 u}{\partial y^2} = \frac{1}{c^2}\frac{\partial^2 u}{\partial t^2}$$

and obtained various special solutions by what is now known as the method of separation of variables. Finally, in a series of definitive papers on the propagation of sound, Euler obtained cylindrical and spherical wave solutions of the wave equation in two and three space variables. Progress, however, was limited by the lack of knowledge of Fourier series and of the behavior of the special functions arising from the application of the method of separation of variables.

Research into the theory of gravitational attraction led to the formulation of Laplace's equation

$$\frac{\partial^2 u}{\partial x^2} + \frac{\partial^2 u}{\partial y^2} + \frac{\partial^2 u}{\partial z^2} = 0$$

for the potential function $u(x,y,z)$. The first significant work on potential theory was done by Legendre (1752–1833) in his 1782 study of the gravitational attraction of spheroids, which introduced what are now called Legendre polynomials. This work was continued by Laplace (1749–1827) in 1785 (although Laplace never mentioned Legendre!). In a series of papers continuing through the 1780s, Legendre and Laplace continued their investigations of potential theory and the use of Legendre polynomials, associated Legendre polynomials, and spherical harmonics, laying the foundation for the vast work in the nineteenth century on the theory of harmonic functions. However, no general method for solving Laplace's equation was developed in the eighteenth century, nor were the full potentialities of the use of special functions appreciated.

The study of partial differential equations experienced phenomenal growth in the nineteenth century. This growth not only illuminated new areas of physics, but created the need for mathematical developments in such diverse areas as analytic function theory, the calculus of variations, ordinary differential equations, and differential geometry. In this brief history, we can only highlight a few of the developments that are relevant to the material covered in this book.

The first major step was taken by Fourier (1768–1830) in 1807 when he submitted a basic paper on heat conduction and trigonometric series to the Academy of Sciences of Paris. His paper was rejected; however, when the Academy made the subject of heat conduction the topic of a grand prize in 1812, Fourier submitted a revised copy and this time won the prize. He continued to work in the area and in 1822 published his classic *Théorie analytique de la chaleur* in which, following his paper of 1807, he derived the equation of heat conduction

$$\frac{\partial^2 u}{\partial x^2} = \frac{1}{\alpha^2}\frac{\partial u}{\partial t}$$

and solved specific heat conduction problems by what is now known as the method of separation of variables and Fourier series. All of Fourier's work was purely formal, and the convergence properties of Fourier series were left unexamined until later in the century. Later, Poisson (1781–1840) made use of Legendre polynomials and spherical harmonics in addition to trigonometric series to study multi-dimensional problems. At the same time as Fourier series were being developed, Fourier, Cauchy (1789–1857), and Poisson discovered what is now called the Fourier integral and applied it to various problems in heat conduction and water waves. Because all three presented papers orally to the Academy of Sciences and published their results only later, it is not possible to assign priority to the discovery of Fourier integrals and transforms.

Mathematicians of the nineteenth century vigorously investigated problems associated with Laplace's equation, continuing the research initiated by Legendre and Laplace. In a basic paper written in 1813, Poisson showed that the gravitational attraction of a body with density $\rho(x,y,z)$ is given by

$$\frac{\partial^2 u}{\partial x^2} + \frac{\partial^2 u}{\partial y^2} + \frac{\partial^2 u}{\partial z^2} = -4\pi\rho(x,y,z)$$

for points inside the body. Poisson's derivation of this result was not rigorous, even by the standards of his time, and the first rigorous derivation of Poisson's equation was given by Gauss (1777–1855) in 1839. However, despite the work of Legendre, Laplace, Poisson, and Gauss, almost nothing was known about the general properties of solutions of Laplace's equation. In 1828, Green (1793–1841), a self-taught English mathematician, published a privately printed booklet entitled *An Essay on the Application of Mathematical Analysis to the Theories of Electricity and Magnetism*. In this small masterwork Green derived what are now known as Green's formulas and introduced the concept of the Green's function. Unfortunately, his work was neglected for over twenty years until Sir William Thomson (later Lord Kelvin, 1824–1907) discovered it and, recognizing its great value, had it published in the *Journal für Mathematik*.

Until the middle part of the nineteenth century, mathematicians simply assumed that a solution to the Laplace or Poisson equation existed, usually arguing from physical considerations. In particular, Green's proof of the existence of a Green's function was based entirely on a physical argument. However, in the second half of the century extensive work was undertaken on the problem of the existence of solutions to partial differential equations, not only for the Laplace and Poisson equations, but for partial differential equations with variable coefficients. In particular, Riemann (1826–1866) and Hadamard (1865–1963) investigated initial value problems for hyperbolic equations, Picard (1856–1941) and others for elliptic equations, while Cauchy and Kowalewsky (1850–1891) studied the initial value, or Cauchy problem, for general systems of partial differential equations with analytic coefficients. Gradually, mathematicians became aware that different types of equations required different types of boundary and initial conditions, leading to the now-standard classification of partial differential equations into elliptic, hyperbolic, and parabolic types. This classification was introduced by DuBois-Reymond (1831–1889) in 1889. In addition to investigations on existence theorems for general partial differential equations, research continued on the Dirichlet and Neumann problems for Laplace's equation through methods involving analytic function theory, the calculus of variations, and the method of integral equations (using successive approximation techniques). Considerable effort was also made to prove the existence of eigenvalues for

$$\frac{\partial^2 u}{\partial x^2} + \frac{\partial^2 u}{\partial y^2} + k^2 u = 0,$$

particularly by Schwarz (1843–1921) and Poincaré (1854–1912). The systematic treatment of eigenvalue problems for partial differential equations was delayed until the development of the theory of integral equations in this century by Fredholm (1866–1927) and Hilbert (1862–1943). We shall return to this theme shortly.

Throughout the nineteenth century, mathematicians and mathematical physicists were concerned with the theory of wave motion, continuing the tradition established by Euler and d'Alembert in the eighteenth century. In particular, numerous papers were written applying the method of separation of variables in curvilinear coordinates to solve initial-boundary value problems for the wave equation and boundary value problems for the reduced wave equation or Helmholtz equation. Of paramount importance was the theory of Bessel functions, which were first systematically studied by Bessel (1784–1846), a mathematician and director of the astronomical observatory in Königsberg. Although these functions are of central importance in the study of wave propagation, Bessel was in fact led to his study while working on the motion of the planets. In addition to the method of separation of variables for solving initial-boundary value problems in wave propagation, integral representations of solutions to the wave equation and reduced wave equation were established by Poisson, Helmholtz (1821–1894), and Kirchhoff (1824–1887). The most spectacular triumph of these investigations into the theory of wave propagation was Maxwell's derivation in 1864 of the laws of electromagnetism. From his equations, Maxwell (1831–1879) predicted that electromagnetic waves travel through space at the speed of light and that light itself was an electromagnetic phenomenon. Maxwell's research was the highlight of nineteenth century mathematical physics and his monograph *A Treatise on Electricity and Magnetism,* published in 1873, is one of the classics of scientific thought.

Early in this century, a major new era in the theory of partial differential equations began with the development of the theory of integral equations to solve boundary value problems for partial differential equations. Integral equations had already been used by Neumann (1832–1925) in 1870 to solve the Dirichlet problem for Laplace's equation in a convex domain by the method of successive approximations. However, due to the fact that no systematic theory of integral equations was available, Neumann was not able to remove the restrictive condition of convexity from his analysis. The first step toward a general theory of integral equations was taken by Volterra (1860–1940) in 1896 and 1897 when he used the method of successive approximations to solve what is now called the Volterra integral equation of the second kind:

$$\phi(s) - \int_a^s K(s,t)\phi(t)dt = f(s).$$

Volterra's ideas were taken up by Fredholm, a professor at Stockholm, who established what is now known as the Fredholm alternative for Fredholm integral equations of the second kind:

$$\phi(s) - \lambda \int_a^b K(s,t)\phi(t)dt = f(s).$$

Fredholm then proceeded to use his theory to solve the Dirichlet problem for Laplace's equation in domains that were not necessarily convex, his first results appearing in a seminal paper published in 1900. Fredholm's ideas were brought to fruition by Hilbert, a professor at Göttingen and the leading mathematician of the early part of the twentieth century. In a series of six papers published between 1904 and 1910, Hilbert more simply formulated Fredholm's ideas, established the fact that an "arbitrary" function can be expanded in a series of eigenfunctions of the integral equation (now called the Hilbert-Schmidt theory), and applied his results to problems in mathematical physics. The method of integral equations has been applied to an increasing number of problems in mathematical physics, most notably the scattering of acoustic, electromagnetic, and elastic waves by inhomogeneities in the medium.

The work by Volterra, Fredholm, and Hilbert has reverberated through the twentieth century, leading first to Hilbert space theory and functional analysis with applications to distributional solutions of initial value and boundary value problems for partial differential equations, and, in a somewhat different direction, to singular integral operators and the "general" theory of linear partial differential operators. However, these topics are beyond the scope of this brief survey; indeed, as the twentieth century reached middle age the area of partial differential equations became so broad and deep that a short survey of the directions taken and results discovered would require a small monograph! Thus we conclude this section by indicating only three of these directions that are relevant to this book: numerical methods, nonlinear problems, and improperly posed problems.

We recall that nineteenth century research in the theory of partial differential equations was concerned primarily with well posed linear problems—by well posed we mean that a solution exists, is unique, and depends continuously on the boundary or initial data. For such problems, interest was focused on obtaining series or integral representations for the solution. However, as the demands of science increased, it became evident that such representations were often not suitable for numerical computation. For example, in using the series representation of the solution of Maxwell's equations describing the propagation of radio waves around the earth it was discovered that over a thousand terms of the series were needed in order to assure the needed accuracy—a formidable task even for a modern computer! Hence mathematicians were led to derive new methods for the approximate solution of boundary and initial value problems of mathematical physics, leading to a fruitful interplay between the art of computer science and the methods of numerical analysis. At the same time, it has become clear that the real world is in fact nonlinear and that although linear models are useful and valid in certain contexts, many phenomena can be understood only by a nonlinear model. Motivated by an increasing number of apparently intractable problems in fluid and gas dynamics, elasticity, and chemical reactions, mathematicians in the twentieth century have systematically studied nonlinear partial differential equations. This subject has by now reached full maturity and forms one of the major areas of the theory of partial differential equations. Finally, by mid-century, mathematicians realized (after some resistance!) that well posed problems were not the only ones of physical interest. In particular, such problems as the design of shock-free airfoils and the inverse scattering problems associated with radar, sonar, and medical imaging have led mathematicians seriously to consider improperly posed problems and to derive methods for their "solution." Although the subject areas of study in partial differential equations in the twentieth century are significantly different from those of the last, the words of Fourier still provide the appropriate guidelines: "The profound study of nature is the most fertile source of mathematical discoveries."

Exercises

1. Find the general solution of

   (a) $\dfrac{\partial u}{\partial x} + \dfrac{\partial u}{\partial y} - u = 0.$

   (b) $y \dfrac{\partial u}{\partial x} + x \dfrac{\partial u}{\partial y} = y.$

   (Answers: (a) $u(x,y) = e^x f(x - y)$ **(b)** $u(x,y) = x + f(x^2 - y^2)$.)

2. Find the general solution of

   (a) $xy \dfrac{\partial u}{\partial x} - x^2 \dfrac{\partial u}{\partial y} - yu = xy.$

   (b) $x^2 \dfrac{\partial u}{\partial x} - xy \dfrac{\partial u}{\partial y} + 2yu = 0.$

3. Show that a continuously differentiable solution of $x \dfrac{\partial u}{\partial x} + y \dfrac{\partial u}{\partial y} = 1$ does not exist in any neighborhood of the origin.

4. Obtain the general solution of the equation

   $$y \frac{\partial u}{\partial x} + x \frac{\partial u}{\partial y} - yu = xe^x.$$

   If we prescribe $u(x,y) = \phi(x,y)$ on the upper portion of the hyperbola $y^2 - x^2 = 1$, $y \geq 1$, show that no solution exists unless $\phi(x,y)$ is of a special form. Find this form and show that in such a case there are infinitely many solutions.

5. Solve the following initial value problems:

   (a) $\dfrac{\partial u}{\partial x} + \dfrac{\partial u}{\partial y} = 1$, $u(x,0) = e^x.$

   (b) $x \dfrac{\partial u}{\partial x} - y \dfrac{\partial u}{\partial y} + u = x$, $u(x,x) = x^2$

   (Answers: (a) $u(x,y) = y + e^{x-y}$. **(b)** $u(x,y) = \dfrac{x}{2} + \sqrt{x}y^{3/2} - \dfrac{y}{2}$.)

6. Solve the initial value problem

   $$x \frac{\partial u}{\partial x} + y \frac{\partial u}{\partial y} = u$$

   $$u(x,1) = f(x),$$

   where $f(x)$ is continuously differentiable.

7. Determine the type of each of the following equations and reduce them to canonical form.

(a) $\dfrac{\partial^2 u}{\partial x^2} + 2\dfrac{\partial^2 u}{\partial x \partial y} + \dfrac{\partial^2 u}{\partial y^2} + \dfrac{\partial u}{\partial x} - \dfrac{\partial u}{\partial y} = 0.$

(b) $\dfrac{\partial^2 u}{\partial x^2} + 2\dfrac{\partial^2 u}{\partial x \partial y} + 5\dfrac{\partial^2 u}{\partial y^2} + 3\dfrac{\partial u}{\partial x} + u = 0.$

(c) $3\dfrac{\partial^2 u}{\partial x^2} + 10\dfrac{\partial^2 u}{\partial x \partial y} + 3\dfrac{\partial^2 u}{\partial y^2} = 0.$

8. Find the general solution of

(a) $\dfrac{\partial^2 u}{\partial x^2} + 6\dfrac{\partial^2 u}{\partial x \partial y} - 16\dfrac{\partial^2 u}{\partial y^2} = 0.$

(b) $\dfrac{\partial^2 u}{\partial x^2} + 2\dfrac{\partial^2 u}{\partial x \partial y} + \dfrac{\partial^2 u}{\partial y^2} = 0.$

9. Reduce the following equations to canonical form:

(a) $y^2\dfrac{\partial^2 u}{\partial x^2} + 2xy\dfrac{\partial^2 u}{\partial x \partial y} + x^2\dfrac{\partial^2 u}{\partial y^2} = 0.$

(b) $\dfrac{\partial^2 u}{\partial x^2} - 2x\dfrac{\partial^2 u}{\partial x \partial y} = 0 \qquad (x \neq 0).$

(**Answers:** **(a)** $\dfrac{\partial^2 v}{\partial \eta^2} - \dfrac{\xi}{2\eta(\xi + \eta)}\dfrac{\partial v}{\partial \xi} + \dfrac{1}{2\eta}\dfrac{\partial v}{\partial \eta} = 0;\ \xi = y^2 - x^2,\ \eta = x^2.$

(b) $\dfrac{\partial^2 v}{\partial \xi \partial \eta} - \dfrac{1}{2(\xi - \eta)}\dfrac{\partial v}{\partial \xi} = 0;\ \xi = x^2 + y,\ \eta = y.$)

10. Reduce the following equations to canonical form.

(a) $(1 + x^2)^2\dfrac{\partial^2 u}{\partial x^2} + \dfrac{\partial^2 u}{\partial y^2} + 2x(1 + x^2)\dfrac{\partial u}{\partial x} = 0.$

(b) $y\dfrac{\partial^2 u}{\partial x^2} + \dfrac{\partial^2 u}{\partial y^2} = 0 \qquad (y < 0).$

11. Using the norm defined in Section 1.4 prove the parallelogram law

$$\|f + g\|^2 + \|f - g\|^2 = 2\|f\|^2 + 2\|g\|^2.$$

12. Find the Fourier series of the following functions:

(a) $f(x) = \begin{cases} 0 & (-\pi \le x \le 0) \\ x & (0 < x \le \pi). \end{cases}$

(b) $f(x) = |x| \qquad (-\pi \le x \le \pi).$

(**Answers:** **(a)** $\dfrac{\pi}{4} - \displaystyle\sum_{n=1}^{\infty}\left(\dfrac{2}{\pi(2n-1)^2}\cos(2n-1)x + \dfrac{(-1)^n}{n}\sin nx\right).$

(b) $\dfrac{\pi}{2} - \dfrac{4}{\pi}\displaystyle\sum_{n=1}^{\infty}\dfrac{\cos(2n-1)x}{(2n-1)^2}$ y'.)

13. Find the Fourier series of the given functions on the interval indicated and describe graphically the periodic function to which the series converges.

(a) $f(x) = \begin{cases} -\pi & (-\pi < x < 0) \\ x & (0 \leq x < \pi). \end{cases}$

(b) $f(x) = \begin{cases} x + \dfrac{1}{2} & (-1 \leq x \leq 0) \\ \dfrac{1}{2} - x & (0 \leq x \leq 1). \end{cases}$

(c) $f(x) = e^{ax} \qquad (-L < x < L).$

(d) $f(x) = \begin{cases} 0 & (-\pi \leq x \leq 0) \\ \sin x & (0 \leq x \leq \pi). \end{cases}$

(Answers: (a) $-\dfrac{\pi}{4} + \displaystyle\sum_{n=1}^{\infty} \left(\frac{(-1)^n - 1}{\pi n^2} \cos nx + \frac{1 + 2(-1)^{n+1}}{n} \sin nx \right).$

(b) $\dfrac{4}{\pi^2} \displaystyle\sum_{n=1}^{\infty} \frac{\cos (2n-1)\pi x}{(2n-1)^2}.$

(c) $\dfrac{\sinh aL}{aL} + 2 \sinh aL \displaystyle\sum_{n=1}^{\infty} \frac{(-1)^n}{a^2L^2 + n^2\pi^2} \left(aL \cos \frac{n\pi x}{L} - n\pi \sin \frac{n\pi x}{L} \right).$

(d) $\dfrac{1}{\pi} - \dfrac{2}{\pi} \displaystyle\sum_{n=1}^{\infty} \frac{\cos 2nx}{4n^2 - 1} + \frac{1}{2} \sin x.$**)**

14. Find the Fourier cosine and sine series of the given functions:

(a) $f(x) = x \qquad (0 \leq x \leq \pi).$

(b) $f(x) = \pi^2 - x^2 \qquad (0 \leq x < \pi).$

(Answers: (a) $\dfrac{\pi}{2} + \dfrac{2}{\pi} \displaystyle\sum_{n=1}^{\infty} \frac{(-1)^n - 1}{n^2} \cos nx.$

$2 \displaystyle\sum_{n=1}^{\infty} \frac{(-1)^{n+1}}{n} \sin nx.$

(b) $\dfrac{2\pi^2}{3} + 4 \displaystyle\sum_{n=1}^{\infty} \frac{(-1)^{n+1}}{n^2} \cos nx.$

$\displaystyle\sum_{n=1}^{\infty} \left(\frac{2\pi}{n} + \frac{4}{\pi} \frac{1 - (-1)^n}{n^3} \right) \sin nx.$**)**

15. Find the Fourier cosine and sine series of the given function and graph the periodic function to which the series converges.

(a) $f(x) = \begin{cases} x & \left(0 \le x \le \dfrac{1}{2}\right) \\ 1 - x & \left(\dfrac{1}{2} \le x \le 1\right). \end{cases}$

(b) $f(x) = 1 - x \qquad (0 < x < 2).$

(c) $f(x) = e^x \qquad (0 < x < \pi).$

(Answers: (a) $\dfrac{1}{4} - \dfrac{2}{\pi^2}\sum_{n=1}^{\infty}\dfrac{\cos 2(2n-1)\pi x}{(2n-1)^2}$

$\dfrac{4}{\pi^2}\sum_{n=1}^{\infty}\dfrac{(-1)^{n+1}}{(2n-1)^2}\sin(2n-1)\pi x.$

(b) $\dfrac{8}{\pi^2}\sum_{n=1}^{\infty}\dfrac{\cos\frac{1}{2}(2n-1)\pi x}{(2n-1)^2}$

$\dfrac{2}{\pi}\sum_{n=1}^{\infty}\dfrac{\sin n\pi x}{n}.$

(c) $\dfrac{e^{\pi}-1}{\pi} + \dfrac{2}{\pi}\sum_{n=1}^{\infty}\dfrac{(-1)^n e^{\pi}-1}{1+n^2}\cos nx$

$\dfrac{2}{\pi}\sum_{n=1}^{\infty}\dfrac{n}{n^2+1}(e^{\pi}(-1)^{n+1}-1)\sin nx.$**)**

16. Let $f(x) = x^2$ when $-\pi \le x \le \pi$ and $f(x + 2\pi) = f(x)$ for all x. Show that

$$f(x) = \frac{\pi^2}{3} + 4\sum_{n=1}^{\infty}(-1)^n\frac{\cos nx}{n^2}$$

for all x and deduce that

$$\sum_{n=1}^{\infty}\frac{(-1)^{n-1}}{n^2} = \frac{\pi^2}{12}$$

and

$$\sum_{n=1}^{\infty}\frac{1}{n^2} = \frac{\pi^2}{6}.$$

17. Show that for $-\pi \le x \le \pi$ and a not an integer

$$\cos ax = \frac{\sin \pi a}{\pi a} + \sum_{n=1}^{\infty}(-1)^n\frac{2a\sin \pi a}{\pi(a^2-n^2)}\cos nx.$$

Deduce that

$$\cot \pi a = \frac{1}{\pi}\left(\frac{1}{a} - \sum_{n=1}^{\infty}\frac{2a}{n^2-a^2}\right).$$

18. Define the **Fourier transform** of a function $f(x)$ by

$$\hat{f}(\lambda) = F[f] = \int_{-\infty}^{\infty} f(x)e^{-i\lambda x}\,dx,$$

where $f(x)$ satisfies the hypothesis of Lemma 2 of Section 4. Assuming that $\hat{f}(x)$ has compact support on $(-\infty,\infty)$ derive **Parseval's equality**

$$\int_{-\infty}^{\infty} |\hat{f}(\lambda)|^2\,d\lambda = 2\pi \int_{-\infty}^{\infty} |f(x)|^2\,dx.$$

19. Let $\hat{f}(\lambda)$ be the Fourier transform of $f(x)$. Show that
(a) the Fourier transform of $f(x)e^{iax}$ is $\hat{f}(\lambda - a)$.
(b) the Fourier transform of $f(ax)$ is $\dfrac{1}{a}\hat{f}(s/a)$.

20. Let $f(x)$, $g(x)$, and $h(x)$ satisfy the hypothesis of Lemma 2 of Section 4 and let their Fourier transforms (cf. problem 18) be $\hat{f}(\lambda)$, $\hat{g}(\lambda)$, and $\hat{h}(\lambda)$ respectively. Assume that $g(x)$ and $f(x)$ have compact support and let $h(x)$ be the **convolution** of $f(x)$ and $g(x)$ defined by

$$h(x) = f * g = \int_{-\infty}^{\infty} f(\xi)g(x - \xi)\,d\xi.$$

Show that $f * g = g * f$ and prove the **convolution theorem**

$$\hat{h}(\lambda) = \hat{f}(\lambda)\hat{g}(\lambda).$$

21. Use the residue theorem and Jordan's lemma when necessary to compute the Fourier transforms of the following functions:
(a) $f(x) = e^{-|x|}$.
(b) $f(x) = \dfrac{1}{x^2 + 1}$.
(c) $f(x) = \dfrac{x}{1 + x^2}$.

(Answers: **(a)** $\hat{f}(\lambda) = \dfrac{2}{1 + \lambda^2}$.
(b) $\hat{f}(\lambda) = \pi e^{-|\lambda|}$.
(c) $\hat{f}(\lambda) = \begin{cases} -\pi i e^{-\lambda} & (\lambda > 0) \\ \pi i e^{\lambda} & (\lambda < 0).) \end{cases}$

Note that in (c), $f(x)$ is not absolutely integrable.

22. Use the convolution theorem for Fourier transforms and Fourier's integral theorem to solve the integral equation

$$\int_{-\infty}^{\infty} e^{-|x-\xi|} f(\xi)\, d\xi = -\frac{1}{4} f(x) + e^{-|x|} \qquad (-\infty < x < \infty).$$

(Answer: $f(x) = \frac{4}{3} e^{-3|x|}$.**)**

23. Solve the integral equation

$$\int_0^x f(x-y)\, e^{-y}\, dy = \begin{cases} 0 & (x < 0) \\ x^2 e^{-x} & (x \geq 0). \end{cases}$$

(Answer: $f(x) = 2xe^{-x}$, $x \geq 0$.**)**

24. Use Cauchy's formula to prove **Liouville's theorem:** A bounded entire function must be a constant.

25. Apply Cauchy's theorem to the function $f(z) = e^{-z^2}$ and the sector of the circle $|z| = R$ bounded by a section of the real axis and a linear segment making the angle $\pi/4$ with the real axis. Show that the integral over the circular boundary tends to zero as R tends to infinity and show that

$$\int_0^{\infty} \cos(x^2)\, dx = \int_0^{\infty} \sin(x^2)\, dx = \frac{1}{\sqrt{2}} \int_0^{\infty} e^{-x^2}\, dx = \frac{1}{2}\sqrt{\frac{\pi}{2}}.$$

26. Prove the fundamental theorem of algebra (i.e., every polynomial has a zero) by applying the maximum principle to the function

$$f(z) = \frac{1}{a_0 + a_1 z + \ldots + a_n z^n}$$

(which would be analytic if the theorem were false) in the region $|z| \leq R$ and letting R tend to infinity. (Hint: Use exercise 24.)

27. With the help of Cauchy's formula show that

$$\left(\frac{z^n}{n!}\right)^2 = \frac{1}{2\pi i} \int_C \frac{z^n e^{z\zeta}}{n!\zeta^n} \frac{d\zeta}{\zeta},$$

where C is a simple closed wave surrounding the origin. Use this identity to show that

$$\sum_{n=0}^{\infty} \left(\frac{z^n}{n!}\right)^2 = \frac{1}{2\pi} \int_0^{2\pi} e^{2z\cos\theta}\, d\theta.$$

28. Use the residue theorem to establish the following identities.

(a) $$\int_0^{\infty} \frac{dx}{1+x^4} = \frac{\pi}{2\sqrt{2}}.$$

(b) $$\int_0^{\infty} \frac{\cos ax}{x^2+1}\, dx = \frac{\pi}{2} e^{-a} \qquad (a \geq 0).$$

(c) $$\int_0^{2\pi} \frac{d\theta}{1 + a\cos\theta} = \frac{2\pi}{\sqrt{1-a^2}} \qquad (-1 < a < 1).$$

(d) $$\int_0^{\pi} \sin^{2n}\theta\, d\theta = \frac{(2n)!\pi}{2^{2n}(n!)^2} \qquad (n = 1, 2, \ldots).$$

(e) $$\int_{-\infty}^{\infty} \frac{x \sin ax}{x^4 + 4}\, dx = \frac{\pi}{2} e^{-a} \sin a \qquad (a > 0).$$

(Hint: If a function is even, its integral over $(-\infty,\infty)$ is twice its integral over $(0,\infty)$. In (b) and (e) replace $\cos ax$ and $\sin ax$ by e^{iax} and take the real or imaginary part.)

29. Evaluate the following integrals:

(a) $\displaystyle\int_0^{\infty} \frac{\sin^2 x}{x^2}\, dx.$ (Hint: Integrate $\dfrac{1}{z^2}(e^{2iz} - 1 - 2iz)$ around a large semicircle.)

(b) $\displaystyle\int_0^{\infty} \frac{dx}{1 + x^n}$, where $n \geq 2$ is an integer. (Hint: Consider a contour bounding the wedge $z = t$, $z = te^{2\pi i/n}$, $0 \leq t < \infty$.)

30. Show that the binomial coefficient $\binom{n}{k}$ has the representation

$$\binom{n}{k} = \frac{1}{2\pi i} \int_C \frac{(1+z)^n}{z^{k+1}}\, dz,$$

where C is any simple closed curve surrounding the origin. Choosing C to be the unit circle, deduce that

$$\binom{2n}{n} \leq 4^n.$$

31. Prove Morera's theorem. (Hint: For $z_0 \in \Omega$ consider $F(z) = \displaystyle\int_{z_0}^{z} f(w)\, dw$. Show that for z in a neighborhood of z_0, $F(z)$ is independent of the path of integration and $F'(z)$ exists and equals $f(z)$. Now apply the holomorphy theorem.)

Chapter **2**

The Wave Equation

In this chapter we are primarily concerned with two basic problems for the wave equation: the Cauchy problem and the initial-boundary value problem. In the introductory spirit of this book, we consider the initial-boundary value problem only for domains of simple shape for which we can apply the method of separation of variables; however, the reader should not get the impression that somewhere down the mathematical road he or she will meet a simple analytic method that will yield the solution of initial-boundary value problems for the wave equation in arbitrary domains. Solving initial-boundary value problems is hard, and more often than not the only recourse is to some numerical method such as finite difference approximations. (The method of finite differences will be touched on in Chapter 4 for the case of Laplace's equation; the same idea can also be applied to initial-boundary value problems for the wave equation.) In our study of these two basic problems for the wave equation we also take the opportunity to investigate some aspects of hyperbolic equations different from the wave equation, in particular the method of successive approximations for solving the Cauchy problem for hyperbolic equations with variable coefficients, and shock formation arising in the initial value problem for the nonlinear equations of gas dynamics. We also draw the reader's attention to Section 4.6 and

Chapter 6 of this book where exterior boundary value problems for the reduced wave equation are considered.

2.1 THE WAVE EQUATION IN TWO INDEPENDENT VARIABLES

We consider the wave equation in two independent variables

$$\frac{\partial^2 u}{\partial x^2} = \frac{\partial^2 u}{\partial t^2}. \tag{2.1}$$

We begin by finding the general solution to (2.1), an approach which in general is not very useful in the study of second order partial differential equations, but which bears fruit here. To do so we make the change of variables

$$\begin{aligned} \xi &= x - t \\ \eta &= x + t \end{aligned} \tag{2.2}$$

and arrive at

$$\frac{\partial^2 U}{\partial \xi \partial \eta} = 0.$$

From Section 1.3.3, the general solution of this equation is

$$U(\xi,\eta) = \phi(\xi) + \psi(\eta),$$

where we assume that $\phi(\xi)$ and $\psi(\eta)$ are twice continuously differentiable functions of their independent variable. Then from (2.2) we see that the general solution of (2.1) that is twice continuously differentiable with respect to x and t is given by

$$u(x,t) = \phi(x - t) + \psi(x + t) \tag{2.3}$$

where $\phi(\xi)$ and $\psi(\eta)$ are arbitrary functions having two continuous derivatives. The solution $\phi(x - t)$ can be interpreted as a solution of the wave equation moving to the right (i.e., if at $x = x_0$ and $t = t_0$ we have $u(x_0,t_0) = \phi(x_0 - t_0)$, then this value is unchanged as x and t increase such that $x - t = x_0 - t_0$). Similarly $\psi(x + t)$ can be interpreted as a solution of the wave equation moving to the left. The speed of propagation of these traveling waves is

$$\left| \frac{x - x_0}{t - t_0} \right| = 1.$$

Note that if we had considered the unnormalized wave equation

$$\frac{\partial^2 u}{\partial x^2} = \frac{1}{c^2} \frac{\partial^2 u}{\partial t^2}$$

then the corresponding general solution would be

$$u(x,t) = \phi(x - ct) + \psi(x + ct),$$

which represents the sum of waves traveling to the right and left with speed c. The solution (2.3) of the wave equation (2.1) is known as **d'Alembert's solution.**

We now want to use d'Alembert's solution to solve the Cauchy problem for the two-dimensional wave equation, i.e., to find a twice continuously differentiable function $u(x,t)$ such that

$$\frac{\partial^2 u}{\partial x^2} = \frac{\partial^2 u}{\partial t^2} \qquad (-\infty < x < \infty,\ 0 \le t < \infty)$$

$$u(x,0) = \phi_0(x), \qquad \frac{\partial u}{\partial t}(x,0) = \phi_1(x) \qquad (-\infty < x < \infty),$$

where $\phi_0 \in C^2(-\infty,\infty)$ and $\phi_1 \in C^1(-\infty,\infty)$. (For the definition of the notation see the beginning of Chapter 1.) Representing $u(x,t)$ in the form (2.3) we see that

$$\phi_0(x) = \phi(x) + \psi(x)$$
$$\phi_1(x) = -\phi'(x) + \psi'(x)$$

or

$$\phi(x) + \psi(x) = \phi_0(x)$$
$$\phi(x) - \psi(x) = -\int_0^x \phi_1(x)\,dx + c,$$

where c is an arbitrary constant of integration. Solving this system of equations gives

$$\phi(x) = \frac{1}{2}\phi_0(x) - \frac{1}{2}\int_0^x \phi_1(x)\,dx + \frac{c}{2}$$
$$\psi(x) = \frac{1}{2}\phi_0(x) + \frac{1}{2}\int_0^x \phi_1(x)\,dx - \frac{c}{2}.$$

Hence from (2.3) we see that the solution of Cauchy's problem is

$$u(x,t) = \frac{\phi_0(x - t) + \phi_0(x + t)}{2} + \frac{1}{2}\int_{x-t}^{x+t} \phi_1(x)\,dx. \tag{2.4}$$

Equation (2.4) is known as **d'Alembert's formula.**

By construction, the solution of the Cauchy problem as given by (2.4) is unique. Furthermore, small perturbations of the initial data $\phi_0(x)$ and $\phi_1(x)$, measured with respect to the maximum norm

$$\|\phi\| = \max_{a \le x \le b} |\phi(x)|,$$

lead to small perturbations of the solution $u(x,t)$ as given by (2.4). If $\phi_1(x) = 0$, then perturbations of $\phi_0(x)$ for $a \le x \le b$ lead to perturbations of $u(x,t)$ in the region

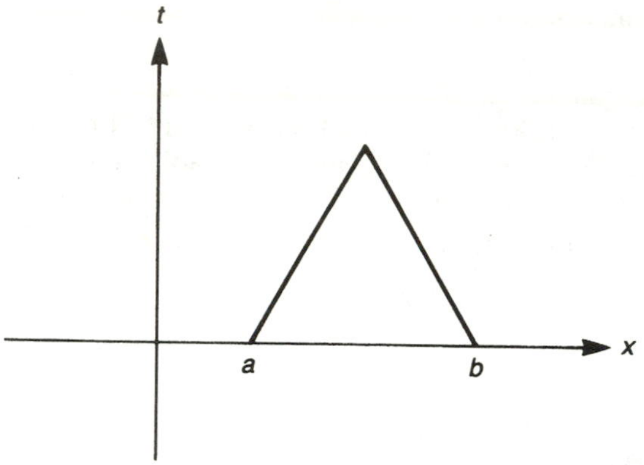

Figure 2.1

bounded by the triangle $x - t = a$, $x + t = b$, $t = 0$ (see Figure 2.1). Note that the lines $x - t = a$ and $x + t = b$ are characteristics for the wave equation (2.1) (see Section 1.3.3). In particular, the solution of Cauchy's problem at a point (x,t) depends only on the initial data on an interval of the x axis cut out by the characteristics passing through the point (x,t). This interval is known as the **domain of dependence** of the solution $u(x,t)$ at the point (x,t).

We have established above that the Cauchy problem for the wave equation in two independent variables is **well posed,** that is to say:

(1) a solution exists,

(2) the solution is unique,

(3) the solution depends continuously on the given data, measured with respect to some norm.

These are the basic conditions that we shall seek for the various initial value and boundary value problems considered in this book; indeed, one would expect that any reasonable mathematical model of a physical phenomenon should satisfy these three conditions. This line of thinking led mathematicians to say for many years that well posed (or "properly" posed) problems were the *only* ones of interest in mathematical physics; however, in recent years this rigid view has been eroded by confrontation with physical reality. In particular, the need to solve various inverse problems has shown mathematicians that well posed problems are not the only ones to appropriately reflect real phenomena. We shall later study two such inverse problems, the inverse Stefan problem (Section 3.6) and the inverse scattering problem (Section 6.6), and show how meaningful answers can be found where one or more of the above three conditions is violated. Such problems are called **improperly posed;** for an introduction to this rapidly growing area of partial differential equations we refer the reader to Payne.

Before leaving the wave equation in two independent variables, we want to observe what happens when the initial data to the Cauchy problem fails to be smooth enough so that $u(x,t)$ as defined by (2.4) is a twice continuously differentiable solution of the wave equation (2.1). For the sake of simplicity, assume that $\phi_1(x)$ is identically zero and $\phi_0(x)$ is given by

$$\phi_0(x) = \begin{cases} x^2 & (x > 0) \\ 0 & (x = 0) \\ -x^2 & (x < 0). \end{cases}$$

Then $\phi_0''(x)$ does not exist and the solution (2.4) is easily seen to be

$$u(x,t) = \begin{cases} x^2 + t^2 & (x > t) \\ 2tx & (|x| \leq t) \\ -(x^2 + t^2) & (x < -t). \end{cases}$$

Hence, $u(x,t)$ is continuously differentiable, whereas the second derivatives of $u(x,t)$ fail to exist along the lines $x - t = 0$, $x + t = 0$. We note that these lines are in fact the characteristics of the wave equation (2.1) passing through the point (0,0) at which the initial data $\phi_0(x)$ fails to be twice continuously differentiable; we can say that the discontinuities in the initial data are propagated along the characteristics. This behavior is true in general for initial value problems for hyperbolic equations with variable coefficients.

2.2 THE CAUCHY PROBLEM FOR HYPERBOLIC EQUATIONS IN TWO INDEPENDENT VARIABLES

In the previous section we solved the Cauchy problem for the wave equation in two independent variables. We now want to give a method for solving Cauchy's problem for linear hyperbolic equations in two independent variables with variable coefficients. To this end, consider the hyperbolic equation (written in canonical form)

$$\frac{\partial^2 u}{\partial x \partial y} + a(x,y)\frac{\partial u}{\partial x} + b(x,y)\frac{\partial u}{\partial y} + c(x,y)u = f(x,y) \tag{2.5}$$

where $a(x,y)$, $b(x,y)$, $c(x,y)$, and $f(x,y)$ are real valued continuous functions. Note that the characteristics of the hyperbolic equation (2.5) are given by $x =$ constant and $y =$ constant. Let l be an arc described by both $y = g(x)$ and $x = h(y)$, where $g(x)$ and $h(y)$ are continuously differentiable such that the derivatives $g'(x)$ and $h'(y)$ do not vanish and note that such a parameterization of l implies that l is a noncharacteristic curve. The Cauchy problem for (2.5) is to find a solution in a neighborhood of l such that

$$u(x,g(x)) = \phi_0(x) \tag{2.6}$$

$$\frac{\partial u}{\partial y}(x,g(x)) = \phi_1(x),$$

where $\phi_0(x)$ and $\phi_1(x)$ are given functions such that $\phi_0(x)$ is twice continuously differentiable and $\phi_1(x)$ is continuously differentiable. Note that from (2.6) we have

$$\frac{\partial u}{\partial x}(x,g(x)) + \frac{\partial u}{\partial y}(x,g(x))g'(x) = \phi_0'(x),$$

so

$$\begin{aligned}\frac{\partial u}{\partial x}(x,g(x)) &= \phi_0'(x) - \phi_1(x)g'(x)\\ &= \phi_2(x).\end{aligned} \tag{2.7}$$

In order to solve the Cauchy problem for (2.5) we first reduce the second order equation (2.5) to a first order system of equations. This is accomplished by setting

$$\begin{aligned}v &= \frac{\partial u}{\partial x}\\ w &= \frac{\partial u}{\partial y},\end{aligned}$$

allowing us to produce the system

$$\begin{aligned}\frac{\partial v}{\partial y} &= f(x,y) - a(x,y)v - b(x,y)w - c(x,y)u\\ \frac{\partial w}{\partial x} &= f(x,y) - a(x,y)v - b(x,y)w - c(x,y)u\\ \frac{\partial u}{\partial y} &= w.\end{aligned} \tag{2.8}$$

Now let $ABCD$ be a rectangle containing l and $N(x,y)$ be an arbitrary point in $ABCD$. Let NP and NQ be characteristics through N intersecting l (see Figure 2.2). Then from (2.8) we have

$$\begin{aligned}v(x,y) &= \phi_2(x) + \int_{g(x)}^{y} [f(x,y) - a(x,y)v - b(x,y)w - c(x,y)u]\,dy\\ w(x,y) &= \phi_1(h(y)) + \int_{h(y)}^{x} [f(x,y) - a(x,y)v - b(x,y)w - c(x,y)u]\,dx\\ u(x,y) &= \phi_0(x) + \int_{g(x)}^{y} w(x,y)\,dy.\end{aligned} \tag{2.9}$$

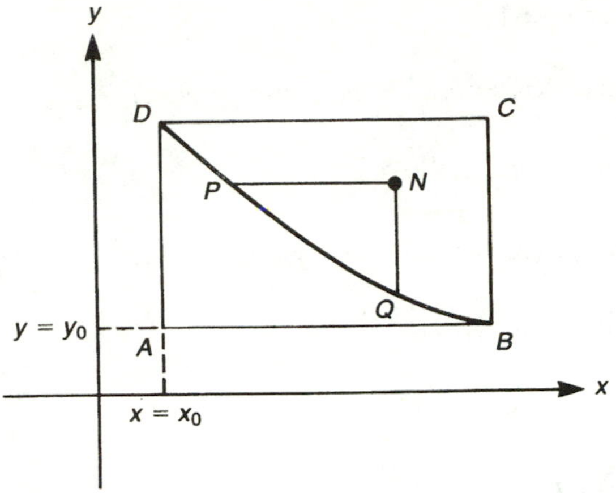

Figure 2.2

Any solution of the Cauchy problem as described by (2.5) and (2.6) will satisfy the system of integral equations (2.9). Conversely, if (u,v,w) is a continuous solution of (2.9), then $u(x,y)$ will be a solution of (2.5) and (2.6). Hence, we have reduced our problem to that of showing the existence of a continuous solution of the system of integral equations (2.9). This approach of converting an initial value or boundary value problem for a partial differential equation to the problem of solving an integral equation is a theme that will reoccur throughout this book and shows how the theories of partial differential equations and integral equations are so intimately related. Note that the integral equation (2.9) is a system of Volterra integral equations of the second kind (cf. Section 1.6).

We shall now solve the system of integral equations (2.9) by the method of **successive approximations.** Let $v_0(x,y) = \phi_2(x)$, $w_0(x,y) = \phi_1(h(y))$, $u_0(x,y) = \phi_0(x)$, and define the functions $v_n(x,y)$, $w_n(x,y)$, and $u_n(x,y)$ recursively for $n \geq 1$ by

$$
\begin{aligned}
v_n(x,y) &= \phi_2(x) + \int_{g(x)}^{y} [f(x,y) - a(x,y)v_{n-1} - b(x,y)w_{n-1} - c(x,y)u_{n-1}]\,dy \\
w_n(x,y) &= \phi_1(h(y)) + \int_{h(y)}^{x} [f(x,y) - a(x,y)v_{n-1} - b(x,y)w_{n-1} - c(x,y)u_{n-1}]\,dx \\
u_n(x,y) &= \phi_0(x) + \int_{g(x)}^{y} w_{n-1}(x,y)\,dy.
\end{aligned}
\tag{2.10}
$$

We shall first show that the sequence $\{v_n,w_n,u_n\}$ converges uniformly in the rectangle

$ABCD$. From (2.10) we know that for $n \geq 1$,

$$\begin{aligned} v_{n+1}(x,y) - v_n(x,y) &= -\int_{g(x)}^{y} [a(x,y)(v_n - v_{n-1}) + b(x,y)(w_n - w_{n-1}) \\ &\qquad + c(x,y)(u_n - u_{n-1})]\,dy, \\ w_{n+1}(x,y) - w_n(x,y) &= -\int_{h(y)}^{y} [a(x,y)(v_n - v_{n-1}) + b(x,y)(w_n - w_{n-1}) \\ &\qquad + c(x,y)(u_n - u_{n-1})]\,dx, \\ u_{n+1}(x,y) - u_n(x,y) &= \int_{g(x)}^{y} (w_n - w_{n-1})\,dy. \end{aligned} \tag{2.11}$$

We first show that in the rectangle $ABCD$

$$\begin{aligned} |v_n(x,y) - v_{n-1}(x,y)| &\leq M^{n-1}A\,\frac{(x + y - x_0 - y_0)^{n-1}}{(n-1)!} \\ |w_n(x,y) - w_{n-1}(x,y)| &\leq M^{n-1}A\,\frac{(x + y - x_0 - y_0)^{n-1}}{(n-1)!} \\ |u_n(x,y) - u_{n-1}(x,y)| &\leq M^{n-1}A\,\frac{(x + y - x_0 - y_0)^{n-1}}{(n-1)!} \end{aligned} \tag{2.12}$$

for $n \geq 1$ ($0! = 1$ by definition—see Section 6.1) where

$$M = \max_{(x,y)\in ABCD} (|a(x,y)| + |b(x,y)| + |c(x,y)|)$$

and A is a constant. The inequalities (2.12) are clearly true when $n = 1$ if A is large enough. In order to establish (2.12) by induction, assume that (2.12) is true for $n = j$. Then from (2.11) and (2.12), we have, in the rectangle $ABCD$,

$$\begin{aligned} |v_{j+1}(x,y) - v_j(x,y)| &\leq \int_{g(x)}^{y} (|a(x,y)| + |b(x,y)| + |c(x,y)|)|v_j - v_{j-1}|\,dy \\ &\leq \int_{g(x)}^{y} (|a(x,y)| + |b(x,y)| + |c(x,y)|)M^{j-1} \\ &\qquad \cdot A\,\frac{(x + y - x_0 - y_0)^{j-1}}{(j-1)!}\,dy \\ &\leq M^{j}A \int_{y_0}^{y} \frac{(x + y - x_0 - y_0)^{j-1}}{(j-1)!}\,dy \\ &= M^{j}A\left(\frac{(x + y - x_0 - y_0)^{j}}{j!} - \frac{(x - x_0)^{j}}{j!}\right) \\ &\leq M^{j}A\,\frac{(x + y - x_0 - y_0)^{j}}{j!}. \end{aligned}$$

The remaining two inequalities can be shown to be true for $n = j + 1$ in the same way, so (2.12) is valid for $n = j + 1$, completing the induction step. Thus, by induction, (2.12) is valid for all integers $n \geq 1$. Hence, each of the series

$$v_0 + \sum_{n=1}^{\infty} (v_n - v_{n-1})$$

$$w_0 + \sum_{n=1}^{\infty} (w_n - w_{n-1})$$

$$u_0 + \sum_{n=1}^{\infty} (u_n - u_{n-1})$$

is dominated in *ABCD* by the uniformly convergent series

$$|\phi_0(x)| + |\phi_1(h(y))| + |\phi_2(x)| + A \sum_{n=1}^{\infty} M^{n-1} \frac{(x + y - x_0 - y_0)^{n-1}}{(n-1)!}$$

and therefore are absolutely and uniformly convergent in *ABCD*. Since the uniform limit of continuous functions is continuous we can conclude that

$$\begin{aligned} v(x,y) &= \lim_{n\to\infty} v_n(x,y) \\ w(x,y) &= \lim_{n\to\infty} w_n(x,y) \\ u(x,y) &= \lim_{n\to\infty} u_n(x,y) \end{aligned} \tag{2.13}$$

exist and are continuous in *ABCD*.

The existence of a solution to the system of integral equations (2.9), and hence the Cauchy problem (2.5), (2.6), now follows from our discussion by letting n tend to infinity in (2.10), and (u,v,w) as defined by (2.13) is the solution of (2.9). Note that the interchange of limit and integration is justified by the fact that the limits in (2.13) are uniform. All that remains is to show uniqueness. Suppose there existed two solutions $u_1(x,y)$ and $u_2(x,y)$ of this Cauchy problem. Then $u_1 - u_2$ would be a solution of (2.5) and (2.6) with $f(x,y)$, $\phi_0(x)$, and $\phi_1(x)$ identically equal to zero. The solution of this problem is equivalent to the system of integral equations (2.9) with $f = \phi_0 = \phi_1 = \phi_2 = 0$ and $v = v_1 - v_2$, $w = w_1 - w_2$, $u = u_1 - u_2$ where $v_1 = \partial u_1/\partial x$, $v_2 = \partial u_2/\partial x$, $w_1 = \partial u_1/\partial y$, $w_2 = \partial u_2/\partial y$. The above induction analysis applied to (2.9) instead of (2.11) now shows that $|u(x,y)|$, $|v(x,y)|$, and $|w(x,y)|$ are bounded by

$$M^n B \frac{(x + y - x_0 - y_0)^n}{n!}$$

where $B = \max_{(x,y)\in ABCD} (|u(x,y)| + |v(x,y)| + |w(x,y)|)$, and n is an arbitrary integer. Letting n tend to infinity now shows that $v(x,y)$, $w(x,y)$, and $u(x,y)$ are identically zero. Hence $u_1(x,y) = u_2(x,y)$ in *ABCD*, so the solution of (2.5), (2.6) is unique.

We note that, in contrast to the wave equation in two independent variables, a closed form solution of (2.5) and (2.6) is in general impossible. However, analytic approximations with error estimates can be obtained through the recursion formula (2.10) and inequalities (2.12).

2.3 THE CAUCHY PROBLEM FOR THE WAVE EQUATION IN MORE THAN TWO INDEPENDENT VARIABLES

We now want to solve the Cauchy problem for the wave equation in three and four independent variables. We begin with the case of four independent variables and will later use the solution to this problem to obtain the corresponding solution for the case of three independent variables. Our present aim, therefore, is to solve

$$\frac{\partial^2 u}{\partial t^2} = \frac{\partial^2 u}{\partial x^2} + \frac{\partial^2 u}{\partial y^2} + \frac{\partial^2 u}{\partial z^2} \qquad (\mathbf{x} = (x,y,z) \in R^3,\ t \geq 0), \tag{2.13}$$

$$\left.\begin{aligned} u(\mathbf{x},0) &= \phi_0(\mathbf{x}) \\ \frac{\partial u}{\partial t}(\mathbf{x},0) &= \phi_1(\mathbf{x}) \end{aligned}\right\} \qquad (\mathbf{x} \in R^3), \tag{2.14}$$

where $\phi_0 \in C^3(R^3)$ and $\phi_1 \in C^2(R^3)$. To this end, we first show that if $\phi \in C^2(R^3)$ then

$$u(\mathbf{x},t) = \frac{1}{4\pi t}\int_{|\boldsymbol{\xi}-\mathbf{x}|=t} \phi(\boldsymbol{\xi})\,d\sigma \tag{2.15}$$

where $\mathbf{x} \in R^3$ is a solution of the wave equation (2.13). The integral in (2.15) is the surface integral of $\phi(\boldsymbol{\xi})$ over the sphere of radius t centered at $\mathbf{x}$. The right hand side of (2.15) is called the **spherical mean** of $\phi(\boldsymbol{\xi})$ and plays an important role in the theory of the wave equation and of more general hyperbolic equations. For a lucid exposition of the use of spherical means in the theory of partial differential equations we refer the reader to John 1955.

We see that in spherical coordinates (2.15) can be rewritten as

$$u(\mathbf{x},t) = \frac{t}{4\pi}\int_{|\boldsymbol{\xi}|=1} \phi(\mathbf{x} + \boldsymbol{\xi}t)\,d\sigma, \tag{2.16}$$

so

$$\begin{aligned} \Delta_3 u = \frac{\partial^2 u}{\partial x^2} + \frac{\partial^2 u}{\partial y^2} + \frac{\partial^2 u}{\partial z^2} &= \frac{t}{4\pi}\int_{|\boldsymbol{\xi}|=1} \Delta_3\phi(\mathbf{x} + \boldsymbol{\xi}t)\,d\sigma \\ &= \frac{1}{4\pi t}\int_{|\boldsymbol{\xi}-\mathbf{x}|=t} \Delta_3\phi(\boldsymbol{\xi})\,d\sigma. \end{aligned} \tag{2.17}$$

On the other hand, from (2.16) we have

$$\begin{aligned}\frac{\partial u}{\partial t} &= \frac{1}{4\pi}\int_{|\boldsymbol{\xi}|=1} \phi(\mathbf{x} + \boldsymbol{\xi}t)\, d\sigma + \frac{t}{4\pi}\int_{|\boldsymbol{\xi}|=1} \nabla\phi(\mathbf{x} + \boldsymbol{\xi}t)\cdot\boldsymbol{\xi}\, d\sigma \\ &= \frac{u(\mathbf{x},t)}{t} + \frac{1}{4\pi t}\int_{|\boldsymbol{\xi}-\mathbf{x}|=t} \nabla\phi(\boldsymbol{\xi})\cdot\boldsymbol{\nu}\, d\sigma\end{aligned} \tag{2.18}$$

where $\boldsymbol{\nu}$ is the unit outward normal to the sphere $|\boldsymbol{\xi} - \mathbf{x}| = t$. By the divergence theorem,

$$\iint_D \operatorname{div} \mathbf{F}(\mathbf{x})\, d\mathbf{x} = \int_{\partial D} \mathbf{F}(\mathbf{x})\cdot\boldsymbol{\nu}(\mathbf{x})\, d\sigma,$$

where D is a region in R^3 with smooth boundary ∂D having unit outward normal $\boldsymbol{\nu}(\mathbf{x})$, and $\mathbf{F}(\mathbf{x})$ is a vector valued continuously differentiable function defined on $\overline{D}$, we have from (2.18) that

$$\frac{\partial u}{\partial t} = \frac{u(\mathbf{x},t)}{t} + \frac{1}{4\pi t}\iint_{|\boldsymbol{\xi}-\mathbf{x}|\le t} \Delta_3\phi(\boldsymbol{\xi})\, d\boldsymbol{\xi}. \tag{2.19}$$

Setting

$$I(\mathbf{x},t) = \iint_{|\boldsymbol{\xi}-\mathbf{x}|\le t} \Delta_3\phi(\boldsymbol{\xi})\, d\boldsymbol{\xi}$$

we see that

$$\begin{aligned}\frac{\partial^2 u}{\partial t^2} &= -\frac{u}{t^2} + \frac{1}{t}\frac{\partial u}{\partial t} + \frac{1}{4\pi t}\frac{\partial I}{\partial t} - \frac{1}{4\pi t^2} I \\ &= \frac{1}{t}\left(\frac{\partial u}{\partial t} - \frac{u}{t} - \frac{1}{4\pi t} I\right) + \frac{1}{4\pi t}\frac{\partial I}{\partial t} = \frac{1}{4\pi t}\frac{\partial I}{\partial t}\,.\end{aligned} \tag{2.20}$$

Furthermore,

$$\frac{\partial I}{\partial t} = \int_{|\boldsymbol{\xi}-\mathbf{x}|=t} \Delta_3\phi(\boldsymbol{\xi})\, d\sigma, \tag{2.21}$$

and hence from (2.17), (2.20), and (2.21) we see that the spherical mean (2.15) is a solution of the wave equation (2.13).

With the above result we can now easily obtain the solution of Cauchy's problem (2.13), (2.14). To see this, we note that if $u(\mathbf{x},t)$ is defined by (2.15) then from (2.16) and (2.18) we have

$$u(\mathbf{x},0) = 0 \qquad \frac{\partial u}{\partial t}(\mathbf{x},0) = \phi(\mathbf{x}). \tag{2.22}$$

On the other hand, if $u(\mathbf{x},t)$ is a solution of the wave equation satisfying (2.22), then it is easily seen that $v = \partial u/\partial t$ is a solution of the wave equation satisfying

$$v(\mathbf{x},0) = \phi(\mathbf{x})$$

$$\frac{\partial v}{\partial t}(\mathbf{x},0) = 0.$$

Under the regularity assumptions on $\phi_0(\mathbf{x})$ and $\phi_1(\mathbf{x})$, we can now conclude that a solution of the Cauchy problem (2.13), (2.14) is given by

$$u(\mathbf{x},t) = \frac{1}{4\pi}\frac{\partial}{\partial t}\left[\frac{1}{t}\int_{|\boldsymbol{\xi}-\mathbf{x}|=t} \phi_0(\boldsymbol{\xi})\, d\sigma\right] + \frac{1}{4\pi t}\int_{|\boldsymbol{\xi}-\mathbf{x}|=t} \phi_1(\boldsymbol{\xi})\, d\sigma. \qquad \textbf{(2.23)}$$

Equation (2.23) is known as **Poisson's formula**. Note that, in contrast to d'Alembert's formula for the wave equation in two independent variables, we cannot conclude from our analysis that the solution of (2.13), (2.14) is unique, for we have not shown that every solution of (2.13), (2.22) can be represented in the form (2.15). We shall later show that the solution of (2.13), (2.14) is in fact unique. However, assuming this fact for the moment, we see from Poisson's formula that the solution of (2.13), (2.14) depends continuously (with respect to the maximum norm) on the data $\phi_1(\mathbf{x})$, $\phi_0(\mathbf{x})$, and $\nabla\phi_0(\mathbf{x})$.

For a given point $(\mathbf{x},t)$ the domain of dependence of $u(\mathbf{x},t)$ is the *surface* of a ball of radius t centered at $\mathbf{x}$. In particular, initial disturbances that vanish outside a ball centered at the origin in R^3 reach a point $\mathbf{x}$ outside this ball in a finite time, are then "heard" for a finite length of time, and then sharply disappear. This phenomenon is known as **Huygen's principle** and, as we shall soon see, is not true for the solution of Cauchy's problem for the wave equation in three independent variables.

We shall now use (2.23) and Hadamard's **method of descent** to solve the Cauchy problem for the wave equation in three independent variables, i.e., the problem

$$\frac{\partial^2 u}{\partial t^2} = \frac{\partial^2 u}{\partial x^2} + \frac{\partial^2 u}{\partial y^2} \qquad (\mathbf{x} = (x,y) \in R^2,\ t \geq 0) \qquad \textbf{(2.24)}$$

$$\left.\begin{aligned} u(\mathbf{x},0) &= \phi_0(\mathbf{x}) \\ \frac{\partial u}{\partial t}(\mathbf{x},0) &= \phi_1(\mathbf{x}) \end{aligned}\right\} \qquad (\mathbf{x} \in R^2), \qquad \textbf{(2.25)}$$

where $\phi_0 \in C^3(R^2)$ and $\phi_1 \in C^2(R^2)$. Observe that if in Poisson's formula $\phi_0(\mathbf{x})$ and $\phi_1(\mathbf{x})$ ($\mathbf{x} = (x,y,z)$) are independent of z, then (2.23) will provide the solution of (2.24), (2.25). This follows from the fact that here the integrals in (2.23) are invariant with respect to translations along the z axis. In order to simplify this expression, let $\boldsymbol{\nu}$ be the unit outward normal to the sphere $|\boldsymbol{\xi} - \mathbf{x}| = t$ and let $\mathbf{k}$ be the unit vector in

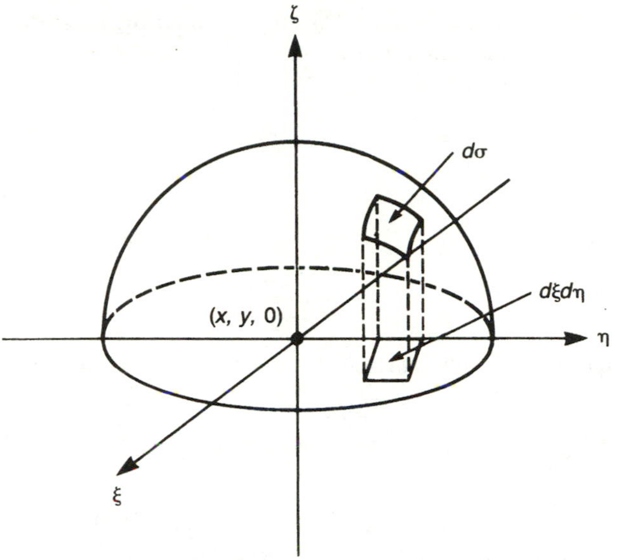

Figure 2.3

the ζ direction, $\boldsymbol{\xi} = (\xi,\eta,\zeta)$. Then, projecting $d\sigma$ onto the plane $\zeta = 0$, we have

$$\begin{aligned} d\xi d\eta &= \boldsymbol{\nu}\cdot\mathbf{k}\, d\sigma \\ &= \frac{1}{t}\sqrt{t^2 - (\xi - x)^2 - (\eta - y)^2}\, d\sigma \end{aligned}$$

where, without loss of generality, we have set $z = 0$ in (2.23) (see Figure 2.3). Remembering that a sphere has both an upper and lower hemisphere, we see that

$$\int_{|\boldsymbol{\xi}-\mathbf{x}|=t} \phi(\boldsymbol{\xi})\, d\sigma = 2t \iint_{\sqrt{(\xi-x)^2+(\eta-y)^2}\le t} \frac{\phi(\xi,\eta)\, d\xi d\eta}{\sqrt{t^2 - (\xi - x)^2 - (\eta - y)^2}} \, .$$

Letting $\boldsymbol{\xi} = (\xi,\eta)$, $\mathbf{x} = (x,y)$, we now see from (2.23) that a solution of the Cauchy problem (2.24), (2.25) is given by

$$u(\mathbf{x},t) = \frac{1}{2\pi}\frac{\partial}{\partial t}\iint_{|\boldsymbol{\xi}-\mathbf{x}|\le t} \frac{\phi_0(\boldsymbol{\xi})\, d\boldsymbol{\xi}}{\sqrt{t^2 - |\boldsymbol{\xi} - \mathbf{x}|^2}} + \frac{1}{2\pi}\iint_{|\boldsymbol{\xi}-\mathbf{x}|\le t} \frac{\phi_1(\boldsymbol{\xi})\, d\boldsymbol{\xi}}{\sqrt{t^2 - |\boldsymbol{\xi} - \mathbf{x}|^2}} \, . \qquad \textbf{(2.26)}$$

Note that for Cauchy's problem in three independent variables Huygen's principle is *not* valid—once a disturbance reaches $\mathbf{x} \in R^2$ (at time $t = t_0$, say) then it can be heard (although with diminishing amplitude) for all times $t \ge t_0$.

It still remains for us to prove the uniqueness of the solution of each of the Cauchy problems (2.13), (2.14) and (2.24), (2.25). We shall do this only for the problem (2.24), (2.25); the proof for (2.13), (2.14) proceeds in an identical manner.

Assuming that there are two solutions of (2.24), (2.25) and considering their difference shows that it suffices to prove that the solution of

$$\frac{\partial^2 u}{\partial t^2} = \frac{\partial^2 u}{\partial x^2} + \frac{\partial^2 u}{\partial y^2} \qquad (\mathbf{x} = (x,y) \in R^2,\ t \geq 0) \tag{2.27}$$

$$\left.\begin{aligned} u(\mathbf{x},0) &= 0 \\ \frac{\partial u}{\partial t}(\mathbf{x},0) &= 0 \end{aligned}\right\} \qquad (\mathbf{x} \in R^2) \tag{2.28}$$

is identically zero for $t > 0$. To this end, let $(x_0, y_0, t_0) \in R^3$, $t_0 > 0$, and consider the **characteristic cone**

$$(t - t_0)^2 - (x - x_0)^2 - (y - y_0)^2 = 0$$

which corresponds to the characteristic lines $(x - x_0) = \pm(t - t_0)$ for the wave equation in two independent variables. We want to show that $u(x_0, y_0, t_0) = 0$. In order to do this, let D be the region bounded by this cone and the plane $t = 0$ where Γ is the cone (with unit outward normal $\boldsymbol{\nu}$) and σ_0 is the disk cut out of the plane $t = 0$ by this cone (see Figure 2.4). Then from the identity

$$2\frac{\partial u}{\partial t}\left(\frac{\partial^2 u}{\partial t^2} - \frac{\partial^2 u}{\partial x^2} - \frac{\partial^2 u}{\partial y^2}\right) = \frac{\partial}{\partial t}\left(\left(\frac{\partial u}{\partial x}\right)^2 + \left(\frac{\partial u}{\partial y}\right)^2 + \left(\frac{\partial u}{\partial t}\right)^2\right) - 2\frac{\partial}{\partial x}\left(\frac{\partial u}{\partial t}\frac{\partial u}{\partial x}\right) - 2\frac{\partial}{\partial y}\left(\frac{\partial u}{\partial t}\frac{\partial u}{\partial y}\right)$$

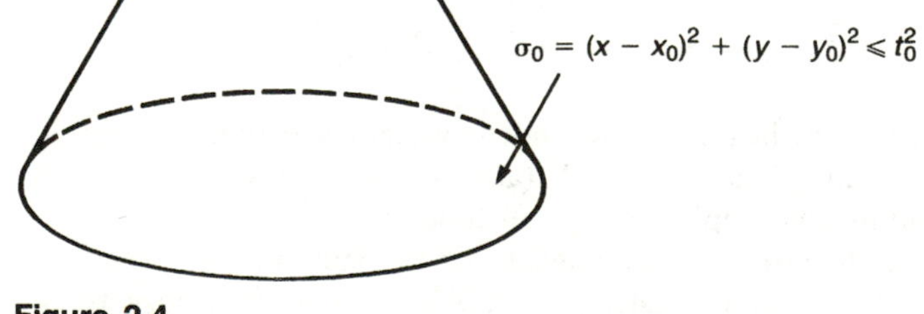

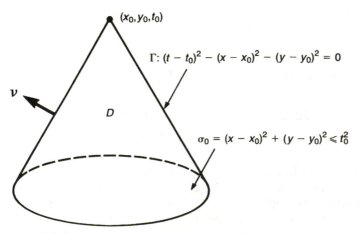

Figure 2.4

we see that if $u(\mathbf{x},t)$ is a solution of the wave equation (2.27) then

$$0 = \iint_D \left\{ \frac{\partial}{\partial t}\left(\left(\frac{\partial u}{\partial x}\right)^2 + \left(\frac{\partial u}{\partial y}\right)^2 + \left(\frac{\partial u}{\partial t}\right)^2 \right) - 2\frac{\partial}{\partial x}\left(\frac{\partial u}{\partial t}\frac{\partial u}{\partial x}\right) - 2\frac{\partial}{\partial y}\left(\frac{\partial u}{\partial t}\frac{\partial u}{\partial y}\right) \right\} d\mathbf{x}\, dt.$$

But from (2.28) we have

$$\frac{\partial u}{\partial x}(\mathbf{x},0) = \frac{\partial u}{\partial y}(\mathbf{x},0) = \frac{\partial u}{\partial t}(\mathbf{x},0) = 0,$$

and hence from the divergence theorem,

$$0 = \int_\Gamma \left\{ \left(\left(\frac{\partial u}{\partial x}\right)^2 + \left(\frac{\partial u}{\partial y}\right)^2 + \left(\frac{\partial u}{\partial t}\right)^2 \right)\mathbf{e}_t - 2\frac{\partial u}{\partial t}\frac{\partial u}{\partial x}\mathbf{e}_x - 2\frac{\partial u}{\partial t}\frac{\partial u}{\partial y}\mathbf{e}_y \right\} \cdot \boldsymbol{\nu}\, d\sigma \tag{2.29}$$

where $\mathbf{e}_t$, $\mathbf{e}_x$, $\mathbf{e}_y$ denote the unit vectors along the positive t, x, and y axes respectively. However, $\boldsymbol{\nu}$ is parallel to the vector

$$(t - t_0)\mathbf{e}_t - (x - x_0)\mathbf{e}_x - (y - y_0)\mathbf{e}_y,$$

which implies (using the equation for Γ) that

$$(\mathbf{e}_t \cdot \boldsymbol{\nu})^2 - (\mathbf{e}_x \cdot \boldsymbol{\nu})^2 - (\mathbf{e}_y \cdot \boldsymbol{\nu})^2 = 0.$$

Using the fact that $\mathbf{e}_t \cdot \boldsymbol{\nu} = \sqrt{2}/2$, we have from (2.29) that

$$\int_\Gamma \left\{ \left(\frac{\partial u}{\partial x}\mathbf{e}_t \cdot \boldsymbol{\nu} - \frac{\partial u}{\partial t}\mathbf{e}_x \cdot \boldsymbol{\nu} \right)^2 + \left(\frac{\partial u}{\partial y}\mathbf{e}_t \cdot \boldsymbol{\nu} - \frac{\partial u}{\partial t}\mathbf{e}_y \cdot \boldsymbol{\nu} \right)^2 \right\} d\sigma = 0,$$

i.e.,

$$\frac{\partial u}{\partial x}\mathbf{e}_t \cdot \boldsymbol{\nu} = \frac{\partial u}{\partial t}\mathbf{e}_x \cdot \boldsymbol{\nu}$$

$$\frac{\partial u}{\partial y}\mathbf{e}_t \cdot \boldsymbol{\nu} = \frac{\partial u}{\partial t}\mathbf{e}_y \cdot \boldsymbol{\nu}$$

or

$$\frac{\dfrac{\partial u}{\partial x}}{\mathbf{e}_x \cdot \boldsymbol{\nu}} = \frac{\dfrac{\partial u}{\partial y}}{\mathbf{e}_y \cdot \boldsymbol{\nu}} = \frac{\dfrac{\partial u}{\partial t}}{\mathbf{e}_t \cdot \boldsymbol{\nu}} = \lambda$$

where $\lambda = \lambda(\mathbf{x},t)$. Now let $\mathbf{l}$ be a unit vector lying along the generator of the characteristic cone. Then

$$\begin{aligned}\frac{\partial u}{\partial \mathbf{l}} &= \frac{\partial u}{\partial x}\mathbf{e}_x \cdot \mathbf{l} + \frac{\partial u}{\partial y}\mathbf{e}_y \cdot \mathbf{l} + \frac{\partial u}{\partial t}\mathbf{e}_t \cdot \mathbf{l} \\ &= \lambda((\mathbf{e}_x \cdot \boldsymbol{\nu})(\mathbf{e}_x \cdot \mathbf{l}) + (\mathbf{e}_y \cdot \boldsymbol{\nu})(\mathbf{e}_y \cdot \mathbf{l}) + (\mathbf{e}_t \cdot \boldsymbol{\nu})(\mathbf{e}_t \cdot \mathbf{l})) \\ &= \lambda \mathbf{l} \cdot \boldsymbol{\nu} \\ &= 0.\end{aligned}$$

Thus $u(\mathbf{x},t)$ is constant along the generator, and since $u(\mathbf{x},0) = 0$ we have that $u(\mathbf{x},t) = 0$ along this generator. It follows that $u(x_0,y_0,t_0) = 0$ and our uniqueness proof is finished.

We conclude this section by considering the nonhomogeneous wave equation in four independent variables. By using identical methods, we can derive analogous results for the wave equation in two and three independent variables, and we leave the derivation of such results to the Exercises. Our present aim is to construct a solution to the nonhomogeneous problem

$$\frac{\partial^2 u}{\partial t^2} = \frac{\partial^2 u}{\partial x^2} + \frac{\partial^2 u}{\partial y^2} + \frac{\partial^2 u}{\partial z^2} + f(x,y,z,t) \qquad (\mathbf{x} = (x,y,z) \in R^3,\ t \geq 0) \tag{2.30}$$

$$\left.\begin{aligned} u(\mathbf{x},0) &= 0 \\ \frac{\partial u}{\partial t}(\mathbf{x},0) &= 0 \end{aligned}\right\} \qquad (\mathbf{x} \in R^3) \tag{2.31}$$

where $f(x,y,z,t)$ is assumed to be twice continuously differentiable for $\mathbf{x} \in R^3$ and $t \geq 0$. The solution of the nonhomogeneous Cauchy problem (2.30), (2.14) can be obtained by adding the solution of (2.13), (2.14) to the solution of (2.30), (2.31). In order to construct a solution of (2.30), (2.31), we first consider the Cauchy problem

$$\frac{\partial^2 v}{\partial t^2} = \frac{\partial^2 v}{\partial x^2} + \frac{\partial^2 v}{\partial y^2} + \frac{\partial^2 v}{\partial z^2} \qquad (\mathbf{x} \in R^3,\ t \geq \tau) \tag{2.32}$$

$$\left.\begin{aligned} v(\mathbf{x},\tau) &= 0 \\ \frac{\partial v}{\partial t}(\mathbf{x},\tau) &= f(\mathbf{x},\tau) \end{aligned}\right\} \qquad (\mathbf{x} \in R^3) \tag{2.33}$$

where τ is a parameter. Then from Poisson's formula (making the change of variables $t \to t - \tau$) we have

$$v(\mathbf{x},t;\tau) = \frac{t-\tau}{4\pi}\int_{|\boldsymbol{\xi}|=1} f(\mathbf{x} + \boldsymbol{\xi}(t-\tau),\tau)\,d\sigma. \tag{2.34}$$

We now show that the solution of (2.30), (2.31) is given by

$$u(\mathbf{x},t) = \int_0^t v(\mathbf{x},t;\tau)\,d\tau. \tag{2.35}$$

Note that from (2.32), (2.33), and (2.34) we have (where $\Delta_3 u = \partial^2 u/\partial x^2 + \partial^2 u/\partial y^2 + \partial^2 u/\partial z^2$)

$$\Delta_3 u = \int_0^t \Delta_3 v(\mathbf{x},t;\tau)\,d\tau$$

$$\frac{\partial u}{\partial t} = \int_0^t \frac{\partial v}{\partial t}(\mathbf{x},t;\tau)\,d\tau + v(\mathbf{x},t;t) = \int_0^t \frac{\partial v}{\partial t}(\mathbf{x},t;\tau)\,d\tau$$

$$\frac{\partial^2 u}{\partial t^2} = \int_0^t \frac{\partial^2 v}{\partial t^2}(\mathbf{x},t;\tau)\,d\tau + \frac{\partial v}{\partial t}(\mathbf{x},t;t) = \int_0^t \frac{\partial^2 v}{\partial t^2}(\mathbf{x},t;\tau)\,d\tau + f(\mathbf{x},t),$$

and hence $u(\mathbf{x},t)$ as defined by (2.35) is a solution of (2.30), (2.31). Substituting (2.34) into (2.35) now gives

$$\begin{aligned} u(\mathbf{x},t) &= \frac{1}{4\pi}\int_0^t (t-\tau)\int_{|\xi|=1} f(\mathbf{x}+\xi(t-\tau),\tau)\,d\sigma\,d\tau \\ &= \frac{1}{4\pi}\int_0^t r\int_{|\xi|=1} f(\mathbf{x}+\xi r, t-r)\,d\sigma\,dr \qquad \textbf{(2.36)} \\ &= \frac{1}{4\pi}\iiint_{|\xi-\mathbf{x}|\le t} \frac{f(\xi, t-|\xi-\mathbf{x}|)}{|\xi-\mathbf{x}|}\,d\xi. \end{aligned}$$

Equation (2.36) is called the **retarded potential.**

In the same manner (cf. Exercise 15), it can be shown that the solution of

$$\frac{\partial^2 u}{\partial t^2} = \frac{\partial^2 u}{\partial x^2} + \frac{\partial^2 u}{\partial y^2} + f(\mathbf{x},t) \qquad (\mathbf{x}=(x,y)\in R^2,\ t\ge 0)$$

$$\left.\begin{aligned} u(\mathbf{x},0) &= 0 \\ \frac{\partial u}{\partial t}(\mathbf{x},0) &= 0 \end{aligned}\right\} \qquad (\mathbf{x}\in R^2)$$

is given by

$$u(\mathbf{x},t) = \frac{1}{2\pi}\int_0^t \iint_{|\xi-\mathbf{x}|\le t-\tau} \frac{f(\xi,\tau)\,d\xi\,d\tau}{\sqrt{(t-\tau)^2-|\xi-\mathbf{x}|^2}} \qquad \textbf{(2.37)}$$

and the solution of

$$\frac{\partial^2 u}{\partial t^2} = \frac{\partial^2 u}{\partial x^2} + f(x,t) \qquad (-\infty < x < \infty,\ t\ge 0)$$

$$\left.\begin{aligned} u(x,0) &= 0 \\ \frac{\partial u}{\partial t}(x,0) &= 0 \end{aligned}\right\} \qquad (-\infty < x < \infty)$$

is given by

$$u(x,t) = \frac{1}{2}\int_0^t \int_{x-(t-\tau)}^{x+(t-\tau)} f(\xi,\tau)\,d\xi d\tau, \tag{2.38}$$

subject to appropriate smoothness assumptions on the nonhomogeneous term.

2.4 THE INITIAL-BOUNDARY VALUE PROBLEM FOR THE WAVE EQUATION IN TWO INDEPENDENT VARIABLES

We now leave our study of the Cauchy problem for the wave equation and begin to examine initial-boundary value problems. The physical applications of such problems have already been indicated in Section 1.1; the boundary condition represents the situation in which there are non-penetrable inhomogeneities in a medium that reflect waves generated by the source terms prescribed at time $t = 0$. We shall study problems of time harmonic wave motion in some detail in Chapter 6, where because of the assumption of time harmonic wave motion we will no longer need to prescribe initial conditions. In this and the following section we shall not assume that the wave motion is time harmonic; instead, we show how the initial-boundary value problem for the wave equation can be solved for simple geometries by the method of **separation of variables.** The physical manifestations of the problems we are considering include the mathematical modeling of vibrating strings and vibrating membranes; however, we choose not to emphasize this physical motivation, preferring instead to view the initial-boundary value problems considered in this chapter as models for those mentioned in Section 1.1.

Our aim in this section is to study the simple initial-boundary value problem

$$\frac{\partial^2 u}{\partial t^2} = \frac{\partial^2 u}{\partial x^2} \qquad (0 < x < l,\ t > 0) \tag{2.39}$$

$$u(0,t) = u(l,t) = 0 \qquad (t \geq 0) \tag{2.40}$$

$$\left.\begin{aligned} u(x,0) &= \phi_0(x) \\ \frac{\partial u}{\partial t}(x,0) &= \phi_1(x) \end{aligned}\right\} \qquad (0 \leq x \leq l) \tag{2.41}$$

where $u(x,t)$ continuously assumes its initial-boundary data, l is a positive constant, and the functions $\phi_0(x)$ and $\phi_1(x)$ satisfy the compatibility condition $\phi_0(0) = \phi_0(l) = \phi_1(0) = \phi_1(l) = 0$. We first need to show that there is at most one solution to the initial-boundary value problem (2.39)–(2.41), i.e. (considering the difference of two solutions, assuming that they exist), to show that if $\phi_0(x)$ and $\phi_1(x)$ are identically zero then $u(x,t)$ is identically zero in the rectangle

$$R = \{(x,t):\ \ 0 < x < l,\ \ 0 < t < T\}$$

where T is an arbitrary, but fixed, positive number. Assume that $u \in C^2(R) \cap C^1(\overline{R})$

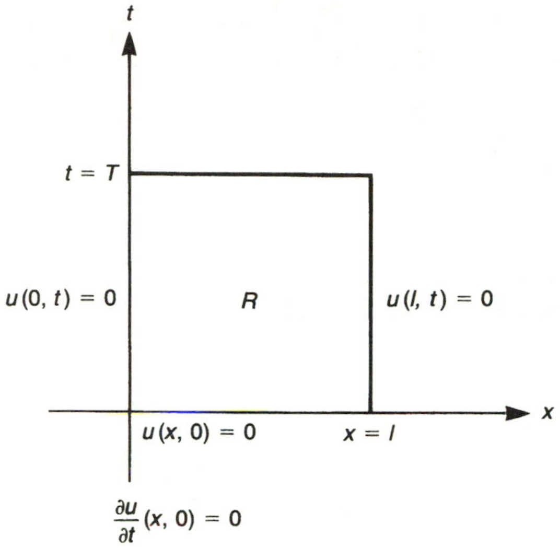

Figure 2.5

and consider the **energy integral**

$$E(t) = \frac{1}{2}\int_0^l \left\{\left(\frac{\partial u}{\partial t}\right)^2 + \left(\frac{\partial u}{\partial x}\right)^2\right\} dx,$$

which can be interpreted as the sum of the potential and kinetic energies of the physical problem modeled by (2.39)–(2.41) (e.g., a vibrating string, with $u(x,t)$ representing the vertical displacement of the string at position x and time t).

Then, since $\phi_0(x)$ and $\phi_1(x)$ are identically zero, we have $E(0) = 0$. Furthermore,

$$\frac{dE(t)}{dt} = \int_0^l \left\{\frac{\partial u}{\partial t}\frac{\partial^2 u}{\partial t^2} + \frac{\partial u}{\partial x}\frac{\partial^2 u}{\partial x \partial t}\right\} dx$$

$$= \int_0^l \left(\frac{\partial^2 u}{\partial t^2} - \frac{\partial^2 u}{\partial x^2}\right)\frac{\partial u}{\partial t}\, dx + \left.\frac{\partial u}{\partial x}\frac{\partial u}{\partial t}\right|_{x=0}^{x=l} = 0$$

from (2.39) and (2.40). Hence $E(t)$ is independent of t, and since $E(0) = 0$, $E(t) = 0$ for $0 \le t \le T$. Thus $\partial u(x,t)/\partial t = \partial u(x,t)/\partial x = 0$ in Q, and $u(x,t)$ equals a constant in R. But since $u \in C^1(\overline{R})$ and $u(x,0) = 0$, we know that $u(x,t)$ is identically zero in R.

Assuming again that $u \in C^2(R) \cap C^1(\overline{R})$, we now want to show that the solution $u(x,t)$ of (2.39)–(2.41) depends continuously on $\phi_0'(x)$ and $\phi_1(x)$ with respect to the maximum norm. The above analysis implies that even if $\phi_0(x)$ and $\phi_1(x)$ are not identically zero, we still have $E(t) = E(0)$, i.e., $E(t)$ is a constant. Hence, if $u(x,t)$ is a solution of (2.39)–(2.41),

$$\int_0^l \left\{\left(\frac{\partial u}{\partial t}\right)^2 + \left(\frac{\partial u}{\partial x}\right)^2\right\} dx = \int_0^l \{(\phi_1(x))^2 + (\phi_0'(x))^2\} dx$$

for $0 \leq t \leq T$. Therefore, if $u^{(1)}(x,t)$ is the solution with initial data $\phi_0^{(1)}(x)$ and $\phi_1^{(1)}(x)$, and $u^{(2)}(x,t)$ is the solution with initial data $\phi_0^{(2)}(x)$ and $\phi_1^{(2)}(x)$ such that for some positive constant ε,

$$\max_{0 \leq x \leq l} \left| \frac{d}{dx}\phi_0^{(1)}(x) - \frac{d}{dx}\phi_0^{(2)}(x) \right| < \varepsilon$$

$$\max_{0 \leq x \leq l} |\phi_1^{(1)}(x) - \phi_1^{(2)}(x)| < \varepsilon,$$

then for $u(x,t) = u^{(1)}(x,t) - u^{(2)}(x,t)$ the above equality implies that

$$\int_0^l \left\{ \left(\frac{\partial u}{\partial t}\right)^2 + \left(\frac{\partial u}{\partial x}\right)^2 \right\} dx \leq 2\varepsilon^2 l \qquad (0 \leq t \leq T).$$

Because

$$\begin{aligned} u(x,t) &= \int_0^x \frac{\partial u}{\partial x}\, dx + u(0,t) \\ &= \int_0^x \frac{\partial u}{\partial x}\, dx, \end{aligned}$$

the Schwarz inequality implies that for $(x,t) \in R$

$$\begin{aligned} |u(x,t)| &\leq \int_0^l \left|\frac{\partial u}{\partial x}\right| dx \\ &\leq \left[\int_0^l \left(\frac{\partial u}{\partial x}\right)^2 dx \int_0^l dx \right]^{1/2} \\ &\leq \sqrt{2}\,\varepsilon l, \end{aligned}$$

which establishes the desired result on continuous dependence of the solution of (2.39)–(2.41) on the initial data.

We now use the method of separation of variables, or **Fourier's method,** to construct a solution to our initial-boundary value problem. We begin by looking for a nontrivial solution of (2.39) in the form

$$u(x,t) = X(x)T(t)$$

where $X(x)$ is a function only of x and $T(t)$ is a function only of t such that (2.40) is satisfied. Substituting into (2.39) gives

$$T''X = TX''$$

or

$$\frac{T''}{T} = \frac{X''}{X}$$

for points (x,t) where $u(x,t)$ does not equal zero. But the left side of this equation is a function only of t and the right side is a function only of x. Hence

$$\frac{T''}{T} = \frac{X''}{X} = \text{constant}.$$

We set this constant equal to $-\lambda$ (no assumption is made at this point as to whether λ is positive or negative) and arrive at the differential equations

$$T'' + \lambda T = 0 \tag{2.42}$$

$$X'' + \lambda X = 0. \tag{2.43}$$

From the boundary condition (2.40),

$$X(0) = X(l) = 0. \tag{2.44}$$

From the elementary theory of ordinary differential equations it is easily seen that (2.43), (2.44) can have a nontrivial solution only if $\lambda > 0$, and the solution of (2.43) is

$$X(x) = c_1 \cos \sqrt{\lambda} x + c_2 \sin \sqrt{\lambda} x$$

where c_1 and c_2 are arbitrary constants. Since we want $X(0) = 0$, we must have $c_1 = 0$, and $X(l) = 0$ implies that

$$c_2 \sin \sqrt{\lambda} l = 0.$$

In order for $X(x)$ not to vanish identically we now must have that

$$\lambda = \left(\frac{k\pi}{l}\right)^2 \qquad (k = 1,2,\ldots),$$

i.e.,

$$X(x) = X_k(x) = \sin \frac{k\pi x}{l}$$

and we have set $c_2 = 1$. From (2.42),

$$T(x) = T_k(x) = a_k \cos \frac{k\pi t}{l} + b_k \sin \frac{k\pi t}{l},$$

where a_k and b_k are arbitrary constants. Hence, we have constructed a set of solutions to the wave equation (2.39) satisfying the boundary condition (2.40) given by

$$u_k(x,t) = \left(a_k \cos \frac{k\pi t}{l} + b_k \sin \frac{k\pi t}{l}\right) \sin \frac{k\pi x}{l}.$$

In order to satisfy the initial condition (2.41) we consider now the infinite series

$$u(x,t) = \sum_{k=1}^{\infty} \left(a_k \cos \frac{k\pi t}{l} + b_k \sin \frac{k\pi t}{l}\right) \sin \frac{k\pi x}{l}. \tag{2.45}$$

The function $u(x,t)$, as defined here, will be a solution of (2.39), (2.40) provided we can differentiate the series termwise with respect to x and t twice. In order to satisfy (2.41) we must have

$$\phi_0(x) = u(x,0) = \sum_{k=1}^{\infty} a_k \sin\frac{k\pi x}{l}$$

$$\phi_1(x) = \frac{\partial u}{\partial t}(x,0) = \frac{\pi}{l}\sum_{k=1}^{\infty} kb_k \sin\frac{k\pi x}{l}$$

and from the theory of Fourier series (cf. Section 1.4),

$$\begin{aligned} a_k &= \frac{2}{l}\int_0^l \phi_0(x)\sin\frac{k\pi x}{l}\,dx \\ b_k &= \frac{2}{k\pi}\int_0^l \phi_1(x)\sin\frac{k\pi x}{l}\,dx. \end{aligned} \tag{2.46}$$

Hence, the solution of (2.39)–(2.41) is given by (2.45), (2.46), provided that the series can be differentiated termwise twice and continuously assumes the initial data (2.41) having the given Fourier series expansions.

THEOREM 14

Let $\phi_0 \in C^3[0,l]$, $\phi_1 \in C^2[0,l]$, such that $\phi_0(0) = \phi_0(l) = 0$, $\phi_0''(0) = \phi_0''(l) = 0$, and $\phi_1(0) = \phi_1(l) = 0$. Then the function $u(x,t)$ as given by (2.45), (2.46) has continuous second derivatives in $\overline{R} = \{(x,t)\colon 0 \le x \le l,\ 0 \le t \le T\}$ and satisfies (2.39)–(2.41).

■ ***Proof.*** Integrating (2.46) by parts gives

$$a_k = -\frac{2}{l}\left(\frac{l}{\pi k}\right)^3 \int_0^l \phi_0'''(x)\cos\frac{k\pi x}{l}\,dx$$

$$b_k = -\frac{2}{l}\left(\frac{l}{\pi k}\right)^3 \int_0^l \phi_1''(x)\sin\frac{k\pi x}{l}\,dx.$$

Hence for every positive integer N,

$$\frac{2}{l}\int_0^l \left|\phi_0'''(x) + \left(\frac{\pi}{l}\right)^3 \sum_{k=1}^{N} k^3 a_k \cos\frac{k\pi x}{l}\right|^2 dx =$$
$$\frac{2}{l}\int_0^l |\phi_0'''(x)|^2\,dx - 2\left(\frac{\pi}{l}\right)^6 \sum_{k=1}^{N} k^6|a_k|^2 + \left(\frac{\pi}{l}\right)^6 \sum_{k=1}^{N} k^6|a_k|^2,$$

i.e.,

$$\left(\frac{\pi}{l}\right)^6 \sum_{k=1}^{N} k^6|a_k|^2 \le \frac{2}{l}\int_0^l |\phi_0'''(x)|^2\,dx.$$

We can now conclude that

$$\sum_{k=1}^{\infty} k^6|a_k|^2$$

is convergent. Similarly,

$$\sum_{k=1}^{\infty} k^6|b_k|^2$$

is convergent. But

$$2k^2|a_k| = \left(\frac{1}{k^2} + k^6|a_k|^2\right) - \left(\frac{1}{k} - k^3|a_k|\right)^2,$$

and hence

$$k^2|a_k| \leq \frac{1}{2}\left(\frac{1}{k^2} + k^6|a_k|^2\right).$$

Similarly,

$$k^2|b_k| \leq \frac{1}{2}\left(\frac{1}{k^2} + k^6|b_k|^2\right).$$

The above inequalities imply that the series

$$\sum_{k=1}^{\infty} k^2|a_k|$$

$$\sum_{k=1}^{\infty} k^3|b_k|$$

are convergent, and we can now conclude that the series (2.45), (2.46) converges absolutely and uniformly and can be differentiated termwise twice with respect to x and t. The assumptions on $\phi_0(x)$ and $\phi_1(x)$ also guarantee that $u(x,t)$ as defined by (2.45) and (2.46) continuously assumes the initial data (2.41), since $\phi_0(x)$ and $\phi_1(x)$ equal their Fourier series expansion for $0 \leq x \leq l$ (cf. Section 1.4). The theorem is now proved. ■

The method of separation of variables can also be applied in a straightforward manner to problems of the form (2.39)–(2.41) where instead of prescribing the boundary data $u(0,t) = u(l,t) = 0$ we prescribe the conditions $\partial u(0,t)/\partial x = \partial u(l,t)/\partial x = 0$.

2.5 FOURIER'S METHOD FOR THE WAVE EQUATION IN THREE INDEPENDENT VARIABLES

We now want to generalize the method of separation of variables described above to the case of the wave equation in three independent variables. Such a generalization only works for cylindrical domains whose cross section is of a particularly simple shape, e.g., a rectangle or a disk. We begin with the case of a rectangle. Consider the problem of finding the solution of the following initial-boundary value problem for the wave equation in a rectangular cylinder: find a function $u(x,y,t)$ such that

$$\frac{\partial^2 u}{\partial t^2} = \frac{\partial^2 u}{\partial x^2} + \frac{\partial^2 u}{\partial y^2} \qquad ((x,y) \in R,\ t > 0) \tag{2.47}$$

$$\begin{aligned} u(0,y,t) &= u(a,y,t) = 0 \qquad (0 \le y \le b,\ t \ge 0) \\ u(x,0,t) &= u(x,b,t) = 0 \qquad (0 \le x \le a,\ t \ge 0) \end{aligned} \tag{2.48}$$

$$\left.\begin{aligned} u(x,y,0) &= \phi_0(x,y) \\ \frac{\partial u}{\partial t}(x,y,0) &= \phi_1(x,y) \end{aligned}\right\} \qquad ((x,y) \in R) \tag{2.49}$$

when $u(x,y,t)$ continuously assumes its initial-boundary data, $R = \{(x,y): 0 < x < a,\ 0 < y < b\}$, and a and b are positive constants. Assuming that $\phi_0(x,y) = \phi_1(x,y)$ are smooth enough, we shall formally construct a solution to (2.47)–(2.49) by Fourier's method.

We first look for a nontrivial solution of (2.47) in the form

$$u(x,y,t) = T(t)v(x,y)$$

such that (2.48) is satisfied. Substituting into (2.47) shows that (cf. Section 2.4)

$$\frac{T''}{T} = \frac{\dfrac{\partial^2 v}{\partial x^2} + \dfrac{\partial^2 v}{\partial y^2}}{v} = -k^2,$$

where k is a constant (we are anticipating that, as in the case of the wave equation in two independent variables, the separation constant is a nonpositive number). Hence,

$$T'' + k^2 T = 0 \tag{2.50}$$

$$\frac{\partial^2 v}{\partial x^2} + \frac{\partial^2 v}{\partial y^2} + k^2 v = 0 \qquad ((x,y) \in R) \tag{2.51}$$

$$\begin{aligned} v(0,y) &= v(a,y) = 0 \qquad (0 \le y \le b) \\ v(x,0) &= v(x,b) = 0 \qquad (0 \le x \le a). \end{aligned} \tag{2.52}$$

Instead of an eigenvalue problem for an ordinary differential equation as in Section 2.4 we have an eigenvalue problem (2.51), (2.52) for a partial differential equation. Such problems will be discussed in Chapter 5 for domains of arbitrary shape; however, for the moment, we shall show how the eigenvalue problem (2.51), (2.52) can be solved by the method of separation of variables. (This is possible because the domain of definition of $v(x,y)$ is particularly simple, in this case a rectangle.) Let

$$v(x,y) = X(x)Y(y).$$

Then, substituting into (2.51), we have (after dividing by $X(x)Y(y)$)

$$\frac{Y''}{Y} + k^2 = -\frac{X''}{X} = k_1^2,$$

where k_1 is a constant, or

$$X'' + k_1^2 X = 0$$

$$Y'' + k_2^2 Y = 0$$

where $k^2 = k_1^2 + k_2^2$. Solving these differential equations gives

$$X(x) = c_1 \cos k_1 x + c_2 \sin k_1 x$$

$$Y(y) = c_3 \cos k_2 y + c_4 \sin k_2 y,$$

where c_1, c_2, c_3, and c_4 are arbitrary constants. From (2.52) we have that

$$X(0) = X(a) = 0$$

$$Y(0) = Y(b) = 0$$

and setting $c_2 = c_4 = 1$,

$$X(x) = \sin k_1 x$$

$$Y(y) = \sin k_2 y$$

where

$$\sin k_1 a = 0$$

$$\sin k_2 b = 0.$$

Hence

$$k_1 = k_{1m} = \frac{m\pi}{a} \qquad (m = 1,2,\ldots)$$

$$k_2 = k_{2n} = \frac{n\pi}{b} \qquad (n = 1,2,\ldots)$$

i.e., the eigenvalues of (2.51), (2.52) are given by

$$k_{mn}^2 = k_{1m}^2 + k_{2n}^2$$

$$= \pi^2\left(\frac{m^2}{a^2} + \frac{n^2}{b^2}\right) \qquad (m,n = 1,2,3,\ldots)$$

with corresponding eigenfunctions

$$v_{mn}(x,y) = \sin\frac{m\pi x}{a}\sin\frac{n\pi y}{b}.$$

The solution of (2.50) is now given by

$$T_{mn}(t) = a_{mn} \cos k_{mn}t + b_{mn} \sin k_{mn}t$$

and hence formally

$$u(x,y,t) = \sum_{n=1}^{\infty} \sum_{m=1}^{\infty} (a_{mn} \cos k_{mn}t + b_{mn} \sin k_{mn}t) \sin \frac{m\pi x}{a} \sin \frac{n\pi y}{b} \tag{2.53}$$

will be a solution of (2.47), (2.48). In order to satisfy (5.3) we must have

$$u(x,y,0) = \phi_0(x,y) = \sum_{n=1}^{\infty} \sum_{m=1}^{\infty} a_{mn} \sin \frac{m\pi x}{a} \sin \frac{n\pi y}{b}$$

$$\frac{\partial u}{\partial t}(x,y,0) = \phi_1(x,y) = \sum_{n=1}^{\infty} \sum_{m=1}^{\infty} b_{mn}k_{mn} \sin \frac{m\pi x}{a} \sin \frac{n\pi y}{b} .$$

From Fourier's theorem we should choose the coefficients a_{mn}, b_{mn} to be

$$\begin{aligned} a_{mn} &= \frac{4}{ab} \int_0^b \int_0^a \phi_0(x,y) \sin \frac{m\pi x}{a} \sin \frac{n\pi y}{b}\, dxdy \\ b_{mn} &= \frac{4}{abk_{mn}} \int_0^b \int_0^a \phi_1(x,y) \sin \frac{m\pi x}{a} \sin \frac{n\pi y}{b}\, dxdy. \end{aligned} \tag{2.54}$$

Equations (2.53) and (2.54) define the formal solution of (2.47)–(2.49). In order to verify this as a solution, we must make smoothness assumptions on $\phi_0(x,y) = \phi_1(x,y)$ and argue as in the previous section.

We conclude this section by considering an initial-boundary value problem for the wave equation in a circular cylinder. This will lead us to the need to study certain properties of Bessel functions. (We postpone a more complete study of these important special functions until Chapter 6 where we will need them in order to discuss time harmonic wave propagation in an exterior domain.) In polar coordinates $x = r \cos \theta$, $y = r \sin \theta$, the wave equation

$$\frac{\partial^2 u}{\partial t^2} = \frac{\partial^2 u}{\partial x^2} + \frac{\partial^2 u}{\partial y^2}$$

becomes

$$\frac{\partial^2 u}{\partial t^2} = \frac{\partial^2 u}{\partial r^2} + \frac{1}{r}\frac{\partial u}{\partial r} + \frac{1}{r^2}\frac{\partial^2 u}{\partial \theta^2} .$$

If we consider the wave equation in the circular cylinder $\{(r,\theta,t): 0 < r < R, 0 \leq \theta \leq 2\pi, t > 0\}$ it makes sense to prescribe the initial-boundary data

$$u(R,\theta,t) = 0 \qquad (0 \leq \theta \leq 2\pi, t \geq 0)$$

$$\left.\begin{aligned} u(r,\theta,0) &= \phi_0(r,\theta) \\ \frac{\partial u}{\partial t}(r,\theta,0) &= \phi_1(r,\theta) \end{aligned}\right\} \qquad (0 < r < R, 0 \leq \theta \leq 2\pi).$$

We shall consider only the special case when $\phi_0(r,\theta)$ and $\phi_1(r,\theta)$ are independent of θ and look for a solution $u(r,\theta,t)$ of the above initial-boundary value problem that is also independent of θ. In this case, the above equations become

$$\frac{\partial^2 u}{\partial t^2} = \frac{\partial^2 u}{\partial r^2} + \frac{1}{r}\frac{\partial u}{\partial r} \qquad (0 < r < R,\ t > 0) \tag{2.55}$$

$$u(R,t) = 0 \qquad (t \geq 0) \tag{2.56}$$

$$\left.\begin{aligned} u(r,0) &= \phi_0(r) \\ \frac{\partial u}{\partial t}(r,0) &= \phi_1(r) \end{aligned}\right\} \qquad (0 \leq r \leq R). \tag{2.57}$$

Our aim is to apply Fourier's method to (2.55)–(2.57). To this end we look for a nontrivial solution of (2.55), (2.56) in the form

$$u(r,t) = T(t)W(r).$$

Substituting into (2.55) and dividing by $T(t)W(r)$ gives

$$\frac{T''}{T} = \frac{W'' + \frac{1}{r}W'}{W} = -k^2$$

where k is a constant (anticipating as usual that the separation constant is nonpositive). Hence we have that

$$T'' + k^2 T = 0 \tag{2.58}$$

$$W'' + \frac{1}{r}W' + k^2 W = 0. \tag{2.59}$$

Equation (2.59) is called Bessel's equation. The indicial equation is $m^2 = 0$, so there is only one solution of (2.59) that is bounded at the origin. This solution can be found by power series methods and, if normalized such that it equals 1 at $r = 0$, is called the **Bessel function (of order zero),** denoted by $J_0(kr)$:

$$J_0(kr) = \sum_{n=0}^{\infty} \frac{(-1)^n}{n!n!}\left(\frac{kr}{2}\right)^{2n}. \tag{2.60}$$

By the ratio test, the series (2.60) converges for all values of r. In order to satisfy the boundary condition (2.56) we must have that

$$J_0(kR) = 0.$$

We shall show later (in Section 6.2.5) that $J_0(\mu)$ has an infinite number of real zeros.

Assuming this fact for now, let $\mu_1, \mu_2, \ldots$ be the positive roots of $J_0(\mu) = 0$ and set

$$k_n = \frac{\mu_n}{R}.$$

Then $J_0(k_n R) = 0$. From (2.58),

$$T_n(t) = a_n \cos k_n t + b_n \sin k_n t$$

where a_n and b_n are arbitrary constants. Hence

$$u(r,t) = \sum_{n=1}^{\infty} \left(a_n \cos \frac{\mu_n t}{R} + b_n \sin \frac{\mu_n t}{R} \right) J_0 \left(\frac{\mu_n r}{R} \right) \tag{2.61}$$

is a formal solution of (2.55), (2.56). In order to satisfy (2.57) we must have

$$\phi_0(r) = \sum_{n=1}^{\infty} a_n J_0 \left(\frac{\mu_n r}{R} \right) \tag{2.62}$$

$$\phi_1(r) = \sum_{n=1}^{\infty} \frac{\mu_n}{R} b_n J_0 \left(\frac{\mu_n r}{R} \right). \tag{2.63}$$

We shall assume that these expansions are possible and determine what the coefficients a_n and b_n must be.

In order to determine the coefficients a_n and b_n in (2.62), (2.63), we first introduce the **Bessel function of order *m***, defined by

$$J_m(kr) = \sum_{n=0}^{\infty} \frac{(-1)^n}{n!(n+m)!} \left(\frac{kr}{2} \right)^{m+2n} \tag{2.64}$$

and note that $J_m(kr)$ is a solution of

$$W'' + \frac{1}{r} W' + \left(k^2 - \frac{m^2}{r^2} \right) W = 0.$$

Using the fact (see Section 6.1) that $m! = \Gamma(m+1)$ where $\Gamma(z)$ is the gamma function, we see that $m! = \infty$ for m a negative integer, so $J_m(kr)$ is well defined for both positive and negative integers. From (2.64),

$$\frac{d}{dt} [t^{-m} J_m(t)] = -t^{-m} J_{m+1}(t)$$

$$\frac{d}{dt} [t^m J_m(t)] = t^m J_{m-1}(t),$$

so

$$\frac{d}{dt} [t^2 \{ (J_m(t))^2 - J_{m+1}(t) J_{m-1}(t) \}] = 2t (J_m(t))^2.$$

Setting $m = 0$, we now see that

$$\begin{aligned}\int_0^R r\left[J_0\left(\frac{\mu_n r}{R}\right)\right]^2 dr &= \frac{R^2}{\mu_n^2}\int_0^{\mu_n} tJ_0^2(t)\,dt \\ &= \frac{R^2}{2\mu_n^2}\left[t^2\{(J_0(t))^2 - J_1(t)J_{-1}(t)\}\right]_{t=0}^{t=\mu_n} \\ &= \frac{R^2}{2\mu_n^2}\left[t^2\{(J_0(t))^2 + (J_1(t))^2\}\right]_{t=0}^{t=\mu_n} \\ &= \frac{R^2}{2}(J_1(\mu_n))^2,\end{aligned} \tag{2.65}$$

which is nonzero, because the integral on the left side is positive. On the other hand, setting

$$y_k(r) = J_0\left(\frac{\mu_k r}{R}\right),$$

we have

$$(ry_k')' + \frac{\mu_k^2}{R^2} ry_k = 0$$

$$(ry_l')' + \frac{\mu_l^2}{R^2} ry_l = 0,$$

and multiplying the first equation by $y_l(r)$, the second by $y_k(r)$, subtracting, and integrating the resulting expression by parts from 0 to R gives

$$r[y_k'(r)y_l(r) - y'(r)y_k(r)]_{r=0}^{r=R} = \frac{\mu_k^2 - \mu_l^2}{R^2}\int_0^R ry_k(r)y_l(r)\,dr.$$

But $y_l(R) = y_k(R) = 0$ and from (2.60) we have $y_l'(0) = y_k'(0) = 0$. Hence the above equation implies that

$$\int_0^R rJ_0\left(\frac{\mu_k r}{R}\right)J_0\left(\frac{\mu_l r}{R}\right) dr = 0, \quad l \neq k. \tag{2.66}$$

From (2.65) and (2.66), we can now conclude that the coefficients a_n and b_n in (2.62) and (2.63) respectively are given by

$$\begin{aligned} a_n &= \frac{2}{R^2 J_1^2(\mu_n)}\int_0^R r\phi_0(r)J_0\left(\frac{\mu_n r}{R}\right) dr \\ b_n &= \frac{2}{\mu_n R J_1^2(\mu_n)}\int_0^R r\phi_1(r)J_0\left(\frac{\mu_n r}{R}\right) dr. \end{aligned} \tag{2.67}$$

Hence, the formal solution to (2.55)–(2.57) is given by (2.61), (2.67). In order to verify that this is in fact a solution we must verify that (2.62) and (2.63) are in fact valid with the coefficients given by (2.67), and that (2.61) is indeed a solution of (2.55) that continuously assumes the initial-boundary data. This requires making smoothness assumptions on $\phi_0(r)$ and $\phi_1(r)$. We shall not pursue this task here, but instead refer the reader to Weinberger where further examples of the method of separation of variables can also be found.

2.6 THE EQUATIONS OF GAS DYNAMICS

Since this book is designed to be an introductory text, our primary concern necessarily must be with linear problems. However, we would hate for the student to leave this book believing that all problems of physical interest are either linear or can be adequately treated by linearized models. Thus, we shall introduce at appropriate places some examples of physical problems for which linearization would mean "throwing the baby out with the bath water" but which nevertheless can be treated by relatively elementary methods. We now consider the one-dimensional flow of an isentropic gas, which is basically a nonlinear problem that is not amenable to study by linear analysis.

The one-dimensional isentropic (the entropy is assumed to be constant) flow of an inviscid gas is governed by the equations

$$\begin{aligned} &\frac{\partial u}{\partial t} + u\frac{\partial u}{\partial x} + \frac{c^2}{\rho}\frac{\partial \rho}{\partial x} = 0 \\ &\frac{\partial \rho}{\partial t} + u\frac{\partial \rho}{\partial x} + \rho\frac{\partial u}{\partial x} = 0 \end{aligned} \tag{2.68}$$

where $u(x,t)$ and $\rho(x,t)$ are the velocity and density of the gas at position x and time t, and $c = c(\rho)$ is the known local speed of sound. We can write (2.68) in matrix form as

$$\frac{\partial}{\partial t}\begin{bmatrix} u \\ \rho \end{bmatrix} + \begin{bmatrix} u & \dfrac{c^2}{\rho} \\ \rho & u \end{bmatrix} \frac{\partial}{\partial x}\begin{bmatrix} u \\ \rho \end{bmatrix} = \begin{bmatrix} 0 \\ 0 \end{bmatrix}. \tag{2.69}$$

This system is a special case of the more general $n \times n$ quasilinear system

$$\frac{\partial \mathbf{u}}{\partial t} + A(x,t,\mathbf{u})\frac{\partial \mathbf{u}}{\partial x} + \mathbf{b}(x,t,\mathbf{u}) = 0 \tag{2.70}$$

where $\mathbf{u}(x,t)$ and $\mathbf{b}(x,t,\mathbf{u})$ are vectors in R^n and $A(x,t,\mathbf{u})$ is a $n \times n$ matrix. Recall that a linear system of the form (2.70) was previously considered in Section 2.2.

□ EXAMPLE 18

The wave equation

$$\frac{\partial^2 u}{\partial x^2} = \frac{\partial^2 u}{\partial t^2}$$

can be written as the system

$$\frac{\partial u}{\partial x} = \frac{\partial v}{\partial t}$$

$$\frac{\partial u}{\partial t} = \frac{\partial v}{\partial x},$$

or, in matrix form,

$$\frac{\partial}{\partial t}\begin{bmatrix} u \\ v \end{bmatrix} = \begin{bmatrix} 0 & 1 \\ 1 & 0 \end{bmatrix} \frac{\partial}{\partial x}\begin{bmatrix} u \\ v \end{bmatrix}.$$

Then

$$A = \begin{bmatrix} 0 & 1 \\ 1 & 0 \end{bmatrix}.$$

The matrix A has eigenvalues $\lambda = \pm 1$ (i.e., the roots of det $|A - \lambda I| = 0$) which correspond to the characteristic equation

$$\frac{dx}{dt} = \pm 1.$$ □

Motivated by this example, we have the following definition:

DEFINITION 7

A curve $x = x(t)$ is said to be a characteristic curve for the system (2.70) and corresponding solution $\mathbf{u}(x,t)$ if

$$\frac{dx}{dt} = \lambda(x,t,\mathbf{u}(x,t)),$$

where $\lambda(x,t,\mathbf{u})$ is an eigenvalue of the matrix $A(x,t,\mathbf{u})$.

□ EXAMPLE 19

The matrix

$$A(x,t,u,\rho) = \begin{bmatrix} u & \dfrac{c^2}{\rho} \\ \rho & u \end{bmatrix}$$

corresponding to the quasilinear system (2.69) of gas dynamics has two real and distinct eigenvalues

$$\lambda_+ = u + c(\rho)$$
$$\lambda_- = u - c(\rho).$$

There are two characteristic curves C_+ and C_- defined by the solutions of

$$\begin{aligned} \frac{dx}{dt} &= u(x,t) + c(\rho(x,t)) \\ \frac{dx}{dt} &= u(x,t) - c(\rho(x,t)) \end{aligned} \qquad (2.71)$$

respectively. □

Note that if the system (2.70) is genuinely quasilinear (A depends on $\mathbf{u}$) then a solution of the system (2.70) must be known in advance in order for determination of the corresponding characteristic curves.

DEFINITION 8

If the $n \times n$ matrix A has n real and distinct eigenvalues for all possible solutions of the system (2.70), then (2.70) is said to be of **hyperbolic type.**

By Example 19, we see that the system (2.69) of gas dynamics is of hyperbolic type.

We now examine the system (2.68) (or (2.69)) in more detail. It takes on a more symmetric form if in place of $\rho(x,t)$ we define the new unknown

$$l(x,t) = \int_{\rho_0}^{\rho(x,t)} \frac{c(\rho')}{\rho'} d\rho' \qquad (2.72)$$

where ρ_0 is the density of the gas at rest. Since

$$\frac{dl}{d\rho} = \frac{c(\rho)}{\rho} > 0,$$

(2.72) can be solved for $\rho(x,t)$ in terms of $l(x,t)$. We can now rewrite the system (2.68) in the form

$$\begin{aligned} \frac{\partial u}{\partial t} + u\frac{\partial u}{\partial x} + c\frac{\partial l}{\partial x} &= 0 \\ \frac{\partial l}{\partial t} + c\frac{\partial u}{\partial x} + u\frac{\partial l}{\partial x} &= 0, \end{aligned} \qquad (2.73)$$

or, adding and subtracting,

$$\frac{\partial}{\partial t}(l + u) + (u + c)\frac{\partial}{\partial x}(l + u) = 0$$

$$\frac{\partial}{\partial t}(l - u) + (u - c)\frac{\partial}{\partial x}(l - u) = 0.$$

In particular, if we define $r(x,t)$ and $s(x,t)$ by

$$r = \frac{1}{2}(l + u)$$

$$s = \frac{1}{2}(l - u),$$

then $r(x,t)$ and $s(x,t)$ satisfy the equations

$$\begin{aligned} \frac{\partial r}{\partial t} + (u + c)\frac{\partial r}{\partial x} &= 0 \\ \frac{\partial s}{\partial t} + (u - c)\frac{\partial s}{\partial x} &= 0. \end{aligned} \tag{2.74}$$

From (2.71) and (2.74) we see that $r(x,t)$ is constant along the C_+ characteristic while $s(x,t)$ is constant along the C_- characteristic.

DEFINITION 9

The functions $r(x,t)$ and $s(x,t)$ are called the **Riemann invariants.**

Now let us consider the initial value problem for the system (2.68) with initial data

$$\begin{aligned} u(x,0) &= \phi_1(x) \qquad (a < x < b) \\ \rho(x,0) &= \phi_2(x) \qquad (a < x < b) \end{aligned} \tag{2.75}$$

where $\phi_1,\phi_2 \in C^1[a,b]$. We shall accept without proof the existence of a unique solution to (2.68), (2.75) in some domain containing the interval (a,b) (cf. Garabedian 1964). However, we want to emphasize that by a solution we mean a pair of continuously differentiable functions satisfying (2.68) and (2.75) and note that the nonlinearity of the system (2.68) may cause the solution to develop discontinuities known as **shocks** (this term will be clarified later).

As a first step in understanding the general nature of solutions to the system (2.68), we consider a linear approximation to this system, assuming that $u(x,t)$ and $\rho(x,t)$ are not very different from their values at time $t = 0$ and that these values and their derivatives are small. Neglecting products of small order terms in (2.68), we arrive at the linear system

$$\frac{\partial u}{\partial t} + \frac{c_0^2 \partial \rho}{\rho_0 \partial x} = 0$$
$$\frac{\partial \rho}{\partial t} + \rho_0 \frac{\partial u}{\partial x} = 0, \tag{2.76}$$

where ρ_0 is the density of the fluid at rest and $c_0 = c(\rho_0)$. This is just the wave equation written as a system and by the analysis of Section 6.1 is easily seen to have the general solution (cf. Exercise 29)

$$u(x,t) = \frac{c_0}{\sqrt{\rho_0}} [f(x - c_0 t) + g(x + c_0 t)]$$
$$\rho(x,t) = \sqrt{\rho_0} [f(x - c_0 t) - g(x + c_0 t)], \tag{2.77}$$

where f and g are arbitrary continuously differentiable functions. If g is identically zero, (2.77) represents an undistorted wave traveling in the $+x$ direction with speed c_0 and is called a **forward wave.** Similarly, if f is identically zero, (2.77) represents an undistorted wave traveling in the $-x$ direction with speed c_0 and is called a **backward wave.** The solution of the initial value problem (2.75), (2.76) is easily obtained from the general solution (2.77) (cf. Exercise 29) and is given explicitly by

$$u(x,t) = \frac{1}{2} [\phi_1(x - c_0 t) + \phi_1(x + c_0 t)] + \frac{c_0}{2\rho_0} [\phi_2(x - c_0 t) - \phi_2(x + c_0 t)]$$
$$\rho(x,t) = \frac{1}{2} [\phi_2(x - c_0 t) + \phi_2(x + c_0 t)] + \frac{\rho_0}{2c_0} [\phi_1(x - c_0 t) - \phi_1(x + c_0 t)]. \tag{2.78}$$

Thus an initial disturbance splits into forward and backward waves which appear to travel undistorted for all time. This is due to the fact that nonlinearities have been completely neglected. As we shall soon see, if the nonlinearities are not neglected, the traveling waves become distorted and as these distortions build up with time shocks can develop. Thus the usefulness of the linear approximation (2.76) is limited to small disturbances over limited time intervals.

To see how the nonlinearities of the system (2.68) can cause distortion of traveling waves, we consider a class of solutions of (2.68) known as **simple waves,** which occur when the velocity $u(x,t)$ can be expressed as a function of the density $\rho(x,t)$, i.e., $u = u(\rho)$. In this case,

$$\frac{\partial u}{\partial x} = \frac{du}{d\rho}\frac{\partial \rho}{\partial x}$$
$$\frac{\partial u}{\partial t} = \frac{du}{d\rho}\frac{\partial \rho}{\partial t}$$

and multiplying the second equation in (2.68) by $du/d\rho$, we have

$$\frac{\partial u}{\partial t} + u\frac{\partial u}{\partial x} + \rho\left(\frac{du}{d\rho}\right)^2\frac{\partial \rho}{\partial x} = 0.$$

Using the first equation in (2.68) now shows that

$$\left(\frac{du}{d\rho}\right)^2 = \frac{c^2(\rho)}{\rho^2},$$

i.e.,

$$u(x,t) = \pm\int_{\rho_0}^{\rho(x,t)} \frac{c(\rho')}{\rho'}\,d\rho' = \pm l(x,t)$$

where $l(x,t)$ is the same function previously introduced in (2.72). This suggests working with the equivalent system (2.73). Solving for $\rho(x,t)$ in terms of $l(x,t)$ allows us to write (2.73) as

$$\begin{aligned}
&\frac{\partial u}{\partial t} + u\frac{\partial u}{\partial x} + c(\rho(l))\frac{\partial l}{\partial x} = 0 \\
&\frac{\partial l}{\partial t} + c(\rho(l))\frac{\partial u}{\partial x} + u\frac{\partial l}{\partial x} = 0.
\end{aligned} \tag{2.79}$$

When $u(x,t) = +l(x,t)$, both equations reduce to

$$\frac{\partial u}{\partial t} + a(u)\frac{\partial u}{\partial x} = 0, \tag{2.80}$$

where

$$a(u) = c(\rho(u)) + u. \tag{2.81}$$

In order to find the general solution of (2.80) we set

$$\begin{aligned}
\eta &= x + a(u)t \\
\xi &= x - a(u)t
\end{aligned}$$

and rewrite (2.80) in the form

$$2a(u)\frac{\partial u}{\partial \eta} = 0,$$

from which we can conclude that the general solution of (2.80) is given implicitly by

$$u(x,t) = F(x - a(u(x,t))t), \tag{2.82}$$

where F is an arbitrary continuously differentiable function. Hence, simple waves for which $u(x,t) = +l(x,t)$ are forward waves traveling with speed $a(u)$, which depends on $u = u(x,t)$. When $u(x,t) = -l(x,t)$, both equations in (2.79) reduce to

$$\frac{\partial u}{\partial t} - a(-u)\frac{\partial u}{\partial x} = 0,$$

i.e.,

$$u(x,t) = G(x + a(-u(x,t))t),$$

where G is an arbitrary continuously differentiable function. Hence, simple waves for which $u(x,t) = -l(x,t)$ are backward waves traveling with speed $a(-u)$, which depends on $u = u(x,t)$.

Let us examine the implications of these formulas, restricting our attention to the forward wave (2.82) and making the physically reasonable assumption that $c(\rho)$ is a positive, monotonically increasing function of ρ. We first note that the traveling wave (2.82) becomes distorted with time, since the speed of propagation depends on $u(x,t)$. Second, writing (2.81) in the form

$$a(u) = c_0 + [c(\rho(u)) - c_0] + u \tag{2.83}$$

shows that for small amplitude flows ($u(x,t)$ and $\rho(x,t) - \rho_0$ are small) the speed of propagation is nearly equal to the constant value $c_0 = c(\rho_0)$. Hence, for short time intervals, simple waves of small amplitude behave almost like the undistorted traveling waves of linear approximation theory. However, the last two terms in (2.83) always cause some distortion, which builds up with time and may eventually lead to the formation of shocks. In particular, by the implicit function theorem, (2.82) is soluble for $u(x,t)$ only if

$$1 + tF'(x - a(u)t)a'(u) \neq 0. \tag{2.84}$$

But, since $u(x,t) = +l(x,t)$, it follows from (2.72) that

$$\frac{du}{d\rho} = \frac{c(\rho)}{\rho} > 0,$$

i.e., $u = u(\rho)$ is a monotonically increasing function of ρ. By (2.81) and our assumptions on $c(\rho)$, $a(u)$ is a monotonically increasing function of u, and

$$a'(u) > 0.$$

Therefore, if the initial velocity profile $u(x,0) = F(x)$ is decreasing for x on some interval of the x axis, i.e.,

$$F'(x) < 0$$

on this interval, then it is possible for (2.84) to be violated for sufficiently large values of the time t. Then $u(x,t)$ ceases to exist as a continuously differentiable function, and shocks are present.

For further information on the equations of gas dynamics and the development of shocks, the reader is referred to Courant and Friedrichs, Lax, and Smoller.

Exercises

1. Use d'Alembert's solution of the wave equation to find the solution of

$$\frac{\partial^2 u}{\partial x^2} = \frac{\partial^2 u}{\partial t^2} \qquad (t \geq 0,\ x \geq 0)$$

$$\frac{\partial u(0,t)}{\partial t} = \alpha \frac{\partial u(0,t)}{\partial x} \qquad (t \geq 0)$$

$$\left.\begin{aligned} u(x,0) &= \phi_0(x) \\ \frac{\partial u}{\partial t}(x,0) &= \phi_1(x) \end{aligned}\right\} \qquad (x \geq 0),$$

where $\alpha \neq -1$ is a constant and $\phi_0(x)$ and $\phi_1(x)$ are twice continuously differentiable for $x > 0$ and vanish near $x = 0$. Show that in general no solution exists when $\alpha = -1$.

2. Find the solution of

$$\frac{\partial^2 u}{\partial x^2} = \frac{\partial^2 u}{\partial t^2} \qquad (t \geq 0,\ -\infty < x < \infty)$$

$$u(x,x) = \phi(x) \qquad (-\infty < x < \infty)$$

$$\frac{\partial u}{\partial x}(x,-x) - \frac{\partial u}{\partial t}(x,-x) = \psi(x) \qquad (-\infty < x < \infty),$$

where $\phi(x)$ and $\psi(x)$ are twice continuously differentiable.

3. Show that

$$\frac{\partial^2 u}{\partial x^2} = \frac{\partial^2 u}{\partial t^2} \qquad (t \geq 0,\ -\infty < x < \infty)$$

$$u(x,x) = \phi(x) \qquad (-\infty < x < \infty)$$

$$\frac{\partial u}{\partial x}(x,x) + \frac{\partial u}{\partial t}(x,x) = \psi(x) \qquad (-\infty < x < \infty)$$

is not properly posed. Determine conditions on $\phi(x)$ and $\psi(x)$ such that it does have a solution.

4. Find the solution of

$$\frac{\partial^2 u}{\partial x^2} = \frac{\partial^2 u}{\partial t^2} \qquad ((x,t) \in D)$$

$$u(x,0) = \phi(x) \qquad (0 \leq x \leq a)$$

$$u(x,x) = \psi(x) \qquad (0 \leq x \leq a),$$

where $\phi(x)$ and $\psi(x)$ are twice continuously differentiable, $\phi(0) = \psi(0)$, and D is the region bounded by the straight lines $t = 0$, $x - t = 0$, $x - t = a$, and $x + t = 2a$.

5. Show that any solution $u(x,y)$ of

$$\frac{\partial^2 u}{\partial x \partial y} = 0$$

satisfies the difference equation $u_1 - u_2 + u_3 - u_4 = 0$ where u_1, u_2, u_3, and u_4 are the values of $u(x,y)$ at the four successive corners of any rectangle whose edges are characteristics.

6. Find the relationship connecting the real constants a, b, and c for which the hyperbolic partial differential equation

$$a^2 \frac{\partial^2 u}{\partial x^2} + b^2 \frac{\partial^2 u}{\partial y^2} = c^2 \frac{\partial^2 u}{\partial t^2}$$

possesses solutions of the form

$$u(x,y,t) = f(\alpha x + \beta y + \gamma t),$$

where α, β, and γ are real constants and f is an arbitrary twice continuously differentiable function. For such an equation, find a solution satisfying the Cauchy data

$$u(x,y,0) = x^2 - y^2$$

$$\frac{\partial u}{\partial t}(x,y,0) = xy.$$

(Answer: $\alpha^2 a^2 + \beta^2 b^2 - \gamma^2 c^2 = 0$; $u(x,y,t) = x^2 - y^2 + \dfrac{(a^2 - b^2)t^2}{c^2} + xyt$.**)**

7. Show that the general solution of the hyperbolic equation

$$\frac{\partial^2 u}{\partial x^2} - y\frac{\partial^2 u}{\partial y^2} - \frac{1}{2}\frac{\partial u}{\partial y} = 0 \qquad (y > 0)$$

has the form

$$u(x,y) = f_1(x + 2\sqrt{y}) + f_2(x - 2\sqrt{y}),$$

where f_1 and f_2 are arbitrary twice continuously differentiable functions.

8. Use the method of successive approximations to solve the characteristic initial value problem

$$\frac{\partial^2 u}{\partial x \partial y} + a(x,y)\frac{\partial u}{\partial x} + b(x,y)\frac{\partial u}{\partial y} + c(x,y)u = f(x,y)$$

$$u(x_0,y) = \phi_0(y) \qquad (y_0 \le y \le b)$$

$$u(x,y_0) = \phi_1(x) \qquad (x_0 \le x \le b),$$

where $a(x,y)$, $b(x,y)$, $c(x,y)$, and $f(x,y)$ are real valued continuous functions, $\phi_0(y)$ and $\phi_1(x)$ are continuously differentiable functions, and $\phi_0(y_0) = \phi_1(x_0)$.

9. Consider the characteristic initial value problem

$$\frac{\partial^2 u}{\partial x \partial y} + \lambda u = 0$$

$$u(x,0) = u(0,y) = 1,$$

where λ is a constant. Show that $u(x,y)$ is a solution of the integral equation

$$u(x,y) = 1 - \lambda \int_0^y \int_0^x u(\xi,\eta)\, d\xi d\eta,$$

and solve this by the method of successive approximations to show that

$$u(x,y) = J_0(2\sqrt{\lambda xy}),$$

where $J_0(z)$ is Bessel's function.

10. Let D be a bounded, simply connected domain in the plane with a continuously differentiable boundary and $u(x,y,t)$ a solution of the wave equation

$$\frac{\partial^2 u}{\partial t^2} = \frac{\partial^2 u}{\partial x^2} + \frac{\partial^2 u}{\partial y^2} \qquad ((x,y) \in D,\ t \ge 0)$$

such that $u(x,y,t) = 0$ on ∂D for all $t \ge 0$. Show that

$$\frac{\partial}{\partial t} \iint_D \left[\left(\frac{\partial u}{\partial t}\right)^2 + \left(\frac{\partial u}{\partial x}\right)^2 + \left(\frac{\partial u}{\partial y}\right)^2 \right] dxdy = 0.$$

Use this result to establish a uniqueness theorem for the initial-boundary value problem

$$\frac{\partial^2 u}{\partial t^2} = \frac{\partial^2 u}{\partial x^2} + \frac{\partial^2 u}{\partial y^2} \qquad ((x,y) \in D,\ t \ge 0)$$

$$u(x,y,t) = f(x,y,t) \qquad ((x,y) \in \partial D,\ t \ge 0)$$

$$\left.\begin{aligned} u(x,y,0) &= \phi_0(x,y) \\ \frac{\partial u}{\partial t}(x,y,0) &= \phi_1(x,y) \end{aligned}\right\} \qquad ((x,y) \in D).$$

11. Show that the general solution of

$$\frac{\partial^2 u}{\partial x^2} + \frac{\partial^2 u}{\partial y^2} + \frac{\partial^2 u}{\partial z^2} = \frac{\partial^2 u}{\partial t^2}$$

which depends only on $r = \sqrt{x^2 + y^2 + z^2}$ and t is of the form

$$u(r,t) = \frac{f_1(r+t)}{r} + \frac{f_2(r-t)}{r},$$

where f_1 and f_2 are arbitrary twice continuously differentiable functions.

12. Use the method of descent to obtain d'Alembert's formula for the wave equation in one space variable from equation (2.26).

13. Let $u(x,y,t)$ be the unique solution of the Cauchy problem

$$\frac{\partial^2 u}{\partial t^2} = \frac{\partial^2 u}{\partial x^2} + \frac{\partial^2 u}{\partial y^2}$$

$$u(x,y,0) = \phi_0(x,y)$$

$$\frac{\partial u}{\partial t}(x,y,0) = \phi_1(x,y),$$

where $\phi_0(x,y)$ and $\phi_1(x,y)$ have compact support, i.e., $\phi_0(x,y)$ and $\phi_1(x,y)$ vanish for $x^2 + y^2 > R_0^2$, where R_0 is sufficiently large. Show that for any $R > 0$ there is a $T > 0$ such that for all $(x,y,t) \in \{(x,y,t)\colon x^2 + y^2 \le R^2,\ t \ge T\}$, $u(x,y,t)$ can be represented in the form

$$u(x,y,t) = \sum_{n=0} \frac{c_n(x,y)}{t^{n+1}}$$

and find $c_0(x,y)$ and $c_1(x,y)$.

14. Consider the initial-boundary value problem

$$\frac{\partial^2 u}{\partial t^2} = \frac{\partial^2 u}{\partial x^2} + \frac{\partial^2 u}{\partial y^2} \qquad ((x,y) \in D,\ t \in [0,T])$$

$$\frac{\partial u}{\partial \nu} + u = 0 \qquad ((x,y) \in \partial D,\ t \in [0,T])$$

$$\left.\begin{aligned} u(x,y,0) &= \phi(x,y) \\ \frac{\partial u}{\partial t}(x,y,0) &= \psi(x,y) \end{aligned}\right\} \qquad ((x,y) \in D),$$

where $\boldsymbol{\nu}$ is the unit outward normal to ∂D.

(a) Show that

$$\iint_D |\nabla u(x,y,T)|^2\,dx\,dy + \int_{\partial D} |u(x,y,T)|^2 dx$$
$$= \iint_D |\nabla u(x,y,0)|^2\,dx\,dy + \int_{\partial D} |u(x,y,0)|^2 ds.$$

(b) Use (a) to derive a uniqueness theorem for the solution to this initial-boundary value problem.

15. Derive equations (2.37) and (2.38) of Section 2.3 (see pp. 77–78).

16. Derive the analogue of Poisson's formula for the solution of Cauchy's problem for the equation

$$\frac{\partial^2 u}{\partial t^2} = \frac{\partial^2 u}{\partial x^2} + \frac{\partial^2 u}{\partial y^2} + \lambda u,$$

where λ is a constant. (Hint: Consider solutions of the wave equation in four independent variables of the form $v(x,y,z,t) = u(x,y,t)e^{kz}$ where $\lambda = k^2$.)

(Answer: $$u(\mathbf{x},t) = \frac{1}{2\pi}\frac{\partial}{\partial t}\iint_{|\xi|\le t} \frac{\phi_0(\mathbf{x}+\xi)\cosh\left(k\sqrt{t^2-|\xi|^2}\right)}{\sqrt{t^2-|\xi|^2}}\,d\xi + \frac{1}{2\pi}\iint_{|\xi|\le t} \frac{\phi_1(\mathbf{x}+\xi)\cosh\left(k\sqrt{t^2-|\xi|^2}\right)}{\sqrt{t^2-|\xi|^2}}\,d\xi.)$$

17. Use Fourier's method to formally solve the initial boundary value problem

$$\frac{\partial^2 u}{\partial t^2} - \frac{\partial^2 u}{\partial x^2} + u = 0 \qquad (0 < x < \pi,\ t > 0)$$

$$\frac{\partial u}{\partial x}(0,t) = \frac{\partial u}{\partial x}(\pi,t) = 0 \qquad (t > 0)$$

$$\left.\begin{aligned} u(x,0) &= 0 \\ \frac{\partial u}{\partial t}(x,0) &= 1 + \cos^3 x \end{aligned}\right\} \qquad (0 \le x \le \pi)$$

(Answer: $u(x,t) = \sin t + \dfrac{3\sqrt{2}}{8}\cos x \sin \sqrt{2}t + \dfrac{\sqrt{10}}{40}\cos 3x \sin \sqrt{10}t$)

18. Use Fourier's method to formally solve the initial-boundary value problem

$$\frac{\partial^2 u}{\partial t^2} + 2a\frac{\partial u}{\partial t} + bu - c^2\frac{\partial^2 u}{\partial x^2} = 0 \qquad (0 < x < \pi,\ t > 0)$$

$$u(0,t) = u(\pi,t) = 0 \qquad (t > 0)$$

$$\left.\begin{aligned} u(x,0) &= 0 \\ \frac{\partial u}{\partial t}(x,0) &= g(x) \end{aligned}\right\} \qquad (0 \le x \le \pi),$$

where $g \in C[0,\pi]$ and a, b, c are real numbers.

19. Use Fourier's method to formally solve the initial-boundary value problem

$$\frac{\partial^2 u}{\partial t^2} - \frac{\partial^2 u}{\partial x^2} + u = 0 \qquad (0 < x < \pi,\ t > 0)$$

$$u(0,t) = u(\pi,t) = 0 \qquad (t \ge 0)$$

$$\left.\begin{aligned} u(x,0) &= f(x) \\ \frac{\partial u}{\partial t}(x,0) &= 0 \end{aligned}\right\} \qquad (0 \le x \le \pi),$$

where

$$f(x) = \begin{cases} x & \left(0 \le x \le \dfrac{\pi}{2}\right) \\ \pi - x & \left(\dfrac{\pi}{2} \le x \le \pi\right). \end{cases}$$

(**Answer:** $u(x,t) = \dfrac{4}{\pi}\displaystyle\sum_{k=0}^{\infty} \frac{(-1)^k}{(2k+1)^2} \cos \sqrt{(2k+1)^2 + 1}\, t \sin (2k+1)x.$)

20. By reducing the problem to a homogeneous equation with homogeneous boundary conditions, formally solve the initial-boundary value problem

$$\frac{\partial^2 u}{\partial t^2} = \frac{\partial^2 u}{\partial x^2} + f(x) \qquad (0 < x < 1,\ t > 0)$$

$$u(0,t) = u(1,t) = 0 \qquad (t \ge 0)$$

$$\left.\begin{aligned} u(x,0) &= 0 \\ \frac{\partial u}{\partial t}(x,0) &= 0 \end{aligned}\right\} \qquad (0 \le x \le 1),$$

where $f(x)$ is a continuous function.

21. Use Fourier's method to formally solve the initial-boundary value problem

$$\frac{\partial^2 u}{\partial t^2} = \frac{\partial^2 u}{\partial x^2} + \frac{\partial^2 u}{\partial y^2} + \frac{\partial^2 u}{\partial z^2} \qquad (0 < x < \pi, 0 < y < \pi, 0 < z < \pi, t > 0)$$

$$\left.\begin{aligned} u(0,y,z,t) = u(x,0,z,t) = u(x,y,0,t) = 0 \\ u(\pi,y,z,t) = u(x,\pi,z,t) = u(x,y,\pi,t) = 0 \end{aligned}\right\} \qquad \begin{cases} (0 \le x \le \pi) \\ (0 \le y \le \pi, t \ge 0) \\ (0 \le z \le \pi) \end{cases}$$

$$\left.\begin{aligned} u(x,y,z,0) = xyz(\pi - x)(\pi - y)(\pi - z) \\ \frac{\partial u}{\partial t}(x,y,z,0) = 0 \end{aligned}\right\} \qquad \begin{cases} (0 \le x \le \pi) \\ (0 \le y \le \pi) \\ (0 \le z \le \pi). \end{cases}$$

(Answer: $u(x,y,z,t) = \left(\frac{8}{\pi}\right)^3 \sum_{k=1}^{\infty} \sum_{l=1}^{\infty} \sum_{m=1}^{\infty} \frac{\sin(2k-1)x \sin(2l-1)y \sin(2m-1)z}{(2k-1)^3(2l-1)^3(2m-1)^3}$

$\cdot \cos\sqrt{(2k-1)^2 + (2l-1)^2 + (2m-1)^2}\,t$.)

22. Determine the coefficients a_n in the Fourier-Bessel expansion (2.62) where $R = 1$ and $\phi_0(r) = 1$, $0 < r < 1$.

(Answer: $1 = 2\sum_{n=1}^{\infty} \frac{J_0(\mu_n r)}{\mu_n J_1(\mu_n)}, \qquad 0 < r < 1.$)

23. Determine the coefficients a_n in the Fourier-Bessel expansion (2.62) where $R = 1$ and $\phi(r) = 1 - r^2$, $0 < r < 1$.

(Answer: $1 - r^2 = 8\sum_{n=1}^{\infty} \frac{J_0(\mu_n r)}{\mu_n^3 J_1(\mu_n)}, \qquad 0 < r < 1.$)

24. Use Fourier's method to formally solve the initial-boundary value problem

$$\frac{\partial^2 u}{\partial t^2} = \frac{\partial^2 u}{\partial r^2} + \frac{1}{r}\frac{\partial u}{\partial r} + \frac{1}{r^2}\frac{\partial^2 u}{\partial \theta^2} \qquad (0 < r < 1, 0 \le \theta \le 2\pi, t > 0)$$

$$u(1,\theta,t) = 0 \qquad (0 \le \theta \le 2\pi, t \ge 0)$$

$$\left.\begin{aligned} u(r,\theta,0) = 0 \\ \frac{\partial u}{\partial t}(r,\theta,0) = 1 - r^2 \end{aligned}\right\} \qquad (0 \le r \le 1, 0 \le \theta \le 2\pi).$$

(Hint: See Exercise 23.)

25. Show that

$$J_0(x) > 1 - \frac{x^2}{4} \qquad (0 \le x \le 2)$$

and

$$J_0(x) < 1 - \frac{x^2}{4} + \frac{x^4}{64} \qquad (0 \le x \le 3).$$

Deduce that $J_0(2) > 0 > J_0(2\sqrt{2})$, and hence if μ_1 is the first positive root of $J_0(\mu) = 0$, then

$$2 < \mu_1 < 2\sqrt{2}.$$

26. Let $P_0(r) = 1/2$ and $P_{2n}(r)$ be the unique polynomial of degree $2n$ satisfying

$$(rP'_{2n})' = -rP_{2n-2}$$

$$P_{2n}(1) = 0.$$

Let μ_n be the nth positive root of $J_0(\mu) = 0$. Show that

$$\mu_n^2 \int_0^1 rP_{2n}(r)\,J_0(\mu_n r)dr = \int_0^1 rP_{2n-2}(r)\,J_0(\mu_n r)dr.$$

27. Use Exercises 23 and 26 to show that

$$P_4(r) = \frac{1}{128}(3 - 4r^2 + r^4) = \sum_{n=1}^{\infty} \frac{J_0(\mu_n r)}{\mu_n^5 J_1(\mu_n)} \qquad (0 < r < 1).$$

28. (a) Let μ_n^m be the positive roots of $J_m(\mu) = 0$ where $J_m(\mu)$ is Bessel's function of order m. Show that

$$\int_0^1 rJ_m(\mu_k^m r)J_m(\mu_l^m r)dr = \begin{cases} 0 & \text{if} \quad l \ne k \\ \dfrac{1}{2}\,[J_{m+1}(\mu_l^m)]^2 & \text{if} \quad l = k. \end{cases}$$

(b) Use Fourier's method to formally solve the initial-boundary value problem

$$\frac{\partial^2 u}{\partial t^2} = \frac{\partial^2 u}{\partial r^2} + \frac{1}{r}\frac{\partial u}{\partial r} + \frac{1}{r^2}\frac{\partial^2 u}{\partial \theta^2} \qquad (0 < r < 1,\, 0 \le \theta \le 2\pi,\, t > 0)$$

$$u(1,\theta,t) = 0 \qquad (0 \le \theta \le 2\pi,\, t \ge 0)$$

$$\left.\begin{aligned} u(r,\theta,0) &= f(r,\theta) \\ \frac{\partial u}{\partial t}(r,\theta,0) &= 0 \end{aligned}\right\} \qquad (0 \le r \le 1,\, 0 \le \theta \le 2\pi),$$

where (r,θ) denote polar coordinates and $f(r,\theta)$ is a continuous function.

29. Derive equations (2.77) and (2.78) of Section 2.6 (see p. 94).

30. Let $u(x,t)$ satisfy

$$\frac{\partial u}{\partial t} + u\,\frac{\partial u}{\partial x} = 0 \qquad (-\infty < x < \infty,\ t \geq 0)$$

$$u(x,0) = -x \qquad (-\infty < x < \infty).$$

Show that shocks develop when $t = 1$. Do shocks ever develop for the initial value problem

$$\frac{\partial u}{\partial t} + u\,\frac{\partial u}{\partial x} = 0 \qquad (-\infty < x < \infty,\ t \geq 0)$$

$$u(x,0) = x \qquad (-\infty < x < \infty)?$$

Chapter 3

The Heat Equation

The heat equation is the simplest example of a partial differential equation of parabolic type. In this chapter we consider initial value and initial-boundary value problems for the heat equation (including, briefly, their numerical solution), as well as the qualitative behavior of solutions to the heat equation such as maximum principles, regularity, and analytic continuation. In contrast to our treatment of the wave equation, we shall restrict our attention to the heat equation in two independent variables. Finally, in an effort to remind the reader that linear problems are not the only ones of interest in heat conduction, we consider some heat conduction problems involving nonlinear source terms as well as a simple free boundary problem for the heat equation called the Stefan problem. The analysis of the Stefan problem also provides us with an opportunity to consider the inverse Stefan problem, giving an example of how an improperly posed problem can arise in a realistic physical situation.

3.1 THE WEAK MAXIMUM PRINCIPLE FOR PARABOLIC EQUATIONS

In this section we shall consider the parabolic equation

$$\frac{\partial u}{\partial t} = \frac{\partial^2 u}{\partial x^2} + a(x,t)\frac{\partial u}{\partial x} + b(x,t)u \tag{3.1}$$

where $a(x,t)$ and $b(x,t)$ are real valued continuous functions, noting that the heat equation is the special case when $a(x,t)$ and $b(x,t)$ are identically zero. Motivated by Section 1.1, it is natural to consider solutions of (3.1) defined in a rectangle $0 < x < l$, $t > 0$, which physically corresponds to heat conduction in a rod of length l. In this case, a well posed initial-boundary value problem would consist in prescribing the intitial temperature of the rod $u(x,0)$, and suitable boundary conditions at the ends of the rod, for example, the temperature at $x = 0$ and $x = l$, i.e., $u(0,t)$ and $u(l,t)$. More generally, one could consider the case when ablation is present, and the length of the rod is a function of time. Such considerations lead us to consider solutions of (3.1) in domains D of the form pictured in Figure 3.1, where $\partial D = \partial D_1 + \partial D_2$, ∂D_1 being an open horizontal line segment lying on the line $t = T$ and ∂D_2 a curve lying below ∂D_1. We assume that D lies in the half plane $t > 0$, a situation which can be realized by making a preliminary linear change of variables in (3.1) if necessary. Finally, we assume that the coefficients $a(x,t)$ and $b(x,t)$ of (3.1) are continuous in $\overline{D} = D \cup \partial D$.

Our main aim in this section is to prove a **weak maximum principle** for solutions of (3.1) defined in D. Under appropriate assumptions, this result will state that if the maximum value of a solution $u(x,t)$ of (3.1) in $\overline{D}$ is positive, or the minimum value is negative, then this value is achieved on ∂D_2. If $b(x,t)$ is identically zero, for example in the case of the heat equation, we no longer need to assume that the maximum value is positive or that the minimum value is negative, since in this case constant functions are solutions of (3.1) and hence by adding an appropriate constant to the solution $u(x,t)$ we can always assume that $u(x,t)$ has a positive maximum (or a negative minimum). The maximum principle we shall prove is called *weak* since it does not exclude the possibility that the positive maximum or negative

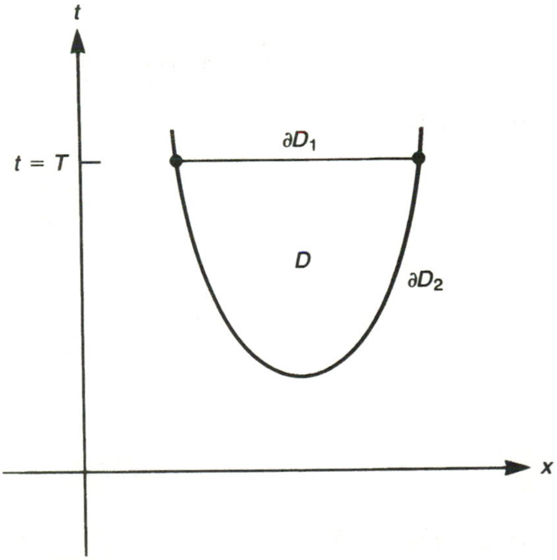

Figure 3.1

minimum (if it exists) is also achieved at an interior point of D. In Section 3.5 we shall prove a *strong* maximum principle for the heat equation which states under what conditions a maximum or minimum value of a solution to the heat equation can be achieved at an interior point. For the heat equation, the physical significance of the maximum principle is clear: the temperature in the interior of the body cannot exceed the initial temperature or the temperature on the boundary.

In what follows, and in the remainder of this chapter, we shall always assume that a solution of (3.1) is twice continuously differentiable with respect both to x and to t in its domain of definition.

THEOREM 15 Weak Maximum Principle

Let $u(x,t)$ be a solution of (3.1) in $D \cup \partial D_1$ that is continuous in $\overline{D}$, and assume that $b(x,t) \leq 0$ in D. Then $u(x,t)$ assumes its positive maximum and negative minimum, if they exist, on ∂D_2.

■ ***Proof.*** For abitrary $\varepsilon > 0$ consider the function $v(x,t)$ defined by $v(x,t) = u(x,t) - \varepsilon t$ and note that $v(x,t)$ is continuous in $\overline{D}$. Suppose a positive maximum of $v(x,t)$ is attained at $(x_0,t_0) \in D \cup \partial D_1$. Then for $k > 0$ sufficiently small, the points (x,t), $x_0 - k \leq x \leq x_0 + k$, $t = t_0$ lie in $D \cup \partial D_1$. Now at (x_0,t_0) we have that

$$\frac{\partial^2 v}{\partial x^2}(x_0,t_0) \leq 0$$

$$\frac{\partial v}{\partial x}(x_0,t_0) = 0,$$

because (x_0,t_0) is a positive maximum. Since $u(x,t)$ is a solution of (3.1) we have

$$\frac{\partial^2 v}{\partial x^2} + a(x,t)\frac{\partial v}{\partial x} + b(x,t)v - \frac{\partial v}{\partial t} = \frac{\partial^2 u}{\partial x^2} + a(x,t)\frac{\partial u}{\partial x} + b(x,t)u - \frac{\partial u}{\partial t}$$
$$- \varepsilon t b(x,t) + \varepsilon = \varepsilon(1 - tb(x,t))$$

and hence at $(x,t) = (x_0,t_0)$,

$$\frac{\partial v}{\partial t}(x_0,t_0) \leq b(x_0,t_0)v(x_0,t_0) + \varepsilon t_0 b(x_0,t_0) - \varepsilon \leq -\varepsilon,$$

since $b(x,t) \leq 0$. Now choose $h > 0$ such that

$$\frac{\partial v}{\partial t}(x_0,t) \leq -\frac{\epsilon}{2}$$

for $t_0 - h \leq t \leq t_0$. Then

$$v(x_0,t_0) - v(x_0,t_0 - h) = \int_{t_0-h}^{t_0} \frac{\partial v}{\partial t}(x_0,t)\,dt \leq -\frac{\varepsilon}{2}h < 0,$$

i.e., $v(x_0,t_0) < v(x_0,t_0 - h)$. But this a contradiction, since $v(x,t)$ has a positive maximum at (x_0,t_0). Hence the positive maximum of $v(x,t)$ lies on ∂D_2.

Now suppose that $u(x,t)$ has a positive maximum in $\overline{D}$. Then for ε sufficiently small $v(x,t)$ has a positive maximum in $\overline{D}$. Since $t \geq 0$ in $\overline{D}$ we have

$$\max_{\overline{D}} u(x,t) = \max_{\overline{D}} (\varepsilon t + v(x,t)) \leq \varepsilon T + \max_{\overline{D}} v(x,t) = \varepsilon T + \max_{\partial D_2} v(x,t)$$

$$= \varepsilon T + \max_{\partial D_2} (u(x,t) - \varepsilon t) \leq \varepsilon T + \max_{\partial D_2} u(x,t).$$

Since ε can be arbitrarily small, we can now conclude that the positive maximum of $u(x,t)$ occurs on ∂D_2.

If $u(x,t)$ has a negative minimum in $\overline{D}$, by considering $-u(x,t)$ and repeating the above argument we can conclude that the negative minimum of $u(x,t)$ occurs on ∂D_2. ■

We now want to use the weak maximum principle to show the uniqueness of the solution (if it exists) to the initial-boundary value problem

$$\frac{\partial u}{\partial t} = \frac{\partial^2 u}{\partial x^2} + a(x,t)\frac{\partial u}{\partial x} + b(x,t)u \qquad ((x,t) \in D) \tag{3.2}$$

$$u(x,t) = \phi(x,t) \qquad ((x,t) \in \partial D_2) \tag{3.3}$$

where $u(x,t)$ continuously assumes the prescribed boundary data $\phi(x,t)$ on ∂D_2. The only assumption made on the coefficients $a(x,t)$ and $b(x,t)$ is that they are real valued and continuous in $\overline{D}$. In particular, no assumption is made on the non-positivity of $b(x,t)$.

THEOREM 16

There exists at most one solution of the initial-boundary value problem (3.2), (3.3).

■ ***Proof.*** If $u_1(x,t)$ and $u_2(x,t)$ are two solutions of (3.2), (3.3), then $v(x,t) = u_1(x,t) - u_2(x,t)$ is a solution of (3.2) such that $v(x,t) = 0$ on ∂D_2. The theorem will be proved if we show that $v(x,t)$ is identically zero in D. If $b(x,t) \leq 0$ in D, then the theorem follows from the weak maximum principle. If $b(x,t) \leq 0$ in D is not satisfied, we choose a constant $\alpha > 0$ such that $\alpha > b(x,t)$ for $(x,t) \in \overline{D}$ and set $v(x,t) = w(x,t)e^{\alpha t}$. Then $w(x,t)$ is a solution of

$$\frac{\partial w}{\partial t} = \frac{\partial^2 w}{\partial x^2} + a(x,t)\frac{\partial w}{\partial x} + (b(x,t) - \alpha)w$$

in D and $w(x,t) = 0$ on ∂D_2. By the weak maximum principle, $w(x,t)$ is identically zero in D, and hence $v(x,t)$ is also identically zero in D. ■

THEOREM 17

The solution of the initial-boundary value problem (3.2), (3.3), if it exists, depends continuously on the boundary data $\phi(x,t)$ with respect to the maximum norm on ∂D_2.

■ ***Proof.*** Note that ∂D_2 is a closed arc, and since $\phi(x,t)$ is continuous on ∂D_2,

$$\|\phi\| = \max_{(x,t)\in\partial D_2} |\phi(x,t)|$$

is well defined. Set α as in the above theorem (if $b(x,t) \leq 0$ set $\alpha = 0$) and let $u_1(x,t)$, $u_2(x,t)$ be two solutions of (3.2) such that their boundary data $\phi_1(x,t)$, $\phi_2(x,t)$, respectively, satisfy

$$\|\phi_1 - \phi_2\| < \varepsilon$$

where ε is a positive constant. Then, setting $v(x,t) = u_1(x,t) - u_2(x,t)$, $v(x,t) = w(x,t)e^{\alpha t}$, we have from the weak maximum principle that

$$\begin{aligned}\max_{(x,t\in\overline{D}}|u_1(x,t) - u_2(x,t)| &= \max_{(x,t)\in\overline{D}}|v(x,t)|\\ &= \max_{(x,t)\in\overline{D}}|w(x,t)e^{\alpha t}|\\ &\leq \varepsilon e^{\alpha T},\end{aligned}$$

where we have again used the fact that $t \geq 0$ for $(x,t) \in \overline{D}$. The theorem is now proved. ■

To conclude this section, we note that we cannot expect to uniquely solve (3.2), (3.3) "backwards in time," i.e., in domains of the form pictured in Figure 3.2 with data prescribed on ∂D_2. For example,

$$u(x,t) = e^{-t}\cos x - \frac{1}{2}e^{-4t}\cos 2x \tag{3.4}$$

is a solution of the heat equation

$$\frac{\partial u}{\partial t} = \frac{\partial^2 u}{\partial x^2}$$

satisfying $u(x,t) = 0$ on the curve

$$t = \frac{1}{3}\log\frac{1}{2}\frac{\cos 2x}{\cos x}.$$

If D is as pictured in Figure 3.3, the initial-boundary value problem

$$\frac{\partial u}{\partial t} = \frac{\partial^2 u}{\partial x^2} \qquad ((x,t) \in D)$$

$$u(x,t) = 0 \qquad \left(\mathrm{t} = \frac{1}{3}\log\frac{1}{2}\frac{\cos 2x}{\cos x}\right)$$

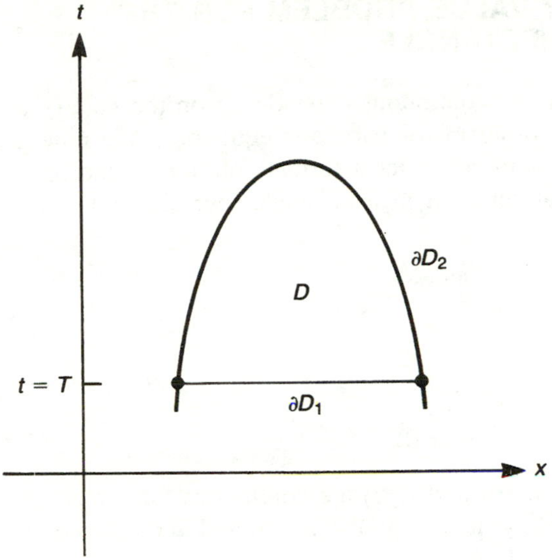

Figure 3.2

has both $u(x,t)$ identically zero and (3.4) as solutions; therefore, uniqueness is violated. In spite of the fact that the solution of the initial-boundary value problem for the heat equation "backwards in time" is improperly posed (in the sense that solutions are not uniquely determined), such problems do appear in practice and can be treated by certain "regularization" methods. For details and further references we refer the reader to Cannon and Payne.

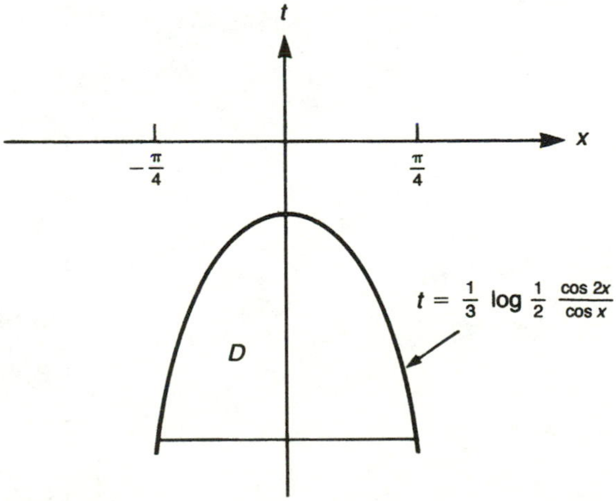

Figure 3.3

3.2 THE INITIAL-BOUNDARY VALUE PROBLEM FOR THE HEAT EQUATION IN A RECTANGLE

We have established the uniqueness and continuous dependence on the data of solutions to the initial-boundary value problem for parabolic equations. We now use the method of separation of variables to construct a formal solution of the initial-boundary value problem for the heat equation in a rectangle; our aim is to find a function $u(x,t)$ such that

$$\frac{\partial u}{\partial t} = \frac{\partial^2 u}{\partial x^2} \qquad (0 < x < l,\ t > 0) \tag{3.5}$$

$$u(x,0) = \phi(x) \qquad (0 \le x \le l) \tag{3.6}$$

$$u(0,t) = \mu_1(t), \qquad u(l,t) = \mu_2(t) \qquad (t \ge 0), \tag{3.7}$$

where l is a positive constant, $\phi(x)$, $\mu_1(t)$, and $\mu_2(t)$ are continuous functions satisfying the compatibility conditions $\phi(0) = \mu_1(0)$, $\phi(l) = \mu_2(0)$, and $u(x,t)$ is required to continuously assume its initial-boundary data.

We first consider the case in which $\mu_1(t)$ and $\mu_2(t)$ are identically zero, i.e., (3.7) is replaced by

$$u(0,t) = u(l,t) = 0 \qquad (t \ge 0) \tag{3.7a}$$

and make the assumption that $\phi(x)$ is continuously differentiable for x on $[0,l]$. Using the method of separation of variables, we look for a solution of the heat equation (3.5) in the form

$$u(x,t) = T(t)X(x).$$

Substitution into (3.5) implies that

$$XT' = X''T$$

or

$$\frac{T'}{T} = \frac{X''}{X} = -\lambda,$$

where λ is a constant, i.e.,

$$T' + \lambda T = 0 \tag{3.8}$$

$$X'' + \lambda X = 0. \tag{3.9}$$

From (3.7a) we have

$$X(0) = X(l) = 0, \tag{3.10}$$

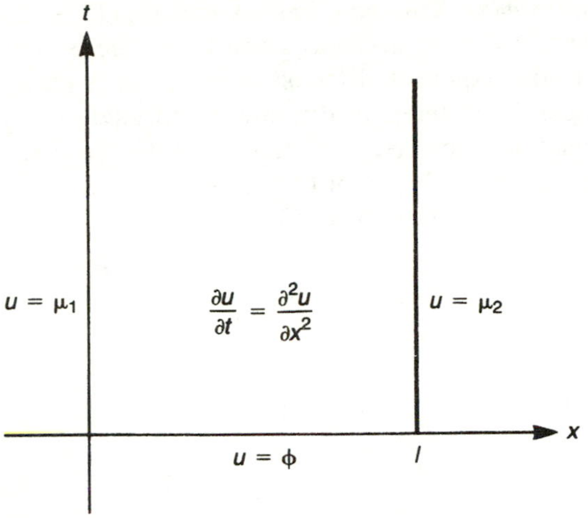

Figure 3.4

and hence (cf. Section 2.4)

$$\lambda = \lambda_n = \left(\frac{n\pi}{l}\right)^2$$

$$X(x) = X_n(x) = \sin\frac{n\pi x}{l}$$

where n is an integer. From (3.8) we now have

$$T(t) = T_n(t) = a_n e^{-(n\pi/l)^2 t}$$

where a_n is a constant (for each integer n). Note that we need only consider positive n, because $X(x)$ and $T(t)$ remain unchanged (except for a possible factor of minus one) if n is replaced by $-n$. The above analysis now implies that the series

$$u(x,t) = \sum_{n=1}^{\infty} a_n e^{-(n\pi/l)^2 t} \sin\frac{n\pi x}{l} \tag{3.11}$$

will be a formal solution of (3.5) and (3.7a). In order to satisfy (3.6) we must have

$$u(x,0) = \phi(x) = \sum_{n=1}^{\infty} a_n \sin\frac{n\pi x}{l}, \tag{3.12}$$

and thus we require that a_n be the Fourier sine coefficient of $\phi(x)$, i.e.,

$$a_n = \frac{2}{l}\int_0^l \phi(x)\sin\frac{n\pi x}{l}\,dx. \tag{3.13}$$

Since $\phi(x)$ is continuously differentiable and $\phi(0) = \phi(l) = 0$ we can conclude from Fourier's theorem and the analysis in Section 2.4 that (3.12) is absolutely and uniformly convergent with sum equal to $\phi(x)$. This now implies that (3.11) is absolutely and uniformly convergent for $t \geq 0$, i.e., $u(x,t)$ as given by (3.11) is continuous for $0 \leq x \leq l$, $t \geq 0$. Due to the exponential factor in (3.11), it is easily seen that for $t > 0$ the series (3.11) can be differentiated termwise infinitely often (note the difference from the wave equation discussed in Section 2.4!). Hence $u(x,t)$ as defined by (3.11) and (3.13) is the desired solution of (3.5), (3.6), (3.7a).

We now consider the initial-boundary value problem (3.5)–(3.7). Let

$$u(x,t) = v(x,t) + w(x,t)$$

where

$$w(x,t) = \mu_1(t) + \frac{1}{l}[\mu_2(t) - \mu_1(t)]x.$$

Then $v(x,t)$ is a solution of

$$\frac{\partial v}{\partial t} = \frac{\partial^2 v}{\partial x^2} - \frac{\partial w}{\partial t} \qquad (0 < x < l,\ t > 0)$$

$$v(x,0) = \phi(x) - w(x,0) \qquad (0 \leq x \leq l)$$

$$v(0,t) = v(l,t) = 0 \qquad (t \geq 0).$$

Subtracting from $v(x,t)$ the solution of (3.5), (3.6), (3.7a) with $\phi(x)$ replaced by $\phi(x) - w(x,0)$, we see that the initial-boundary value problem (3.5), (3.6), (3.7) can be reduced to solving an initial-boundary value problem of the form

$$\frac{\partial u}{\partial t} = \frac{\partial^2 u}{\partial x^2} + f(x,t) \qquad (0 < x < l,\ t > 0) \tag{3.14}$$

$$u(x,0) = 0 \qquad (0 \leq x \leq l) \tag{3.15}$$

$$u(0,t) = u(l,t) = 0 \qquad (t \geq 0), \tag{3.16}$$

where $f(x,t)$ is a known function. We make the assumption that $f(x,t)$ is continuously differentiable for $0 \leq x \leq l$, $t \geq 0$, and look for a solution of (3.14), (3.15), (3.16) in the form

$$u(x,t) = \sum_{n=1}^{\infty} T_n(t) \sin\frac{n\pi x}{l}, \tag{3.17}$$

where the functions $T_n(t)$ are to be determined. Expand $f(x,t)$ in the Fourier sine series

$$f(x,t) = \sum_{n=1}^{\infty} f_n(t) \sin\frac{n\pi x}{l} \tag{3.18}$$

$$f_n(t) = \frac{2}{l}\int_0^l f(x,t) \sin\frac{n\pi x}{l}\,dx. \tag{3.19}$$

Substituting (3.17) into (3.14) (assuming termwise differentiation is permissible) and using (3.18) gives

$$\sum_{n=1}^{\infty}\left[T_n'(t) + \left(\frac{n\pi}{l}\right)^2 T_n(t) - f_n(t)\right]\sin\frac{n\pi x}{l} = 0,$$

i.e.,

$$T_n' + \left(\frac{n\pi}{l}\right)^2 T = f_n(t). \tag{3.20}$$

Since

$$u(x,0) = \sum_{n=1}^{\infty} T_n(0)\sin\frac{n\pi x}{l} = 0,$$

we have

$$T_n(0) = 0. \tag{3.21}$$

Equations (3.20) and (3.21) imply that

$$T_n(t) = \int_0^t f_n(\tau)\exp\left[-\frac{n^2\pi^2}{l^2}(t-\tau)\right]d\tau. \tag{3.22}$$

It can now be verified directly that (3.17), (3.19), (3.22) provides the solution of (3.14), (3.15), (3.16).

We note in closing that we have only considered the initial-boundary value problem for the heat equation where $u(x,t)$ is prescribed at $x = 0$ and $x = l$. The initial-boundary value problem where $\partial u(x,t)/\partial x$ is prescribed at $x = 0$ and $x = l$ also appears in practice and can be handled in a similar manner.

3.3 CAUCHY'S PROBLEM FOR THE HEAT EQUATION

Since the characteristics for the heat equation

$$\frac{\partial u}{\partial t} = \frac{\partial^2 u}{\partial x^2}$$

are the lines t = constant, any non-characteristic curve can be expressed in the form $x = s(t)$. Hence the non-characteristic Cauchy problem for the heat equation is to find a solution of the heat equation such that on the curve $x = s(t)$ we have

$$u(s(t),t) = f(t)$$

$$\frac{\partial u}{\partial x}(s(t),t) = g(t)$$

where $f(t)$ and $g(t)$ are prescribed functions. However, this problem is improperly posed. To see this, consider

$$u_n(x,t) = \frac{1}{n}\,[e^{nx}\sin(2n^2t + nx) + e^{-nx}\sin(2n^2t - nx)],$$

where n is a positive integer. Then $u_n(x,t)$ is a solution of the heat equation such that on the line $x = 0$ we have

$$u_n(0,t) = \frac{2}{n}\sin 2n^2t$$

$$\frac{\partial u_n}{\partial x}(0,t) = 0.$$

But although $u_n(0,t)$ tends to zero as n tends to infinity, $u_n(x,t)$ tends to infinity as n tends to infinity if x is greater than zero. Hence, solutions of the non-characteristic Cauchy problem for the heat equation do not depend continuously on the initial data. Replacing the factor $1/n$ in the definition of $u_n(x,t)$ by $1/n^k$ where k is a positive integer shows that solutions of the non-characteristic Cauchy problem do not depend continuously on the derivatives of the initial data either, in contrast to the case of the wave equation. Nevertheless, the non-characteristic Cauchy problem can arise in certain inverse formulations of free boundary value problems; we shall study such a problem in Section 6 of this chapter. However, for the present, we shall only consider the characteristic Cauchy problem, i.e., to find a function $u(x,t)$ such that

$$\frac{\partial u}{\partial t} = \frac{\partial^2 u}{\partial x^2} \qquad (-\infty < x < \infty,\ t > 0) \tag{3.23}$$

$$u(x,0) = \phi(x) \qquad (-\infty < x < \infty), \tag{3.24}$$

where $\phi(x)$ is a given continuous function that is bounded in absolute value for $-\infty < x < \infty$ and $u(x,t)$ is required to continuously assume the given initial data at $t = 0$. Note that from the results of the previous section the characteristic Cauchy problem where $\phi(x)$ is only prescribed on a finite interval of the x axis is improperly posed since the solution is not uniquely determined.

THEOREM 18

There exists at most one solution of (3.23), (3.24) that is bounded in absolute value for $t \geq 0$, $-\infty < x < \infty$.

■ ***Proof.*** Suppose there exist two solutions $u_1(x,t)$ and $u_2(x,t)$ of (3.23), (3.24) such that for some positive constant M we have $|u_1(x,t)| < M$ and $|u_2(x,t)| < M$ for $t \geq 0$, $-\infty < x < \infty$. Let $w(x,t) = u_1(x,t) - u_2(x,t)$. Then $w(x,t)$ is a solution of the heat equation (3.23) such that $w(x,0) = 0$ for $-\infty < x < \infty$ and $|w(x,t)| \leq 2M$ for

$t \geq 0$, $-\infty < x < \infty$. Consider the bounded region $|x| \leq L$, $0 \leq t \leq T$, where L and T are positive constants and define $v(x,t)$ by

$$v(x,t) = \frac{4M}{L^2}\left(\frac{x^2}{2} + t\right).$$

Then $v(x,t)$ is a solution of the heat equation, and $v(x,0) \geq |w(x,0)| = 0$, $v(\pm L,t) \geq 2M \geq |w(\pm L,t)|$. Hence, by the maximum principle we have that $v(x,t) - w(x,t) \geq 0$ and $v(x,t) + w(x,t) \geq 0$ for $|x| \leq L$, $0 \leq t \leq T$, i.e.,

$$|w(x,t)| \leq v(x,t) = \frac{4M}{L^2}\left(\frac{x^2}{2} + t\right)$$

for $|x| \leq L$, $0 \leq t \leq T$. Fix $(x,t) = (x_0,t_0)$. Then for every $\epsilon > 0$ we have $|w(x_0,t_0)| < \epsilon$ for L sufficiently large. Since $\epsilon > 0$ is arbitrary, $w(x_0,t_0) = 0$; and since (x_0,t_0) is an arbitrary point in the half plane $t > 0$, we can conclude that $w(x,t)$ is identically zero, i.e., $u_1(x,t) = u_2(x,t)$ for $-\infty < x < \infty$, $t \geq 0$. ■

We now want to show the existence of a solution to the Cauchy problem (3.23), (3.24). We begin by proceeding formally. By separation of variables we see that for any number λ

$$u_\lambda(x,t) = g(\lambda) \exp\left[-\lambda^2 t + i\lambda x\right]$$

is a solution of the heat equation (3.23) where $g(\lambda)$ is an arbitrary function of λ. Hence

$$u(x,t) = \int_{-\infty}^{\infty} g(\lambda) \exp\left[-\lambda^2 t + i\lambda x\right] d\lambda$$

will be a solution of the heat equation if the integral exists, and we can differentiate inside the integral sign. Assuming this, we shall choose $g(\lambda)$ such that (3.24) is satisfied, i.e., we want

$$\phi(x) = \int_{-\infty}^{\infty} g(\lambda) e^{i\lambda x}\, d\lambda.$$

By the Fourier integral theorem (see Section 1.4) this will be true (under appropriate assumptions on $\phi(x)$) if we choose $g(\lambda)$ to be

$$g(\lambda) = \frac{1}{2\pi}\int_{-\infty}^{\infty} \phi(\xi) e^{-i\lambda\xi}\, d\xi.$$

Hence,

$$u(x,t) = \frac{1}{2\pi}\int_{-\infty}^{\infty} \phi(\xi)\left[\int_{-\infty}^{\infty} \exp\left[-\lambda^2 t + i\lambda(x-\xi)\right] d\lambda\right] d\xi.$$

But

$$\int_{-\infty}^{\infty} \exp\,[-\lambda^2 t + i\lambda(x-\xi)]\,d\lambda = 2\int_0^{\infty} e^{-\lambda^2 t}\cos\lambda(x-\xi)\,d\lambda$$

$$= \frac{2}{\sqrt{t}}\int_0^{\infty} e^{-\mu^2}\cos\mu z\,d\mu,$$

where we have made the change of variables $\lambda = \mu/\sqrt{t}$, $z = (x-\xi)/\sqrt{t}$. Defining $I(z)$ by

$$I(z) = \int_0^{\infty} e^{-\mu^2}\cos\mu z\,d\mu$$

we see from integration by parts that

$$I'(z) = -\int_0^{\infty} \mu e^{-\mu^2}\sin\mu z\,d\mu = -\frac{1}{2}zI(z)$$

and

$$I(0) = \int_0^{\infty} e^{-\mu^2}\,d\mu = \sqrt{\int_0^{\infty}\int_0^{\infty} e^{-(\mu^2+\nu^2)}\,d\mu d\nu}$$

$$= \sqrt{\frac{1}{4}\int_0^{2\pi}\int_0^{\infty} e^{-r^2} r\,dr d\theta} = \frac{\sqrt{\pi}}{2},$$

where we have changed variables from Cartesian to polar coordinates. Solving the above initial value problem for $I(z)$ gives

$$I(z) = \frac{\sqrt{\pi}}{2}e^{-z^2/4},$$

i.e.,

$$\int_{-\infty}^{\infty} \exp\,[-\lambda^2 t + i\lambda(x-\xi)]\,d\lambda = \sqrt{\frac{\pi}{t}}\exp\left[-\frac{(x-\xi)^2}{4t}\right].$$

Hence

$$u(x,t) = \frac{1}{\sqrt{4\pi t}}\int_{-\infty}^{\infty} \phi(\xi)\exp\left[-\frac{(x-\xi)^2}{4t}\right]d\xi \tag{3.25}$$

provides the formal solution to the Cauchy problem (3.23), (3.24). The kernel

$$K(x,t;\xi) = \frac{1}{\sqrt{4\pi t}}\exp\left[-\frac{(x-\xi)^2}{4t}\right]$$

is known as the **fundamental solution** of the heat equation.

The representation formula (3.25) for the solution of (3.23), (3.24) is often formally obtained by a more direct application of Fourier's integral theorem than that used above. To illustrate this approach, define the **Fourier transform** of $u(x,t)$ with respect to x by

$$\hat{u}(\lambda,t) = \mathrm{F}[u] = \int_{-\infty}^{\infty} u(x,t)e^{-i\lambda x}\,dx$$

and note that formally

$$\mathrm{F}\left[\frac{\partial u}{\partial t}\right] = \frac{\partial \hat{u}}{\partial t}$$

$$\mathrm{F}\left[\frac{\partial^2 u}{\partial x^2}\right] = \int_{-\infty}^{\infty} \frac{\partial^2 u(x,t)}{\partial x^2} e^{-i\lambda x}\,dx = i\lambda \int_{-\infty}^{\infty} \frac{\partial u(x,t)}{\partial x} e^{-i\lambda x}\,dx$$

$$= -\lambda^2 \int_{-\infty}^{\infty} u(x,t)e^{-i\lambda x}\,dx = -\lambda^2 \hat{u}$$

where we have integrated by parts. Hence, taking the Fourier transform of (3.23), (3.24) leads to the initial value problem

$$\frac{\partial \hat{u}}{\partial t} + \lambda^2 \hat{u} = 0$$

$$\hat{u}(\lambda,0) = \hat{\phi}(\lambda)$$

whose solution is

$$\hat{u}(\lambda,t) = \hat{\phi}(\lambda)e^{-\lambda^2 t}.$$

By the Fourier integral theorem,

$$u(x,t) = \frac{1}{2\pi}\int_{-\infty}^{\infty} \hat{\phi}(\lambda)\exp\left[-\lambda^2 t + i\lambda x\right] d\lambda$$

$$= \frac{1}{2\pi}\int_{-\infty}^{\infty} \phi(\xi)\left[\int_{-\infty}^{\infty} \exp\left[-\lambda^2 t + i\lambda(x-\xi)\right] d\lambda\right] d\xi,$$

leading again to (3.25). Further examples of the use of Fourier transforms to solve problems in partial differential equations can be found in Exercise 27 at the end of this chapter.

We shall now prove that (3.25) is the unique bounded solution of (3.23), (3.24) assuming that $\phi(x)$ is continuous for $-\infty < x < \infty$ and satisfies $|\phi(x)| \leq M$ for $-\infty < x < \infty$, where M is a positive constant. It is easily verified that for $t > 0$ we can differentiate inside the integral sign, so (3.25) defines a solution of the heat equation for $t > 0$. We only have to show that $|u(x,t)| \leq M$ for $-\infty < x < \infty$, $t \geq 0$, and

$$\lim_{t\to 0} u(x,t) = \phi(x)$$

uniformly on compact subsets of $-\infty < x < \infty$. Setting $\alpha = (\xi - x)/2\sqrt{t}$ gives

$$u(x,t) = \frac{1}{\sqrt{\pi}} \int_{-\infty}^{\infty} \phi(x + 2\alpha\sqrt{t}) e^{-\alpha^2}\, d\alpha,$$

so

$$|u(x,t)| \leq \frac{M}{\sqrt{\pi}} \int_{-\infty}^{\infty} e^{-\alpha^2}\, d\alpha = \frac{2M}{\sqrt{\pi}} \int_{0}^{\infty} e^{-\alpha^2}\, d\alpha = M$$

by previous calculation. Furthermore,

$$u(x,t) - \phi(x) = \frac{1}{\sqrt{\pi}} \int_{-\infty}^{\infty} [\phi(x + 2\alpha\sqrt{t}) - \phi(x)] e^{-\alpha^2}\, d\alpha. \tag{3.26}$$

For any $\epsilon > 0$ let $N > 0$ be such that

$$\frac{2M}{\sqrt{\pi}} \int_{-\infty}^{-N} e^{-\alpha^2}\, d\alpha \leq \frac{\epsilon}{3}$$

$$\frac{2M}{\sqrt{\pi}} \int_{N}^{\infty} e^{-\alpha^2}\, d\alpha \leq \frac{\epsilon}{3}.$$

Then since $|\phi(x + 2\alpha\sqrt{t}) - \phi(x)| \leq 2M$, (3.26) tells us that

$$\begin{aligned} |u(x,t) - \phi(x)| &\leq \frac{1}{\sqrt{\pi}} \int_{-\infty}^{\infty} |\phi(x + 2\alpha\sqrt{t}) - \phi(x)| e^{-\alpha^2}\, d\alpha \\ &\leq \frac{2\epsilon}{3} + \frac{1}{\sqrt{\pi}} \int_{-N}^{N} |\phi(x + 2\alpha\sqrt{t}) - \phi(x)| e^{-\alpha^2}\, d\alpha. \end{aligned}$$

Let $|x| \leq L$ where L is a positive constant. Since $\phi(x)$ is continuous, there exists a $\delta > 0$ such that for $0 \leq t \leq \delta$, $|x| \leq L$, $|\alpha| \leq N$, we have

$$|\phi(x + 2\alpha\sqrt{t}) - \phi(x)| < \frac{\epsilon}{3};$$

hence from the above,

$$|u(x,t) - \phi(x)| \leq \frac{2\epsilon}{3} + \frac{\epsilon}{3\sqrt{\pi}} \int_{-N}^{N} e^{-\alpha^2}\, d\alpha < \frac{2\epsilon}{3} + \frac{\epsilon}{3\sqrt{\pi}} \int_{-\infty}^{\infty} e^{-\alpha^2}\, d\alpha = \epsilon.$$

This implies that $u(x,t)$ is continuous for $t \geq 0$ and

$$\lim_{t\to 0} u(x,t) = \phi(x).$$

We have now shown that (3.25) is the unique bounded solution of (3.23), (3.24).

An amusing consequence of this analysis is the famous Weierstrass approximation theorem.

THEOREM 19 Weierstrass Approximation Theorem

Let $f(x)$ be a continuous function on the closed interval $[a,b]$. Then for every $\epsilon > 0$ there exists a polynomial $p(x)$ such that

$$\max_{a \le x \le b} |f(x) - p(x)| < \epsilon.$$

■ ***Proof.*** Extend $f(x)$ continuously to $(-\infty,\infty)$ such that $f(x)$ is identically zero for $|x| > L$ where L is a sufficiently large positive number. Then for $t > 0$,

$$u(x,t) = \frac{1}{\sqrt{4\pi t}} \int_{-L}^{L} f(\xi) \exp\left[-\frac{(x-\xi)^2}{4t}\right] d\xi$$

is a solution of the heat equation such that for δ sufficiently small

$$\max_{a \le x \le b} |u(x,\delta) - f(x)| < \frac{\epsilon}{2}. \tag{3.27}$$

Let

$$M = \max_{-L \le \xi \le L} |f(\xi)|$$

and choose n such that

$$\max_{\substack{a \le x \le b \\ -L \le \xi \le L}} \left| \exp\left[-\frac{(x-\xi)^2}{4\delta}\right] - \sum_{k=0}^{n} \frac{(-1)^k}{k!}\left[\frac{(x-\xi)^2}{4\delta}\right]^k \right| < \frac{\epsilon\sqrt{\pi\delta}}{2LM}.$$

This is possible to do by Taylor's theorem. Then

$$p(x) = \frac{1}{\sqrt{4\pi\delta}} \int_{-L}^{L} f(\xi) \sum_{k=0}^{n} \frac{(-1)^k}{k!}\left[\frac{(x-\xi)^2}{4\delta}\right]^k d\xi$$

is a polynomial, and

$$\max_{a \le x \le b} |p(x) - u(x,\delta)| < \frac{\epsilon}{2}. \tag{3.28}$$

Equations (3.27) and (3.28) now imply that

$$\max_{a \le x \le b} |p(x) - f(x)| < \epsilon.$$

This proves the theorem. ■

We close this section by making a few comments and observations. To begin with, it follows easily from (3.25) that the solution of the Cauchy problem (3.23), (3.24) depends continuously on the initial data $\phi(x)$ with respect to the maximum norm over $(-\infty,\infty)$; however, from (3.25) we also see that the initial temperature $\phi(x)$ is propagated instantaneously—any nonzero temperature at time $t = 0$ creates a nonzero temperature $u(x,t)$ for all x, $-\infty < x < \infty$, and $t > 0$. This is obviously a physical impossibility and shows the limitations of the linear equation (3.23) of heat conduction. Even if a mathematical model is well posed such as (3.23), (3.24), the resulting solution may exhibit behavior that is not physically realistic; a well posed mathematical model is not the same thing as a good physical model of the phenomenon being investigated!

One can show by an analysis similar to the one in this section that the unique bounded solution of

$$\frac{\partial u}{\partial t} = \frac{\partial^2 u}{\partial x^2} + \frac{\partial^2 u}{\partial y^2} \qquad ((x,y) \in R^2,\ t > 0)$$

$$u(x,y,0) = \phi(x,y) \qquad ((x,y) \in R^2)$$

where $\phi(x,y)$ is continuous and bounded in absolute value is given by

$$u(x,y,t) = \frac{1}{4\pi t}\int_{-\infty}^{\infty}\int_{-\infty}^{\infty} \phi(\xi,\eta) \exp\left[-\frac{(x-\xi)^2 + (y-\eta)^2}{4t}\right] d\xi d\eta,$$

and the unique bounded solution of

$$\frac{\partial u}{\partial t} = \frac{\partial^2 u}{\partial x^2} + \frac{\partial^2 u}{\partial y^2} + \frac{\partial^2 u}{\partial z^2} \qquad ((x,y,z) \in R^3,\ t > 0)$$

$$u(x,y,z,0) = \phi(x,y,z) \qquad ((x,y,z) \in R^3)$$

where $\phi(x,y,z)$ is continuous and bounded in absolute value is given by

$$u(x,y,z,t) =$$

$$\frac{1}{(4\pi t)^{3/2}}\int_{-\infty}^{\infty}\int_{-\infty}^{\infty}\int_{-\infty}^{\infty} \phi(\xi,\eta,\zeta) \exp\left[-\frac{(x-\xi)^2 + (y-\eta)^2 + (z-\zeta)^2}{4t}\right] d\xi d\eta d\zeta.$$

Our final remark is to note that by imitating the method in Section 2.3 for solving the nonhomogeneous wave equation, it can easily be shown (cf. Exercise 25) that the solution of

$$\frac{\partial u}{\partial t} = \frac{\partial^2 u}{\partial x^2} + f(x,t) \qquad (-\infty < x < \infty,\ t > 0)$$

$$u(x,0) = 0 \qquad (-\infty < x < \infty),$$

where $f(x,t)$ is continuous and bounded with respect to x in absolute value is given by

$$u(x,t) = \int_0^t \int_{-\infty}^{\infty} \frac{f(\xi,\tau)}{\sqrt{4\pi(t-\tau)}} \exp\left[-\frac{(x-\xi)^2}{4(t-\tau)}\right] d\xi d\tau \tag{3.29}$$

with similar expressions holding in higher dimensions.

3.4 REGULARITY OF SOLUTIONS TO THE HEAT EQUATION

We now want to show that solutions of the heat equation in a domain D in the plane are analytic functions of x and infinitely differentiable with respect to t. This strong regularity property of solutions of the heat equation contrasts with the wave equation, where the regularity of solutions is no better than that of the initial data. Since regularity is a local phenomenon, it suffices to consider the heat equation restricted to a small rectangle containing an arbitrary point in D. In particular, we consider the heat equation

$$\frac{\partial u}{\partial t} = \frac{\partial^2 u}{\partial x^2}$$

defined in the open rectangle $R = (0,a) \times (0,T)$ where a and T are positive numbers. Then for (x,t) in R we want to show that $u(x,t)$ can be continued into the complex domain as an analytic function of x for each fixed t and that for real x and t, $u(x,t)$ is infinitely differentiable. In the analysis that follows, R_t denotes the rectangle $(0,a) \times (0,t)$ where $0 < t < T$ (see Figure 3.5).

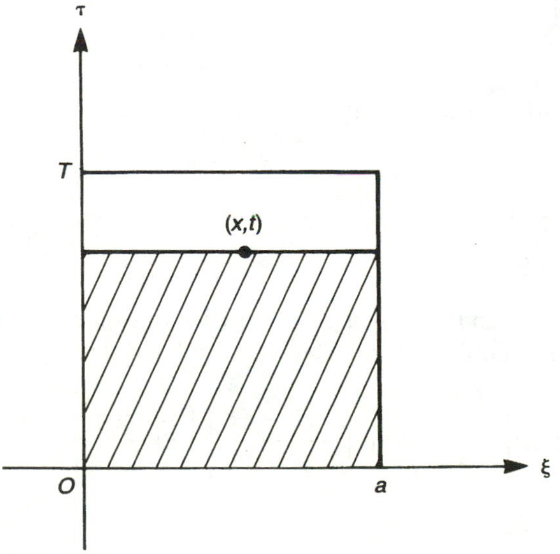

Figure 3.5

We begin by defining the differential operators L and M by

$$L[u] = \frac{\partial^2 u}{\partial \xi^2} - \frac{\partial u}{\partial \tau}$$

$$M[v] = \frac{\partial^2 v}{\partial \xi^2} + \frac{\partial v}{\partial \tau}.$$

Then for $u(\xi,\tau)$ and $v(\xi,\tau)$ twice continuously differentiable with respect to ξ and continuously differentiable with respect to τ,

$$vL[u] - uM[v] = \frac{\partial}{\partial \xi}\left(v\frac{\partial u}{\partial \xi} - u\frac{\partial v}{\partial \xi}\right) - \frac{\partial}{\partial \tau}(uv) \tag{3.30}$$

and integrating (3.30) over the rectangle R_t gives

$$\iint_{R_t} (vL[u] - uM[v])\, d\xi d\tau = \int_0^t \left(v(a,\tau)\frac{\partial u(a,\tau)}{\partial \xi} - u(a,\tau)\frac{\partial v(a,\tau)}{\partial \xi}\right) d\tau$$

$$- \int_0^t \left(v(0,\tau)\frac{\partial u(0,\tau)}{\partial \xi} - u(0,\tau)\frac{\partial v(0,\tau)}{\partial \xi}\right) d\tau \tag{3.31}$$

$$- \int_0^a u(\xi,t)v(\xi,t)\, d\xi + \int_0^a u(\xi,0)v(\xi,0)\, d\xi.$$

Now let $u(\xi,\tau)$ be a solution of $L[u] = 0$ in $\overline{R} = [0,a] \times [0,T]$ and let $v(\xi,\tau)$ be defined by

$$v(\xi,\tau) = \frac{1}{\sqrt{t-\tau}} \exp\left[-\frac{(x-\xi)^2}{4(t-\tau)}\right].$$

Then $M[v] = 0$ for $\tau \neq t$ and $\xi \neq x$ and, following the verification of equation (3.25) for the solution of the Cauchy problem, for $0 < x < a$ we have that

$$\int_0^a u(\xi,t)v(\xi,t)\, d\xi = \lim_{\tau\to t} \int_0^a u(\xi,t)\left[\frac{1}{\sqrt{t-\tau}} \exp\left[-\frac{(x-\xi)^2}{4(t-\tau)}\right]\right] d\xi$$

$$= \sqrt{4\pi}\, u(x,t). \tag{3.32}$$

Hence, from (3.31) and (3.32) we have

$$
\begin{aligned}
u(x,t) = \frac{1}{\sqrt{4\pi}} \int_0^t \Bigg[& \frac{\exp\left(-\dfrac{(x-a)^2}{4(t-\tau)}\right)}{\sqrt{t-\tau}} \frac{\partial u}{\partial \xi}(a,t) \\
& - u(a,t) \frac{\partial}{\partial \xi} \frac{\exp\left(-\dfrac{(x-a)^2}{4(t-\tau)}\right)}{\sqrt{t-\tau}} \Bigg] d\tau \\
- \frac{1}{\sqrt{4\pi}} \int_0^t \Bigg[& \frac{\exp\left(-\dfrac{x^2}{4(t-\tau)}\right)}{\sqrt{t-\tau}} \frac{\partial u(0,t)}{\partial \xi} \\
& - u(0,t) \frac{\partial}{\partial \xi} \frac{\exp\left(-\dfrac{x^2}{4(t-\tau)}\right)}{\sqrt{t-\tau}} \Bigg] d\tau \\
+ \frac{1}{\sqrt{4\pi}} \int_0^a & u(\xi,0) \left[\frac{1}{\sqrt{t}} \exp\left(-\frac{(x-\xi)^2}{4t}\right) \right] d\xi
\end{aligned} \tag{3.33}
$$

where $\partial/\partial\xi \exp(-x^2/4t) = \partial/\partial\xi \exp(-(x-\xi)^2/4t)|_{\xi=0}$, etc. Equation (3.33) now implies the following theorem. (For the reader who is unfamiliar with analytic function theory, omit the phrase "and for each fixed t is an analytic function of x" in the statement of the theorem and in the second paragraph of the proof of the theorem.)

THEOREM 20

Let $u(x,t)$ be a solution of the heat equation

$$\frac{\partial u}{\partial t} = \frac{\partial^2 u}{\partial x^2}$$

in a domain D. Then $u(x,t)$ is infinitely differentiable with respect to x and t and for each fixed t is an analytic function of x for (x,t) in D.

■ ***Proof.*** The fundamental solution

$$K(x,t;\xi) = \frac{1}{\sqrt{4\pi t}} \exp\left[-\frac{(x-\xi)^2}{4t}\right]$$

is infinitely differentiable for $-\infty < x < \infty$, $-\infty < \xi < \infty$, $x \neq \xi$, and $-\infty < t < \infty$ if we define $K(x,t;\xi) = 0$ for $t \leq 0$. Hence we can replace the integrals from zero to t in (3.33) by integrals from zero to infinity, and it is now clear that $u(x,t)$ is infinitely differentiable in D. (For $(x,t) \in D$, let R be a rectangle contained in D such that $(x,t) \in R$.)

To extend $u(x,t)$ to complex values of x, let $x = x_1 + ix_2$ in (3.33), where x_1, x_2, and t are real. Let R be as above, i.e., $R \subset D$, $(x_1,t) \in R$. The last integral in (3.33) is easily seen to be an entire function of x. Furthermore, $K(x,t - \tau;\xi)$ is analytic in x and bounded in absolute value by

$$\frac{1}{\sqrt{4\pi(t-\tau)}} \exp\left[\frac{x_2^2 - (\xi - x_1)^2}{4(t-\tau)}\right].$$

Hence, $K(x,t - \tau;\xi)$ is uniformly bounded for complex x provided $(\xi - x_1)^2 - x_2^2$ is bounded below by a positive constant and $\tau < t$. A similar statement holds for $\partial K/\partial\xi$. In the first two integrals in (3.33) let the range of integration be from zero to $t - \epsilon$ where $\epsilon > 0$, and let ϵ tend to zero. By the above considerations, this limit is uniform, and since the uniform limit of analytic functions is analytic we can conclude that $u(x,t)$ is analytic in x provided $(\xi - x_1)^2 - x_2^2 > 0$ for $\xi = a$ and $\xi = 0$. But this is clearly true provided x_2 is small enough, which completes the proof of the theorem. ■

* 3.5 THE STRONG MAXIMUM PRINCIPLE FOR THE HEAT EQUATION

In Section 3.1 we derived a weak maximum principle for parabolic equations, which states that the maximum value of a solution of the given parabolic equation defined in a domain D is achieved on a distinguished portion of the boundary of D. This theorem says nothing about the possibility of a solution of the parabolic equation also achieving its maximum at an interior point. The strong maximum principle addresses itself to this possibility. The first proof of the strong maximum principle for parabolic equations was given by Nirenberg in 1953 (cf. Protter and Weinberger) and since that time various other mathematicians have given alternate proofs for the special case of the heat equation. The proof given below is original with the author (Colton 1984a) and is chosen for the same reason that a father prefers his own children as well as for the similarity of the method of proof to that to be used in Chapter 4 for the strong maximum principle for Laplace's equation. (In both proofs a "mean value" theorem plays a central role.) Note that a strong minimum principle can be obtained for the heat equation if we replace $u(x,t)$ by $-u(x,t)$ in the theorem below.

THEOREM 21 Strong Maximum Principle

Let $u(x,t)$ satisfy the heat equation

$$\frac{\partial u}{\partial t} = \frac{\partial^2 u}{\partial x^2}$$

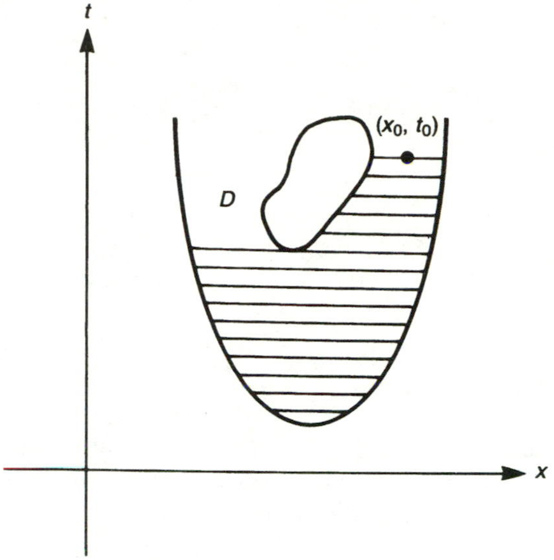

Figure 3.6

in a domain D and suppose that M, the maximum of u in D, is attained at some interior point (x_0,t_0) of D. If (x_1,t_1) is a point in D that can be connected to (x_0,t_0) by a path in D consisting only of horizontal and upward vertical line segments, then $u(x_1,t_1) = M$.

■ ***Proof.*** Suppose $u(x,t)$ achieves its maximum at a point (x_0,t_0) as stated in the theorem, but $u(x,t)$ is not identically equal to M on the horizontal line segment l containing (x_0,t_0) and lying in D. Define the "mean value" $S(x,t)$ by

$$S(x,t) = \frac{1}{2}\,(u(x_0 - x,t) + u(x_0 + x,t)).$$

Then there exists a sequence of positive numbers ρ_n such that ρ_n tends to zero as n tends to infinity and $S(\rho_n,t_0) < M$. This is true, for otherwise $u(x,t)$ is identically equal to M in a neighborhood of (x_0,t_0) on l, so by analyticity and the identity theorem, $u(x,t)$ is identically equal to M on all of l, contrary to assumption. Now note that $S(x,t)$ is an even solution of the heat equation in the rectangle $|x| \leq R_0$, $t_0 - h \leq t \leq t_0$ for R_0 and h sufficiently small and achieves its maximum value M at $(x,t) = (0,t_0)$.

Choose $R_0 = \rho_{n_0}$ for n_0 sufficiently large. Then by the evenness of $S(x,t)$ as a function of x and the weak maximum principle there exists a point (x,t_1) on the base or vertical sides of this rectangle where $0 \leq x_1 \leq \rho_{n_0}$, $t_0 - h \leq t_1 < t_0$, such that $S(x_1,t_1) = M$. Let $\rho_{n_1} < \rho_{n_0}$ be such that $t_1 + \rho_{n_1} < t_0$. Then applying the above argument to the rectangle $|x| \leq \rho_{n_1}$, max $(t_0 - (h/2),\ t_1 + \rho_{n_1}) \leq t \leq t_0$ yields a second point (x_2,t_2) such that $S(x_2,t_2) = M$, and proceeding in this manner we have

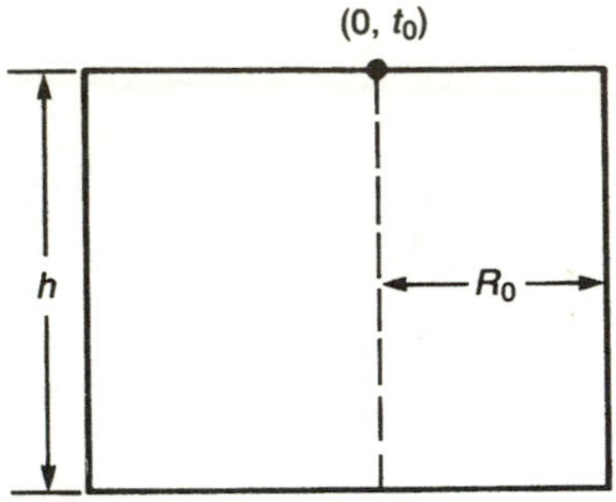

Figure 3.7

an infinite sequence of points (x_n,t_n) such that $S(x_n,t_n) = M$ and whose only accumulation point is $(x,t) = (0,t_0)$. Since the points (x_n,t_n) are maximum points, we have $\partial S(x_n,t_n)/\partial t = 0$ and our choice of concentric rectangles implies that

$$\left| \frac{x_{n+1} - x_n}{t_{n+1} - t_n} \right| \leq 1$$

for all n.

We now want to deduce from the above that all of the derivatives of $u(x,t)$ with respect to t vanish at $(x,t) = (x_0,t_0)$. Since (x_0,t_0) is a maximum point, we immediately have

$$\frac{\partial u(x_0,t_0)}{\partial t} = 0.$$

Applying the mean value theorem to $\partial S(x,t)/\partial t$ and using the above analysis shows that

$$0 = \frac{\dfrac{\partial S}{\partial t}(x_{n+1},t_{n+1}) - \dfrac{\partial S}{\partial t}(x_n,t_n)}{t_{n+1} - t_n} = \frac{\dfrac{\partial^2 S}{\partial x \partial t}(x_n^0,t_n^0)(x_{n+1} - x_n)}{t_{n+1} - t_n} + \frac{\partial^2 S}{\partial t^2}(x_n^0,t_n^0),$$

where (x_n^0,t_n^0) is a point on the line segment joining (x_n,t_n) to (x_{n+1},t_{n+1}). But since $S(x,t)$ is an even function of x we have that $\partial^2 S(0,t_0)/\partial x \partial t = 0$ and letting n tend to infinity, we deduce that

$$\frac{\partial^2 S}{\partial t^2}(0,t_0) = \frac{\partial^2 u(x_0,t_0)}{\partial t^2} = 0.$$

If instead of considering $u(x,t)$ and the point (x_0,t_0) we consider $S(x,t)$ and the point (x_n,t_n) we can immediately conclude from the above analysis that

$$\frac{\partial^2 S(x_n,t_n)}{\partial t^2} = 0$$

for all n. Applying the mean value theorem to $\partial^2 S(x,t)/\partial t^2$ now implies as above that

$$\frac{\partial^3 S}{\partial t^3}(0,t_0) = \frac{\partial^3 u(x_0,t_0)}{\partial t^3} = 0$$

and proceeding in this manner we can conclude that all derivatives of $u(x,t)$ with respect to t vanish at $(x,t) = (x_0,t_0)$.

Since $S(x,t_0)$ is an even analytic function of x, it has a power series expansion of the form

$$S(x,t_0) = \sum_{n=0}^{\infty} \frac{x^{2n}}{(2n)!} \left.\frac{\partial^{2n} S(x,t_0)}{\partial x^{2n}}\right|_{x=0}$$

$$= \sum_{n=0}^{\infty} \frac{x^{2n}}{(2n)!} \frac{\partial^n u(x_0,t_0)}{\partial t^n}$$

valid for x sufficiently small. Hence from the above we see that $S(x,t_0) = u(x_0,t_0) = M$ for all x sufficiently small, contradicting the fact that $S(\rho_n,t_0) < M$ for all n. Thus $u(x,t)$ is identically equal to M on l.

The proof of the theorem will be complete if we can now show that $u(x,t)$ is identically equal to M on every vertical line segment whose upper end point is an interior maximum point of $u(x,t)$. To show this, let (x_0,t_0) be an interior maximum point such that the vertical segment (x_0,t), $t_1 \le t \le t_0$, lies in D. Suppose there is a point (x_0,t_2), $t_1 \le t_2 < t_0$, such that $u(x_0,t_2) < M$ and let τ be the least upper bound of such values of t_2. Then by continuity $u(x_0,\tau) = M$ and $u(x_0,t) < M$ for $t_2 \le t < \tau$. Thus, $u(x,t) < M$ in some rectangle $|x - x_0| \le \epsilon$, $t_2 \le t < \tau$, where $\epsilon > 0$ and $u(x,\tau) = M$ for $|x - x_0| \le \epsilon$. We shall now obtain a contradiction. Let Δ be a disk centered at (x_0,τ) and let Ω be that part of Δ lying below the parabola $(x - x_0)^2 + 3(t - \tau) = 0$. Then $\partial\Omega = C_1 + C_2$ where C_1 is part of $\partial\Delta$, and C_2 is part of the parabola ($\partial\Omega$ and $\partial\Delta$ denote the boundaries of Ω and Δ respectively). Consider the function $v(x,t)$ defined by

$$v(x,t) = u(x,t) - \epsilon((x - x_0)^2 + 3(t - \tau)),$$

where $\epsilon > 0$. Then

$$\frac{\partial v}{\partial t} - \frac{\partial^2 v}{\partial x^2} = -\epsilon,$$

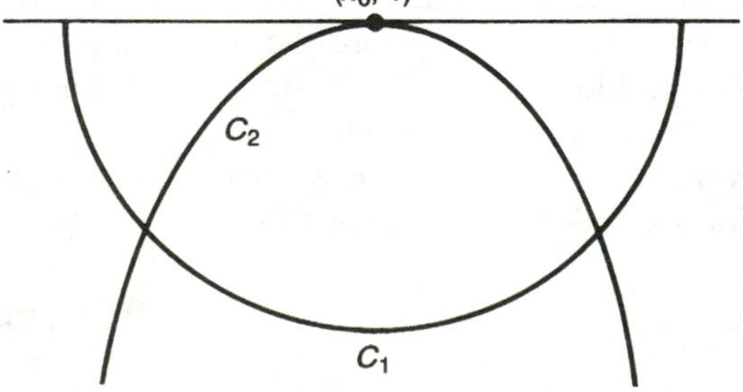

Figure 3.8

and since $v(x,t)$ is continuous, it achieves its maximum in $\overline{\Omega}$. The point at which this maximum is achieved cannot be in Ω since at such a point $\partial v(x,t)/\partial t = 0$ and $\partial^2 v(x,t)/\partial x^2 \le 0$. Choosing ϵ sufficiently small, we see that $v(x,t) < M$ on C_1, $v(x,t) \le M$ on C_2, and since $v(x,t) = u(x,t)$ on C_2 we can conclude that the maximum of $v(x,t)$ is M and it occurs at (x_0,τ). Hence, $\partial v(x_0,\tau)/\partial t \ge 0$. But since $\partial u(x_0,\tau)/\partial t = 0$, we see that $\partial v(x_0,\tau)/\partial t = -3\epsilon < 0$, a contradiction. We conclude that $u(x,t)$ is identically equal to M on every vertical line segment whose upper end point is an interior maximum point of $u(x,t)$. ■

* 3.6 THE STEFAN PROBLEM AND ANALYTIC CONTINUATION

We now want to consider a problem for the heat equation in which the domain of definition of the solution is not known a priori but must be determined. The physical motivation for the problem we have in mind is concerned with the melting of solids. In particular, suppose the half-space $x > 0$ in R^3 is filled with ice at zero degrees Centrigrade, and for time $t \ge 0$ the wall $x = 0$ is kept at a constant positive temperature U. Then the ice will begin to melt, the temperature of the water depending only on x and t. Let the interphase boundary between ice and water be described by $x = s(t)$. The curve $x = s(t)$ is unknown and must be found as part of the solution. If we assume that the ice portion remains at zero degrees, then conservation of energy considerations lead to the following (normalized) equations for the temperature $u(x,t)$ in the water region (cf. Stakgold):

$$\frac{\partial u}{\partial t} = \frac{\partial^2 u}{\partial x^2} \qquad (0 < x < s(t),\ t > 0) \tag{3.34}$$

$$u(0,t) = U \qquad (t \ge 0) \tag{3.35}$$

$$u(s(t),t) = 0 \qquad (t > 0) \tag{3.36}$$

$$\frac{\partial u}{\partial x}(s(t),t) = -\dot{s}(t) \qquad (t > 0). \tag{3.37}$$

Problem (3.34)–(3.37) is a simple version of the **Stefan problem** and is representative of a class of free boundary problems in heat conduction (cf. Cannon). We emphasize that the free boundary $x = s(t)$ is unknown and must be found as part of the solution. Note also that our Stefan problem is nonlinear, since the unknown function $s(t)$ is a nonlinear function of the prescribed temperature U.

Since $u(0,t)$ is constant, we can solve (3.34)–(3.37) in a particularly simple manner. More specifically, we look for a solution $u(x,t)$ in the form

$$u(x,t) = A + B \operatorname{erf}\left(\frac{x}{2\sqrt{t}}\right) \qquad (0 < x < s(t)), \tag{3.38}$$

where A and B are constants to be determined and the **error function** erf x is defined by

$$\operatorname{erf} x = \frac{2}{\sqrt{\pi}} \int_0^x e^{-z^2}\, dz. \tag{3.39}$$

It is easily verified that (3.38) is a solution of the heat equation (3.34). To satisfy (3.35) and (3.36) we must have

$$A = U \tag{3.40}$$

$$A + B \operatorname{erf}\left(\frac{s(t)}{2\sqrt{t}}\right) = 0 \tag{3.41}$$

$$\frac{B}{\sqrt{\pi t}} \exp\left(-\frac{(s(t))^2}{4t}\right) = -\dot{s}(t). \tag{3.42}$$

Equation (3.41) implies that $s(t)$ is proportional to $\sqrt{t}$, i.e.,

$$s(t) = 2\alpha\sqrt{t} \tag{3.43}$$

where α is an unknown constant. From (3.40)–(3.43) we now have

$$\begin{aligned} U + B \operatorname{erf} \alpha &= 0 \\ Be^{-\alpha^2} &= -\alpha\sqrt{\pi} \end{aligned} \tag{3.44}$$

or

$$Ue^{-\alpha^2} = \sqrt{\pi}\, \alpha \operatorname{erf} \alpha. \tag{3.45}$$

It can easily be seen that (3.45) has a unique solution α. Then $s(t)$ is given by (3.43), and

$$u(x,t) = U - \frac{U}{\operatorname{erf} \alpha} \operatorname{erf}\left(\frac{x}{2\sqrt{t}}\right) \qquad (0 < x < s(t)). \tag{3.46}$$

We now want to briefly consider the **inverse Stefan problem**, i.e., given $s(t)$ to determine $u(0,t)$, or, how must ice be heated in order to make it melt along a prescribed curve? Note that if $s(t)$ is not of the form (3.43), we cannot expect $u(0,t)$ to be a constant. The inverse Stefan problem is a non-characteristic Cauchy problem for the heat equation and, as we have already seen, is improperly posed in the real domain. Thus we shall assume that $s(t)$ is analytic and look for a solution in the complex domain. We begin by noting that if $u(x,t)$ is a solution of the heat equation that is an analytic function of both x and t in a neighborhood of the origin, then $u(x,t)$ can be represented in the form

$$\begin{aligned} u(x,t) = &-\frac{1}{2\pi i} \int_{|t-\tau|=\delta} \frac{\partial E}{\partial x}(x, t-\tau) u(0,\tau)\, d\tau \\ &-\frac{1}{2\pi i} \int_{|t-\tau|=\delta} E(x,t-\tau) \frac{\partial u}{\partial x}(0,\tau)\, d\tau, \end{aligned} \tag{3.47}$$

where $\delta > 0$, the circle $|t - \tau| = \delta$ is traversed in a positive sense, and

$$E(x,t) = \sum_{n=0}^{\infty} \frac{(-1)^n n! x^{2n+1}}{(2n+1)! t^{n+1}} . \tag{3.48}$$

This can be verified by direct calculation. An elementary power series analysis shows that analytic solutions of the heat equation are uniquely determined by their Cauchy data on the line $x = 0$, i.e., if

$$u(x,t) = \sum_{n=0}^{\infty} a_n(t) x^n$$

with $a_0(t) = a_1(t) = 0$, then $a_n(t) = 0$ for all n. Since $E(x,t)$ is analytic for $|x| < \infty$, $|t| > 0$, the representation (3.47) immediately implies the following remarkable result on the analytic continuation of analytic solutions of the heat equation:

THEOREM 22

Let $u(x,t)$ be a solution of the heat equation such that $u(x,t)$ is analytic for $|x| < x_0$, $|t| < t_0$. Then $u(x,t)$ can be analytically continued into the infinite strip $|x| < \infty$, $|t| < t_0$.

The representation (3.47) can be viewed as an integral operator mapping analytic functions onto solutions of the heat equation. Such operators are useful in obtaining results on the analytic behavior of solutions to partial differential equations. For the general theory of integral operator methods in partial differential equations, we refer the reader to Bergman, Colton 1980, Gilbert, and Vekua.

As a further application of the representation (3.47), we expand $u(0,t)$ and $(\partial u/\partial x)(0,t)$ in their Taylor series

$$\begin{aligned} u(0,t) &= \sum_{n=0}^{\infty} c_n t^n \\ \frac{\partial u}{\partial x}(0,t) &= \sum_{n=0}^{\infty} d_n t^n \end{aligned} \tag{3.49}$$

and substitute (3.49) into (3.47). Interchanging orders of integration and summation and computing the residue, we find that $u(x,t)$ has the expansion

$$u(x,t) = \sum_{n=0}^{\infty} a_n u_n(x,t) \qquad (|x| < \infty, |t| < t_0) \tag{3.50}$$

where the $u_n(x,t)$ are the **heat polynomials** defined by

$$u_n(x,t) = n! \sum_{k=0}^{[n/2]} \frac{x^{n-2k} t^k}{(n-2k)! k!} \tag{3.51}$$

and the a_n are constants. If we define the **Hermite polynomial** by

$$H_n(x) = n! \sum_{k=0}^{[n/2]} \frac{(-1)^k (2x)^{n-2k}}{(n-2k)! k!} , \tag{3.52}$$

we see that

$$u_n(x,t) = (-t)^{n/2} H_n\left(\frac{x}{\sqrt{-4t}}\right). \tag{3.53}$$

The expansion (3.50) can be viewed as Taylor's theorem for analytic solutions of the heat equation. For an extensive presentation of the role of heat polynomials in the theory of the heat equation, see Widder.

The Hermite polynomials are an important class of polynomials that appear in a variety of areas of mathematical physics. Although we shall not need to discuss their properties in detail, we want to note their recurrence relation

$$H_{n+1}(x) - 2xH_n(x) - 2nH_{n-1}(x) = 0 \qquad (n = 1,2,\ldots)$$
$$H_0(x) = 1,$$

the differential equation satisfied by $y(x) = H_n(x)$

$$y'' - 2xy' + 2ny = 0,$$

and the orthogonality property

$$\int_{-\infty}^{\infty} e^{-x^2} H_n(x)H_m(x)\,dx = \begin{cases} 0 & (m \neq n) \\ 2^n n!\sqrt{\pi} & (m = n). \end{cases}$$

For more information on Hermite polynomials and their applications see Lebedev.

We now return to the inverse Stefan problem. In (3.47) we consider x and t to be complex and deform the circle $|t - \tau| = \delta$ onto the two-dimensional surface $x = s(t)$. Then, since $u(s(t),t) = 0$ and $u_x(s(t),t) = -\dot{s}(t)$, we are led to the following integral representation of the solution to the inverse Stefan problem:

$$u(x,t) = \frac{1}{2\pi i}\int_{|t-\tau|=\delta} E(x - s(\tau), t - \tau)\dot{s}(\tau)\,d\tau. \tag{3.54}$$

Alternatively, it can be verified directly that this is the solution of the inverse Stefan problem. Computing the residue in (3.54) gives

$$u(x,t) = \sum_{n=1}^{\infty} \frac{1}{(2n)!}\frac{\partial^n}{\partial t^n}[x - s(t)]^{2n}, \tag{3.55}$$

so

$$u(0,t) = \sum_{n=1}^{\infty} \frac{1}{(2n)!}\frac{\partial^n}{\partial t^n}[s(t)]^{2n}. \tag{3.56}$$

In particular, we see that in order for the ice to melt along the curve $x = \sqrt{t}$ we must have

$$u(0,t) = \sum_{n=1}^{\infty} \frac{n!}{(2n)!} = \text{constant}, \tag{3.57}$$

which agrees with our previous result. Note that the improperly posed nature of the inverse Stefan problem in the real domain is evident from (3.56), since $u(0,t)$ depends upon all the higher derivatives of $s(t)$; thus, (3.56) is not suitable for numerical computation. For a numerical method for solving the inverse Stefan problem (based on the use of heat polynomials) see Reemtsen and Kirsch. Finally, we note that (3.56) dramatically indicates the nonlinear dependence of $s(t)$ on the initial temperature $u(0,t)$.

3.7 HERMITE POLYNOMIALS AND THE NUMERICAL SOLUTION OF THE HEAT EQUATION IN A RECTANGLE

The polynomial solutions of the heat equation introduced in the previous section provide a convenient method for obtaining approximations to the solution of the initial-boundary value problem for the heat equation in a rectangle. The method, known as **Trefftz's method**, can also be used to approximate the solution of the initial-boundary value problem in domains with moving boundaries (cf. Colton 1980), although we shall not pursue this direction here. The basic ideas can also be applied to other classes of elliptic and parabolic boundary value problems; for a detailed exposition we refer the reader to Colton 1980 and Herrera.

The method is based on the fact that the heat polynomials

$$u_n(x,t) = (-t)^{n/2} H_n\left(\frac{x}{\sqrt{-4t}}\right),$$

where $H_n(x)$ is the Hermite polynomial of degree n, are a complete set of solutions of the heat equation defined in a rectangle

$$R = \{(x,t): 0 < x < l,\ 0 < t < T\}.$$

Thus, if $u \in C^2(R) \cap C(\bar{R})$ is a solution of the heat equation in R, then $u(x,t)$ can be approximated arbitrarily closely with respect to the maximum norm on $\bar{R}$ by a finite linear combination of heat polynomials. Assuming for the moment that this is true, an approximation to the solution of the initial-boundary value problem for the heat equation in R can be obtained as follows. Suppose we seek a solution of the heat equation in R such that $u(x,0) = \phi(x)$ for $0 \le x \le l$ and $u(0,t) = \mu_1(t)$, $u(l,t) = \mu_2(t)$ for $0 \le t \le T$. Then for fixed integer N, choose constants $a_0, a_1, \ldots, a_N$ such that

$$\begin{aligned} F(a_0,\ldots,a_N) = &\int_0^l \left|\phi(x) - \sum_{n=0}^{N} a_n u_n(x,0)\right|^2 dx \\ &+ \int_0^T \left|\mu_1(t) - \sum_{n=0}^{N} a_n u_n(0,t)\right|^2 dt \\ &+ \int_0^T \left|\mu_2(t) - \sum_{n=0}^{N} a_n u_n(l,t)\right|^2 dt \end{aligned} \tag{3.58}$$

is minimized. Since the heat polynomials are complete, $F(a_0,\ldots,a_N)$ can be made arbitrarily small if N is large enough, and it can further be shown (although we shall not prove it here) that if $a_0,\ldots,a_N$ minimizes (3.58), then

$$u^N(x,t) = \sum_{n=0}^{N} a_n u_n(x,t)$$

approximates the solution $u(x,t)$ of the initial-boundary value problem in R. Hence approximations to $u(x,t)$ can be found by minimizing the simple quadratic expression defined by (3.58).

We shall now prove the completeness theorem on which the above approximation scheme is based.

THEOREM 23

Let $u \in C^2(R) \cap C(\bar{R})$ be a solution of the heat equation in R. Then given $\epsilon > 0$ there exists an integer N and constants $a_0,a_1,\ldots,a_N$ such that

$$\max_{(x,t)\in\bar{R}} \left| u(x,t) - \sum_{n=0}^{N} a_n u_n(x,t) \right| < \epsilon.$$

■ ***Proof.*** By the Weierstrass approximation theorem and the maximum principle for the heat equation, there exists a solution $u_1(x,t)$ of the heat equation in R which assumes polynomial initial-boundary data such that

$$\max_{(x,t)\in\bar{R}} |u(x,t) - u_1(x,t)| < \frac{\epsilon}{3}.$$

Let

$$u_1(0,t) = \sum_{m=0}^{M} b_m t^m$$

$$u_1(l,t) = \sum_{m=0}^{M} c_m t^m$$

and look for a solution of the heat equation in the form

$$v(x,t) = \sum_{m=0}^{M} v_m(x) t^m, \tag{3.59}$$

where $v(0,t) = u_1(0,t)$ and $v(l,t) = u_1(l,t)$. Substituting (3.59) into the heat equation

$$\frac{\partial u}{\partial t} = \frac{\partial^2 u}{\partial x^2}$$

leads to the following recursion scheme for the $v_m(x)$:

$$v_M'' = 0 \qquad v_M(0) = b_M \qquad v_M(l) = c_M$$

$$v_{m-1}'' = m v_m \qquad v_{m-1}(0) = b_{m-1} \qquad v_{m-1}(l) = c_{m-1}$$

for $m = 1,2,\ldots,M$. Hence each $v_m(x)$ is a uniquely determined polynomial. Now consider

$$u_2(x,t) = u_1(x,t) - v(x,t).$$

By the method of separation of variables and Fourier's theorem it is seen that there exist constants $d_1,\ldots,d_K$ such that

$$\max_{(x,t)\in\bar{R}} \left| u_2(x,t) - \sum_{k=1}^{K} d_k \sin\frac{k\pi x}{l}\, e^{-(k\pi/l)^2 t} \right| < \frac{\epsilon}{3}.$$

Hence, there exists a solution $u_3(x,t)$ of the heat equation that is analytic for all x and t such that

$$\max_{(x,t)\in\bar{R}} |u(x,t) - u_3(x,t)| < \frac{2\epsilon}{3}.$$

By the expansion (3.50) of the previous section we can write

$$u_3(x,t) = \sum_{n=0}^{\infty} a_n u_n(x,t),$$

and truncating this series now shows that

$$\max_{(x,t)\in\bar{R}} \left| u(x,t) - \sum_{n=0}^{N} a_n u_n(x,t) \right| < \epsilon$$

for N sufficiently large. The theorem is now proved. ∎

3.8 NONLINEAR PROBLEMS IN HEAT CONDUCTION

Within a heat-conducting medium, heat can arise or vanish due to the presence of heat-generating or heat-absorbing processes within the medium. In the simplest case, such an occurrence is governed by the nonhomogeneous heat equation

$$\frac{\partial u}{\partial t} = \frac{\partial^2 u}{\partial x^2} + f(x,t),$$

where $f(x,t)$ is a function of x and t alone. However, in many problems the heat production term depends on the unknown temperature $u(x,t)$. For example, if a chemical reaction that liberates heat takes place, then under certain conditions it is reasonable to assume that the rate at which heat is released is given by

$$f(x,t) = Ae^{-B/u(x,t)},$$

where A and B are constants. Further examples in the area of mathematical biology and diffusion can be found in Jones and Sleeman. Hence we are led to consider nonlinear parabolic equations of the form

$$\frac{\partial u}{\partial t} = \frac{\partial^2 u}{\partial x^2} + f(u). \tag{3.60}$$

In this chapter, we shall consider some simple problems where $f(u)$ is a function only of u and does not depend explicitly on the independent variables x and t.

We begin by considering the special class of solutions that are independent of x. Here, (3.60) becomes

$$\frac{du}{dt} = f(u) \tag{3.61}$$

and if $u(0) = u_0$, then (3.61) has the implicit solution

$$t = \int_{u_0}^{u} \frac{d\eta}{f(\eta)}, \tag{3.62}$$

assuming the integral exists. For example, if $f(u) = u(1 - u)$ and u_0 is not zero, (3.62) becomes

$$\begin{aligned} t &= \int_{u_0}^{u} \frac{d\eta}{\eta(1-\eta)} \\ &= \int_{u_0}^{u} \left(\frac{1}{\eta} + \frac{1}{1-\eta}\right) d\eta \\ &= \log\left[\frac{u(1-u_0)}{(1-u)u_0}\right], \end{aligned}$$

provided

$$\frac{u}{1-u}\,\frac{1-u_0}{u_0} \geq 1.$$

In this case,

$$u(t) = \frac{u_0 e^t}{1 - u_0 + u_0 e^t}.$$

In particular, we note that

$$\lim_{t\to\infty} u(t) = 1.$$

On the other hand, if (3.61) is linear (e.g., $f(u) = \gamma u$ where γ is a constant), then (3.62) becomes (again assuming $u_0 \neq 0$)

$$\gamma t = \log \frac{u}{u_0} \tag{3.63}$$

and hence

$$u(t) = u_0 e^{\gamma t}.$$

In particular,

$$\lim_{t\to\infty} u(t) = \begin{cases} 0 & \text{if} \quad \gamma < 0 \\ \infty & \text{if} \quad \gamma > 0. \end{cases}$$

We now consider the general problem

$$\frac{\partial u}{\partial t} = \frac{\partial^2 u}{\partial x^2} + f(u) \qquad (-1 < x < 1,\, t > 0) \tag{3.64}$$

$$u(x,0) = u_0(x) \qquad (-1 \le x \le 1) \tag{3.65}$$

and, to fix some boundary conditions, set

$$\frac{\partial u}{\partial x}(0,t) = \frac{\partial u}{\partial x}(1,t) = 0 \qquad (t \ge 0). \tag{3.66}$$

Our aim is to determine conditions on $f(u)$ such that $u(x,t)$ tends to a constant value as t tends to infinity. We shall assume that $f(u)$ is a real valued continuously differentiable function such that (3.64)–(3.66) has a (real valued) solution that is bounded for all times $t \ge 0$. (For conditions under which this is true see Smoller.) For similar results in the case of Dirichlet boundary data $u(0,t) = u(1,t) = 0$ we refer the reader to Exercise 31 at the end of this chapter.

We begin by defining the energy function $E(t)$ by

$$E(t) = \frac{1}{2}\int_0^1 \left(\frac{\partial u}{\partial x}\right)^2 dx. \tag{3.67}$$

Differentiating with respect to t gives

$$\begin{aligned} \frac{dE}{dt} &= \int_0^1 \frac{\partial u}{\partial x}\frac{\partial^2 u}{\partial x \partial t}\, dx = \int_0^1 \frac{\partial u}{\partial x}\frac{\partial}{\partial x}\left(\frac{\partial^2 u}{\partial x^2} + f(u)\right) dx \\ &= \int_0^1 \frac{\partial u}{\partial x}\frac{\partial^3 u}{\partial x^3}\, dx + \int_0^1 \left(\frac{\partial u}{\partial x}\right)^2 \frac{df}{du}\, dx. \end{aligned} \tag{3.68}$$

But

$$\int_0^1 \frac{\partial u}{\partial x}\frac{\partial^3 u}{\partial x^3}\, dx = \frac{\partial u}{\partial x}\frac{\partial^2 u}{\partial x^2}\bigg|_{x=0}^{x=1} - \int_0^1 \left(\frac{\partial^2 u}{\partial x^2}\right)^2 dx$$

and hence from (3.66) we can write (3.68) as

$$\frac{dE}{dt} = -\int_0^1 \left(\frac{\partial^2 u}{\partial x^2}\right)^2 dx + \int_0^1 \left(\frac{\partial u}{\partial x}\right)^2 \frac{df}{du}\, dx. \tag{3.69}$$

Let

$$M = \max \left| \frac{df}{du} \right|,$$

where the maximum is taken over all solution values of $u(x,t)$ (which is assumed to be bounded). Then from (3.69) we have

$$\begin{aligned} \frac{dE}{dt} &\leq -\int_0^1 \left(\frac{\partial^2 u}{\partial x^2} \right)^2 dx + M \int_0^1 \left(\frac{\partial u}{\partial x} \right)^2 dx \\ &= -\int_0^1 \left(\frac{\partial^2 u}{\partial x^2} \right)^2 dx + 2ME(t). \end{aligned} \tag{3.70}$$

To proceed further we need the following result, where it is assumed that $w(x)$ is real valued.

THEOREM 24 Poincaré's Inequality

Let $w \in C^2[0,1]$, $w'(0) = w'(1) = 0$. Then

$$\int_0^1 \left(\frac{dw}{dx} \right)^2 dx \leq \frac{1}{\pi^2} \int_0^1 \left(\frac{d^2w}{dx^2} \right)^2 dx.$$

■ ***Proof.*** Interpreting equality in the sense of mean square convergence (cf. Section 1.4), we have

$$\frac{d^2w}{dx^2} = \sum_{n=0}^{\infty} a_n \cos n\pi x,$$

where

$$a_0 = \int_0^1 \frac{d^2w}{dx^2}\, dx = \left. \frac{dw}{dx} \right|_{x=0}^{x=1} = 0 \tag{3.71}$$

and

$$a_n = 2 \int_0^1 \frac{d^2w}{dx^2} \cos n\pi x \qquad (n \geq 1). \tag{3.72}$$

Integrating (3.72) twice by parts shows that

$$a_n = -2n^2\pi^2 \int_0^1 w(x) \cos n\pi x \, dx \qquad (n \geq 1),$$

and

$$w(x) = c - \sum_{n=1}^{\infty} \frac{a_n}{n^2\pi^2} \cos n\pi x,$$

where

$$c = \int_0^1 w(x)\, dx.$$

Hence

$$\int_0^1 \left(\frac{dw}{dx}\right)^2 dx = w\frac{dw}{dx}\bigg|_{x=0}^{x=1} - \int_0^1 w\frac{d^2w}{dx^2}\, dx$$

$$= \int_0^1 \left(\sum_{n=1}^{\infty} \frac{a_n}{n^2\pi^2} \cos n\pi x\right)\left(\sum_{n=1}^{\infty} a_n \cos n\pi x\right) dx$$

$$- c\int_0^1 \left(\sum_{n=1}^{\infty} a_n \cos n\pi x\right) dx$$

$$= \frac{1}{2}\sum_{n=1}^{\infty} \frac{a_n^2}{n^2\pi^2}$$

$$\le \frac{1}{2\pi^2}\sum_{n=1}^{\infty} a_n^2$$

$$= \frac{1}{\pi^2}\int_0^1 \left(\frac{d^2w}{dx^2}\right)^2 dx,$$

where we have used (3.71) and Parseval's equality for Fourier series on the interval [0,1] (see Section 1.4). ■

Returning now to (3.70), we see from Poincaré's inequality that

$$\frac{dE}{dt} \le 2(M - \pi^2)E(t); \tag{3.73}$$

since $E(t) \ge 0$ this equation implies that

$$\frac{d}{dt}\,[e^{-2(M-\pi^2)t}E(t)] \le 0,$$

i.e.,

$$E(t) \le E(0)e^{2(M-\pi^2)t},$$

where $E(0) \ge 0$. If $M < \pi^2$, $E(t)$ decays exponentially to zero as t tends to infinity. Thus, if

$$\max \left|\frac{df}{du}\right| < \pi^2$$

for all solution values of $u(x,t)$, then $u(x,t)$ tends to a constant value in the generalized sense that

$$\lim_{t\to\infty} \int_0^1 \left(\frac{\partial u}{\partial x}\right)^2 dx = 0.$$

Exercises

1. **(a)** Find all solutions of the heat equation $\dfrac{\partial u}{\partial t} = \dfrac{\partial^2 u}{\partial x^2}$ of the form

$$u(x,t) = \frac{1}{\sqrt{t}} f\left(\frac{x}{2\sqrt{t}}\right).$$

(b) Determine a solution of the heat equation $\dfrac{\partial u}{\partial t} = \dfrac{\partial^2 u}{\partial x^2}$ that is bounded for $0 \le x < \infty$, $-\infty < t < \infty$.

2. Show that if $u(x,t)$ is a solution of $\dfrac{\partial u}{\partial t} = \dfrac{\partial^2 u}{\partial x^2}$, then so is

$$v(x,t) = \frac{1}{\sqrt{t}} \exp\left(\frac{-x^2}{4t}\right) u\left(\frac{x}{t}, -\frac{1}{t}\right).$$

3. Let $D = \{(x,t): 0 < x < l, 0 < t < T\}$ and $u(x,t) \in C^2(\overline{D})$ be a solution of

$$\frac{\partial u}{\partial t} = \frac{\partial^2 u}{\partial x^2}$$

in D such that

$$\frac{\partial u(0,t)}{\partial x} = 0.$$

Show that the maximum of $u(x,t)$ in $\overline{D}$ must occur at $t = 0$ or $x = l$. (Hint: Reflect $u(x,t)$ across the t axis.)

4. Let $u(x,t)$ be a solution of

$$\frac{\partial u}{\partial t} - \frac{\partial^2 u}{\partial x^2} = f(x,t)$$

in the rectangle $R = \{(x,t): 0 < x < l, 0 < t < t_0\}$. Show that if $u \in C(\overline{R})$ and $f(x,t) < 0$ for $(x,t) \in \overline{R}$ then the maximum of $u(x,t)$ is attained at $t = 0$ or $x = 0$ or $x = l$.

5. **(a)** Let $u(x,t)$ satisfy

$$\frac{\partial u}{\partial t} = \frac{\partial^2 u}{\partial x^2} \qquad (0 < x < 1, t > 0)$$

$$u(0,t) = u(1,t) = 0 \qquad (t \ge 0)$$

$$u(x,0) = f(x) \qquad (0 \le x \le 1),$$

where $f \in C[0,1]$. Show that for any $T \geq 0$

$$\int_0^1 (u(x,T))^2\, dx \leq \int_0^1 (f(x))^2\, dx.$$

$$\left(\text{Hint: Use the identity } 2u\left(\frac{\partial u}{\partial t} - \frac{\partial^2 u}{\partial x^2}\right) = \frac{\partial u^2}{\partial t} - \frac{\partial}{\partial x}\left(u\frac{\partial u}{\partial x}\right) + 2\left(\frac{\partial u}{\partial x}\right)^2.\right)$$

(b) Use (a) to derive a uniqueness theorem for the above initial-boundary value problem.

6. Let D be a bounded, simply connected domain in the plane with smooth boundary and consider the initial-boundary value problem

$$\frac{\partial u}{\partial t} = \frac{\partial^2 u}{\partial x^2} + \frac{\partial^2 u}{\partial y^2} \quad \text{in} \quad D \times [0,T]$$

$$u(x,y,t) = f(x,y,t) \quad \text{on} \quad \partial D \times [0,T]$$

$$u(x,y,0) = \phi(x,y),$$

where $f(x,y,t)$ and $\phi(x,y)$ are continuous functions. In the case where the data vanishes, derive the two inequalities

$$\frac{\partial}{\partial t}\iint_D \left[\left(\frac{\partial u}{\partial x}\right)^2 + \left(\frac{\partial u}{\partial y}\right)^2\right] dxdy = -2\iint_D \left(\frac{\partial u}{\partial t}\right)^2 dxdy \leq 0$$

$$\frac{\partial}{\partial t}\iint_D u^2\, dxdy = -2\iint_D \left[\left(\frac{\partial u}{\partial x}\right)^2 + \left(\frac{\partial u}{\partial y}\right)^2\right] dxdy \leq 0$$

and use either one to conclude that the solution of the initial-boundary value problem is unique.

7. Let D be a bounded, simply connected domain in the plane with smooth boundary ∂D and unit outward normal $\boldsymbol{\nu}$. Show that the only solution of

$$\frac{\partial^2 u}{\partial x^2} + \frac{\partial^2 u}{\partial y^2} - u = \frac{\partial u}{\partial t} \quad \text{in} \quad D \times [0,T]$$

$$\frac{\partial u}{\partial \nu} + u = 0 \quad \text{on} \quad \partial D \times [0,T]$$

$$u(x,y,0) = 0 \quad \text{in} \quad D$$

is u identically zero.

8. By using Fourier's method and interchanging orders of integration and summation show formally that the solution of the initial-boundary value problem

$$\frac{\partial u}{\partial t} = \frac{\partial^2 u}{\partial x^2} \qquad (0 < x < 1,\ t > 0)$$

$$u(0,t) = u(1,t) = 0 \qquad (t \geq 0)$$

$$u(x,0) = f(x) \qquad (0 \leq x \leq 1)$$

can be written in the form

$$u(x,t) = \int_0^1 G(x,\xi,t)f(\xi)d\xi,$$

where

$$G(x,\xi,t) = -\frac{1}{\sqrt{4\pi t}}\,\theta\left(\frac{i(x+\xi)}{2\pi t}, \frac{i}{\pi t}\right)e^{-(x+\xi)^2/4t} + \frac{1}{\sqrt{4\pi t}}\,\theta\left(\frac{i(x-\xi)}{2\pi t}, \frac{i}{\pi t}\right)e^{-(x-\xi)^2/4t},$$

and $\theta(x,t)$ is the **theta function**

$$\theta(x,t) = \frac{1}{\sqrt{-it}} \sum_{n=-\infty}^{\infty} \exp\left[\frac{-i\pi(x+n)^2}{t}\right].$$

9. Use Fourier's method to formally solve the initial-boundary value problem

$$\frac{\partial u}{\partial t} = \frac{\partial^2 u}{\partial x^2} \qquad (0 < x < 1,\ t > 0)$$

$$u(0,t) = u(1,t) \qquad (t \geq 0)$$

$$\frac{\partial u}{\partial x}(0,t) = \frac{\partial u}{\partial x}(1,t) \qquad (t \geq 0)$$

$$u(x,0) = f(x) \qquad (0 \leq x \leq 1),$$

where $f \in C[0,1]$. This describes heat conduction in an insulated loop made of homogeneous material of negligible diameter.

10. Use Fourier's method to formally solve the initial-boundary value problem

$$\frac{\partial u}{\partial t} - \frac{\partial^2 u}{\partial x^2} - u = 0 \qquad (0 < x < 1,\ t > 0)$$

$$u(0,t) = \frac{\partial u}{\partial x}(1,t) = 0 \qquad (t \geq 0)$$

$$u(x,0) = x(1-x) \qquad (0 \leq x \leq 1).$$

(Answer: $u(x,t) =$

$$\frac{8}{\pi^2}\sum_{n=1}^{\infty}\left[\frac{(-1)^n}{(2n-1)^2}+\frac{4}{(2n-1)^3}\right]\exp\left[\frac{(2n-1)^2\pi^2}{4}-1\right]t \sin\left(\frac{2n-1}{2}\right)\pi x.)$$

11. Use Fourier's method to formally solve the initial-boundary value problem

$$\frac{\partial u}{\partial t}=\frac{\partial^2 u}{\partial x^2} \qquad (0<x<1,\ t>0)$$

$$u(0,t)=\frac{\partial u}{\partial x}(1,t)+u(1,t)=0 \qquad (t\geq 0)$$

$$u(x,0)=f(x) \qquad (0\leq x\leq 1),$$

where $f \in C[0,1]$.

(Answer: $u(x,t)=\sum_{n=1}^{\infty} b_n e^{-\mu_n^2 t}\sin \mu_n x$, where

$$b_n = 2\int_0^1 f(x)\sin \mu_n x\,dx,\ \mu_n = -\tan \mu_n.)$$

12. Use Fourier's method to formally solve the initial-boundary value problem

$$\frac{\partial u}{\partial t}=\frac{\partial^2 u}{\partial x^2} \qquad (0<x<1,\ t>0)$$

$$\frac{\partial u}{\partial x}(0,t)+u(0,t)=u(1,t)=0 \qquad (t\geq 0)$$

$$u(x,0)=f(x) \qquad (0\leq x\leq 1),$$

where $f \in C[0,1]$.

(Answer: $u(x,t)=\sum_{n=1}^{\infty} b_n e^{-\mu_n^2 t}(\sin \mu_n x-\mu_n\cos \mu_n x)$, where

$$b_n = 2\int_0^1 f(x)(\sin \mu_n x-\mu_n\cos \mu_n x)dx,\ \mu_n = \tan \mu_n.)$$

13. Use Fourier's method to solve the initial-boundary value problem

$$\frac{\partial u}{\partial t}=\frac{\partial^2 u}{\partial x^2} \qquad (0<x<\pi,\ t>0)$$

$$\frac{\partial u}{\partial x}(0,t) = \frac{\partial u}{\partial x}(\pi,t) = 0 \qquad (t \geq 0)$$

$$u(x,0) = x \qquad (0 \leq x \leq \pi).$$

(Answer: $u(x,t) = \frac{\pi}{2} - \frac{4}{\pi}\sum_{n=1}^{\infty} \frac{e^{-(2n-1)^2 t}}{(2n-1)^2} \cos(2n-1)x.$**)**

14. Use Fourier's method to formally solve the initial-value boundary problem

$$\frac{\partial u}{\partial t} - \frac{\partial^2 u}{\partial x^2} = f(x) \qquad (0 < x < \pi,\ t > 0)$$

$$u(0,t) = u(\pi,t) = 0 \qquad (t \geq 0)$$

$$u(x,0) = 0 \qquad (0 \leq x \leq \pi),$$

where

$$f(x) = \begin{cases} x & \left(0 \leq x \leq \frac{\pi}{2}\right) \\ \pi - x & \left(\frac{\pi}{2} \leq x \leq \pi\right). \end{cases}$$

(Answer: $u(x,t) = \frac{4}{\pi}\sum_{n=1}^{\infty} \frac{(-1)^{n-1}}{(2n-1)^4}(1 - e^{-(2n-1)^2 t}) \sin(2n-1)x.$**)**

15. Extend the given data to the whole x axis in a proper manner and solve the following problems:

(a) $\frac{\partial u}{\partial t} = \frac{\partial^2 u}{\partial x^2} - u \qquad (0 < x < \infty,\ t > 0)$

$$u(0,t) = 0 \qquad (t > 0)$$

$$u(x,0) = f(x) \qquad (0 < x < \infty);$$

(b) $\frac{\partial u}{\partial t} = \frac{\partial^2 u}{\partial x^2} - u \qquad (0 < x < \infty,\ t > 0)$

$$\frac{\partial u}{\partial x}(0,t) = 0 \qquad (t > 0)$$

$$u(x,0) = f(x) \qquad (0 < x < \infty),$$

where f is continuous and bounded.

16. Let $u(x,t)$ be the solution of the Cauchy problem (3.23), (3.24) and suppose

$$\lim_{t \to \infty} u(0,t) = A,$$

where A is a constant. Show that for any $x \in (-\infty,\infty)$,

$$\lim_{t\to\infty} u(x,t) = A.$$

17. Let $u(x,t)$ be a positive twice continuously differentiable solution of

$$\frac{\partial u}{\partial t} = \mu \frac{\partial^2 u}{\partial x^2}$$

for $-\infty < x < \infty$, $t > 0$ where $\mu > 0$. Show that $\theta(x,t) = -2\mu \dfrac{\partial u}{\partial x}(x,t)/u(x,t)$ satisfies **Burgers' equation**

$$\frac{\partial \theta}{\partial t} + \theta \frac{\partial \theta}{\partial x} = \mu \frac{\partial^2 \theta}{\partial x^2}$$

for $-\infty < x < \infty$, $t > 0$. Find a solution of Burgers' equation such that $\theta(x,0) = \phi(x)$ and

$$\lim_{t\to\infty} \theta(x,t) = 0,$$

where $\phi(x)$ is a twice continuously differentiable function that vanishes for $|x|$ sufficiently large.

18. Let $u(x,t)$ be the solution of the Cauchy problem (3.23), (3.24) such that $\phi(x)$ is bounded and

$$\int_{-\infty}^{\infty} |\phi(x)|^2\, dx < \infty.$$

Show that there exists a positive constant C such that for $-\infty < x < \infty$, $t > 0$,

$$|u(x,t)| \le \frac{C}{t^{1/4}}.$$

19. Let $u(x,t)$ be the solution of the Cauchy problem (3.23), (3.24). Show that

$$\lim_{x\to\infty} \frac{1}{x}\int_{-x}^{x} u(\xi,t)\, d\xi$$

is independent of t for $t > 0$.

20. Let $u(x,t)$ be the solution of the Cauchy problem (3.23), (3.24). Use the substitution

$$t = \int_0^y a(y)dy, \qquad U(x,y) = u(x,t)$$

to solve the Cauchy problem

$$\frac{\partial U}{\partial y} = a(y)\frac{\partial^2 U}{\partial x^2} \qquad (-\infty < x < \infty,\ y > 0)$$

$$U(x,0) = \phi(x) \qquad (-\infty < x < \infty),$$

where $a(y)$ is a positive continuous function for $y \geq 0$.

21. Show that the bounded solution of

$$\frac{\partial u}{\partial t} - \frac{\partial^2 u}{\partial x^2} = h(x,t) \qquad (-\infty < x < \infty,\ t > 0)$$

$$u(x,0) = 0 \qquad (-\infty < x < \infty),$$

where $h(x,t)$ is continuous and bounded for $-\infty < x < \infty$, $t \geq 0$, is given by

$$u(x,t) = \int_0^t \int_{-\infty}^{\infty} \frac{\exp[-(x-\xi)^2/4(t-\tau)]}{\sqrt{4\pi(t-\tau)}} h(\xi,\tau)d\xi d\tau.$$

22. Show that the bounded solution of

$$\frac{\partial u}{\partial t} = \frac{\partial^2 u}{\partial x^2} \qquad (0 < x < \infty,\ t > 0)$$

$$u(0,t) = 0 \qquad (t \geq 0)$$

$$u(x,0) = f(x) \qquad (0 \leq x < \infty),$$

where $f(x)$ is continuous and bounded is given by

$$u(x,t) = \int_0^{\infty} [K(x,t;\xi) - K(x,t;-\xi)]\, f(\xi)\, d\xi$$

where $K(x,t;\xi)$ is the fundamental solution of the heat equation.

23. Find a solution of

$$\frac{\partial u}{\partial t} - \frac{\partial^2 u}{\partial x^2} = 0 \qquad (x > 0,\ t > 0)$$

$$\frac{\partial u}{\partial x}(0,t) = 1 \qquad (t \geq 0)$$

$$u(x,0) = 0 \qquad (x \geq 0)$$

(Hint: Consider $v(x,t) = \dfrac{\partial u}{\partial x}(x,t) - 1$.)

24. **(a)** Let $U(x,t)$ be the solution of the heat equation $\partial U/\partial t = \partial^2 U/\partial x^2$ for $x > 0$, $t \geq 0$ that satisfies $U(0,t) = 1$ for $t > 0$ and $U(x,0) = 0$ for

$x > 0$. Show that

$$u(x,t) = \int_0^t \frac{\partial U}{\partial t}(x,t-\tau)h(\tau)\,d\tau$$

is a solution of

$$\frac{\partial u}{\partial t} = \frac{\partial^2 u}{\partial x^2} \qquad (0 < x < \infty,\ t > 0)$$

$$u(0,t) = h(t) \qquad (t \geq 0)$$

$$u(x,0) = 0 \qquad (x \geq 0),$$

where $h(t)$ is continuous for $t \geq 0$.

(b) By considering solutions of the heat equation of the form $U(x,t) = f(x/2\sqrt{t})$, construct the function $U(x,t)$ of part (a).

25. Derive equation (3.29) of Section 3.3.

26. Let the half-plane $x \geq 0$ be filled with ice at the constant temperature $V < 0$ (instead of $V = 0$ as in Section 3.6). Then the Stefan problem corresponding to (3.34)–(3.36) is

$$\frac{\partial u_1}{\partial t} = \frac{\partial^2 u_1}{\partial x^2} \qquad (0 < x < s(t),\ t > 0)$$

$$\frac{\partial u_2}{\partial t} = a\frac{\partial^2 u_2}{\partial x^2} \qquad (s(t) < x < \infty,\ t > 0)$$

$$u_1(0,t) = U \qquad (t \geq 0)$$

$$u_2(x,0) = V \qquad (x \geq 0)$$

$$u_1(s(t),t) = u_2(s(t),t) = 0 \qquad (t > 0)$$

$$-\frac{\partial u_1}{\partial x}(s(t),t) + k\frac{\partial u_2}{\partial x}(s(t),t) = \dot{s}(t) \qquad (t > 0),$$

where a and k are positive constants, $u_1(x,t)$ is the temperature of the water, and $u_2(x,t)$ is the temperature of the ice. Solve this problem by a method similar to that used to solve (3.34)–(3.36) in Section 3.6.

27. Formally solve the following boundary value problems using Fourier transforms as discussed in Section 3.3.

(a) $$\frac{\partial u}{\partial t} = \frac{\partial^2 u}{\partial x^2} + \frac{\partial u}{\partial x} \qquad (-\infty < x < \infty,\ t > 0)$$

$$u(x,0) = \phi(x) \qquad (-\infty < x < \infty)$$

$u(x,t)$ is bounded for $-\infty < x < \infty$, $t \geq 0$.

$$\left(\textbf{Answer:}\quad u(x,t) = \frac{1}{\sqrt{4\pi t}}\int_{-\infty}^{\infty} \phi(\xi)\exp\left[-\frac{(x-\xi+t)^2}{4t}\right]d\xi.\right)$$

(b) $$\frac{\partial^2 u}{\partial x^2} + \frac{\partial^2 u}{\partial y^2} = 0 \qquad (-\infty < x < \infty,\ 0 < y < 1)$$

$$u(x,0) = e^{-2|x|} \qquad (-\infty < x < \infty)$$

$$u(x,1) = 0 \qquad (-\infty < x < \infty)$$

$u(x,y) \to 0$ uniformly in y as $|x| \to \infty$.

$$\left(\textbf{Answer:}\quad u(x,y) = \frac{2}{\pi}\int_{-\infty}^{\infty} \frac{\sinh \lambda(1-y)}{(4+\lambda^2)\sinh \lambda}\, e^{i\lambda x}\, d\lambda.\right)$$

(c) $$\frac{\partial^2 u}{\partial x^2} + \frac{\partial^2 u}{\partial y^2} - u = 0 \qquad (-\infty < x < \infty,\ 0 < y < 1)$$

$$\frac{\partial u}{\partial y}(x,0) = 0 \qquad (-\infty < x < \infty)$$

$$u(x,1) = e^{-x^2} \qquad (-\infty < x < \infty)$$

$u(x,y) \to 0$ uniformly in y as $|x| \to \infty$.

$$\left(\textbf{Answer:}\quad u(x,y) = \frac{1}{\sqrt{4\pi}}\int_{-\infty}^{\infty} \frac{e^{-\lambda^2/4}\cosh \lambda y}{\cosh \lambda}\, e^{i\lambda x} d\lambda.\right)$$

28. Let $u(x,t)$ be the solution of

$$\Delta_3 u = \frac{\partial u}{\partial t} \qquad (\mathbf{x} \in D,\ t > 0)$$

$$u(\mathbf{x},t) = 0 \qquad (\mathbf{x} \in \partial D,\ t \geq 0)$$

$$u(\mathbf{x},0) = \phi(\mathbf{x}) \qquad (\mathbf{x} \in \overline{D}),$$

where D is a bounded, simply connected domain in R^3 with smooth boundary ∂D. Consider a portion Γ of the boundary surface ∂D. Show that the total amount of heat flowing across Γ,

$$Q = \int_0^{\infty}\left[\int\!\!\int_{\Gamma} \frac{\partial u}{\partial \nu}\, ds\right] dt,$$

where $\boldsymbol{\nu}$ is the unit outward normal to ∂D, is given by

$$Q = -\int\!\!\int_{D} \phi(\mathbf{x}) h(\mathbf{x}) d\mathbf{x},$$

where $h(\mathbf{x})$ is the solution of the Dirichlet problem for Laplace's equation

$$\Delta_3 h = 0 \qquad (\mathbf{x} \in D), \qquad h(\mathbf{x}) = \begin{cases} 1 & (\mathbf{x} \in \Gamma) \\ 0 & (\mathbf{x} \in \partial D \backslash \Gamma). \end{cases}$$

The function $h(\mathbf{x})$ is called the **harmonic measure** of the surface Γ.

29. Derive the following identities for the heat polynomials $u_n(x,t)$ defined by equation (3.51) of Section 3.6.

(a) $\exp[xz + tz^2] = \sum_{n=0}^{\infty} u_n(x,t)\dfrac{z^n}{n!}$.

(b) $u_{n+1}(x,t) = xu_n(x,t) + 2ntu_{n-1}(x,t)$.

30. Show that the solution to

$$\frac{du}{dt} = u^{1+\alpha} \qquad (t > 0)$$

$$u(0) = u_0,$$

where $\alpha > 0$, $u_0 > 0$, tends to infinity as t tends to $\dfrac{1}{\alpha u_0^{\alpha}}$. Conclude that for general nonlinear $f(u)$ we cannot expect to obtain a solution to (3.60) that is defined for arbitrarily large time.

31. Consider the initial-boundary value problem

$$\frac{\partial u}{\partial t} = \frac{\partial^2 u}{\partial x^2} + f(u) \qquad (-1 < x < 1,\ t > 0)$$

$$u(x,0) = u_0(x) \qquad (-1 \leq x \leq 1)$$

$$u(0,t) = u(1,t) = 0 \qquad (t \geq 0),$$

where $f(u)$ is a real valued continuously differentiable function such that the above initial-boundary value problem has a real valued solution that is bounded for all times $t > 0$. Define the energy function $E(t)$ by

$$E(t) = \frac{1}{2}\int_0^1 u^2 dx.$$

Show that if $\max |uf(u)| < \pi^2 u^2$ then

$$\lim_{t\to\infty} \int_0^1 u^2 dx = 0.$$

Chapter 4

Laplace's Equation

The simplest elliptic equation is the Laplace, or potential, equation

$$\Delta_n u \equiv \frac{\partial^2 u}{\partial x_1^2} + \frac{\partial^2 u}{\partial x_2^2} + \ldots + \frac{\partial^2 u}{\partial x_n^2} = 0.$$

For the sake of simplicity, we shall primarily consider the case $n = 2$,

$$\Delta_2 u \equiv \frac{\partial^2 u}{\partial x^2} + \frac{\partial^2 u}{\partial y^2} = 0,$$

but the student should note that with straightforward modifications all our results and proofs are valid in higher dimensions. Solutions of Laplace's equation in a domain D are called **harmonic** in D.

The function

$$\log \frac{1}{|\mathbf{x} - \mathbf{y}|} = \log \frac{1}{\sqrt{(x_1 - y_1)^2 + (x_2 - y_2)^2}}$$

where $\mathbf{x} = (x_1,x_2)$, $\mathbf{y} = (y_1,y_2)$ is the **fundamental solution** of Laplace's equation and is harmonic as a function of either $\mathbf{x}$ or $\mathbf{y}$ for $\mathbf{x} \neq \mathbf{y}$. This function plays the same role as $(z - z_0)^{-1}$ does in analytic function theory (cf. Section 1.5). In R^n, $n \geq 3$, the fundamental solution of Laplace's equation is

$$\frac{1}{|\mathbf{x} - \mathbf{y}|^{n-2}} = \frac{1}{[(x_1 - y_1)^2 + \ldots + (x_n - y_n)^2]^{(n-2)/2}} .$$

This chapter proceeds traditionally, from an analysis of the basic qualitative properties of harmonic functions to the formulation of well posed boundary value problems for Laplace's equation and the use of separation of variables and the fundamental solution in solving these problems. Included here is the use of the fundamental solution in constructing particular solutions to **Poisson's equation**

$$\Delta_2 u = \rho(\mathbf{x})$$

and the application of these ideas to the solution of a problem in time harmonic wave propagation. The full development of the use of the fundamental solution to solve boundary value problems for Laplace's equation is delayed until the next chapter, where we shall present the method of potential theory and integral equations for constructing solutions to the Dirichlet and Neumann boundary value problems. In addition to the method of finite differences (Section 4.7) the method of integral equations also provides a numerical method for solving the classical boundary value problems of potential theory (Sections 5.3 and 5.6). For a detailed exposition of the classical theory of solutions to Laplace's equation, the reader is referred to Kellogg.

4.1 GREEN'S FORMULAS

Let D be a bounded, simply connected domain in R^2 with C^2 boundary ∂D, and let $\mathbf{F}(\mathbf{x})$ be a continuously differentiable vector valued function in $\overline{D}$. Then the divergence theorem says that

$$\iint_D \operatorname{div} \mathbf{F}(\mathbf{x})\, \mathbf{dx} = \int_{\partial D} \mathbf{F}(\mathbf{x}) \cdot \boldsymbol{\nu}(\mathbf{x})\, ds,$$

where $\mathbf{x} = (x,y)$, $\boldsymbol{\nu}(\mathbf{x})$ is the unit outward normal to ∂D, and ds denotes an element of arc length. Let $u \in C^2(\overline{D})$, $v \in C^2(\overline{D})$, and define $\mathbf{F}(\mathbf{x})$ by $\mathbf{F}(\mathbf{x}) = u(\mathbf{x})\nabla v(\mathbf{x})$, where $\nabla v(\mathbf{x})$ denotes the gradient of $v(\mathbf{x})$. Then from the divergence theorem we have **Green's first identity**

$$\iint_D (u\Delta_2 v + \nabla u \cdot \nabla v)\, d\mathbf{x} = \int_{\partial D} u \frac{\partial v}{\partial \nu}\, ds, \qquad \textbf{(4.1)}$$

where $\partial v/\partial \nu = \nabla v \cdot \boldsymbol{\nu}$. Interchanging u and v and subtracting gives us **Green's second identity**

$$\iint_D (u\Delta_2 v - v\Delta_2 u)\, d\mathbf{x} = \int_{\partial D} \left(u\frac{\partial v}{\partial \nu} - v\frac{\partial u}{\partial \nu} \right) ds. \tag{4.2}$$

The above identities are clearly also true for domains D in R^3, where the operator Δ_2 is replaced by Δ_3.

THEOREM 25

Let $u \in C^2(\overline{D})$. Then for $\mathbf{x} \in D$,

$$u(\mathbf{x}) = \frac{1}{2\pi}\int_{\partial D} \left(\frac{\partial u}{\partial \nu}(\mathbf{y}) \log \frac{1}{|\mathbf{x}-\mathbf{y}|} - u(\mathbf{y}) \frac{\partial}{\partial \nu(\mathbf{y})} \log \frac{1}{|\mathbf{x}-\mathbf{y}|} \right) ds(\mathbf{y})$$
$$- \frac{1}{2\pi}\iint_D \log \frac{1}{|\mathbf{x}-\mathbf{y}|}\, \Delta_2 u(\mathbf{y})\, d\mathbf{y}.$$

■ ***Proof.*** Let Ω_ϵ be a disk of radius $\epsilon > 0$ centered at $\mathbf{x}$ and contained in D. Then since $v(\mathbf{y}) = \log (1/|\mathbf{x} - \mathbf{y}|)$ is harmonic in $D\backslash\Omega_\epsilon$, we have from Green's second identity that

$$-\iint_{D\backslash\Omega_\epsilon} \log \frac{1}{|\mathbf{x}-\mathbf{y}|}\, \Delta_2 u(\mathbf{y})\, \mathbf{dy} =$$
$$\int_{\partial D} \left(u(\mathbf{y}) \frac{\partial}{\partial \nu(\mathbf{y})} \log \frac{1}{|\mathbf{x}-\mathbf{y}|} - \log \frac{1}{|\mathbf{x}-\mathbf{y}|} \frac{\partial u}{\partial \nu}(\mathbf{y}) \right) ds(\mathbf{y}) \tag{4.3}$$
$$+ \int_{\partial\Omega_\epsilon} \left(u(\mathbf{y}) \frac{\partial}{\partial \nu(\mathbf{y})} \log \frac{1}{|\mathbf{x}-\mathbf{y}|} - \log \frac{1}{|\mathbf{x}-\mathbf{y}|} \frac{\partial u}{\partial \nu}(\mathbf{y}) \right) ds(\mathbf{y}).$$

On $\partial\Omega_\epsilon$ we have $|\mathbf{x} - \mathbf{y}| = \epsilon$, and $\boldsymbol{\nu}$ is in the negative ϵ direction. Hence

$$\frac{\partial}{\partial \nu(\mathbf{y})} \log \frac{1}{|\mathbf{x}-\mathbf{y}|} = \frac{1}{\epsilon}$$

and

$$\int_{\partial\Omega_\epsilon} u(\mathbf{y}) \frac{\partial}{\partial \nu(\mathbf{y})} \log \frac{1}{|\mathbf{x}-\mathbf{y}|}\, ds(\mathbf{y}) = \frac{1}{\epsilon}\int_{\partial\Omega_\epsilon} u\, ds.$$

Since on $\partial\Omega_\epsilon$ we have $ds = \epsilon\, d\theta$, we now see that

$$\lim_{\epsilon\to 0} \int_{\partial\Omega_\epsilon} u(\mathbf{y}) \frac{\partial}{\partial \nu(\mathbf{y})} \log \frac{1}{|\mathbf{x}-\mathbf{y}|}\, ds(\mathbf{y}) = 2\pi u(\mathbf{x}).$$

On the other hand, there exists a constant K such that

$$\left|\int_{\partial\Omega_\epsilon} \log\frac{1}{|\mathbf{x}-\mathbf{y}|}\frac{\partial u}{\partial\nu}(\mathbf{y})\,ds(\mathbf{y})\right| \le K\left|\log\frac{1}{\epsilon}\right|\int_{\partial\Omega_\epsilon} ds = 2\pi\epsilon K\left|\log\frac{1}{\epsilon}\right|$$

which tends to zero as ϵ tends to zero. Hence, letting ϵ tend to zero in (4.3) gives

$$-\iint_D \log\frac{1}{|\mathbf{x}-\mathbf{y}|}\Delta_2 u(\mathbf{y})d\mathbf{y} =$$

$$\int_{\partial D}\left(u(\mathbf{y})\frac{\partial}{\partial\nu(\mathbf{y})}\log\frac{1}{|\mathbf{x}-\mathbf{y}|} - \log\frac{1}{|\mathbf{x}-\mathbf{y}|}\frac{\partial u}{\partial\nu}(\mathbf{y})\right)ds(\mathbf{y}) + 2\pi u(\mathbf{x}),$$

and this implies the theorem. ■

4.2 BASIC PROPERTIES OF HARMONIC FUNCTIONS

Let $u(\mathbf{x})$ be a harmonic function in a bounded, simply connected domain D in R^2 with C^2 boundary ∂D such that $u \in C^2(\overline{D})$. Setting $v(\mathbf{x}) = 1$ in Green's second identity shows that $u(\mathbf{x})$ satisfies the identity

$$\int_{\partial D}\frac{\partial u}{\partial\nu}ds = 0. \tag{4.4}$$

We also have from the previous theorem that for $\mathbf{x} \in D$

$$u(\mathbf{x}) = \frac{1}{2\pi}\int_{\partial D}\left(\log\frac{1}{|\mathbf{x}-\mathbf{y}|}\frac{\partial u}{\partial\nu}(\mathbf{y}) - u(\mathbf{y})\frac{\partial}{\partial\nu(\mathbf{y})}\log\frac{1}{|\mathbf{x}-\mathbf{y}|}\right)ds(\mathbf{y}). \tag{4.5}$$

We shall now prove a regularity theorem for solutions of Laplace's equation. The reader who is unfamiliar with analytic function theory may omit the following discussion on analytic functions of two complex variables and replace the word "analytic" in the theorem below by the words "infinitely differentiable."

Let

$$f(z_1,z_2) = u(\mathbf{x},\mathbf{y}) + iv(\mathbf{x},\mathbf{y}),$$

where $z_1 = x_1 + iy_1$, $z_2 = x_2 + iy_2$, $\mathbf{x} = (x_1,x_2)$, $\mathbf{y} = (y_1,y_2)$. The function $f(z_1,z_2)$ is said to be analytic in a domain $D \subset \mathfrak{C}^2$, where $\mathfrak{C}^2$ is the space of two complex variables, if $\partial f/\partial z_1$ and $\partial f/\partial z_2$ exist and are continuous, i.e., the Cauchy-Riemann equations

$$\frac{\partial f}{\partial \bar{z}_1} = \frac{\partial f}{\partial \bar{z}_2} = 0$$

are satisfied. Thus $f(z_1,z_2)$ is analytic if $f(z_1,z_2)$ is analytic with respect to each of the variables z_1 and z_2 separately. If $f(z_1,z_2)$ is analytic in a neighborhood of

$(z_1^0,z_2^0) \in D$, then Taylor's theorem is valid, and there exists $r_1 > 0$, $r_2 > 0$ such that

$$f(z_1,z_2) = \sum_{n=0}^{\infty}\sum_{m=0}^{\infty} a_{mn}(z_1 - z_1^0)^m(z_2 - z_2^0)^n \qquad (|z_1 - z_1^0| < r_1,\ |z_2 - z_2^0| < r_2),$$

where the series is absolutely and uniformly convergent on compact subsets of the polycylinder $|z_1 - z_1^0| < r_1$, $|z_2 - z_2^0| < r_2$ with

$$a_{mn} = \frac{1}{m!n!}\frac{\partial^{m+n} f}{\partial z_1^m \partial z_2^n}\bigg|_{\substack{z_1 = z_1^0 \\ z_2 = z_2^0}}.$$

A function $f(x_1,x_2)$ defined in a domain $D \subset R^2$ is said to be analytic in D if $f(z_1,z_2)$ is analytic in a neighborhood of each point $(x_1^0,x_2^0) \in D$, i.e., $f(x_1,x_2)$ can be expanded in a Taylor series at (x_1^0,x_2^0). Finally, if $f(z_1,z_2;t)$ is continuous for $a \leq t \leq b$ and $(z_1,z_2) \in D \subset \mathfrak{C}^2$ and is analytic in z_1 and z_2 for $(z_1,z_2) \in D \subset \mathfrak{C}^2$ and t fixed, then by Morera's theorem,

$$F(z_1,z_2) = \int_a^b f(z_1,z_2;t)\,dt$$

is analytic in D.

THEOREM 26 Regularity Theorem

If $u(\mathbf{x})$ is harmonic in a domain D, then $u(\mathbf{x})$ is analytic in D.

■ ***Proof.*** Let $\mathbf{x} \in D$ and $\Omega \subset D$, Ω being a disk containing $\mathbf{x}$. Then from (4.5),

$$u(\mathbf{x}) = \frac{1}{2\pi}\int_{\partial\Omega}\left(\log\frac{1}{|\mathbf{x} - \mathbf{y}|}\frac{\partial u}{\partial\nu}(\mathbf{y}) - u(\mathbf{y})\frac{\partial}{\partial\nu(\mathbf{y})}\log\frac{1}{|\mathbf{x} - \mathbf{y}|}\right)ds(\mathbf{y})$$

for all $\mathbf{x}$ in Ω. Since $\log(1/|\mathbf{x} - \mathbf{y}|)$ is analytic for $\mathbf{x} \neq \mathbf{y}$, the theorem follows. ■

The reader should note the hierarchy of regularity properties for the wave, heat, and Laplace's equation: solutions of the wave equation are only as smooth as their initial data; solutions of the heat equation are analytic with respect to the space variables but in general only infinitely differentiable with respect to the time variable; solutions of Laplace's equation are analytic functions of all their independent variables.

THEOREM 27 Mean Value Theorem

Let $\Omega = \{\mathbf{x}: |\mathbf{x} - \mathbf{x}_0| < \rho\}$ and $u \in C^2(\Omega) \cap C(\overline{\Omega})$ be harmonic in Ω. Then

$$u(\mathbf{x}_0) = \frac{1}{2\pi\rho}\int_{\partial\Omega} u\,ds.$$

■ ***Proof.*** It suffices to prove the theorem for $u \in C^2(\overline{\Omega})$, for if the theorem is true in this case, a limiting argument shows that the theorem is true for $u(\mathbf{x}) \in$

$C^2(\Omega) \cap C(\overline{\Omega})$. If $u \in C^2(\overline{\Omega})$ is harmonic then from (4.5) we have for $\mathbf{x} = \mathbf{x}_0$,

$$u(\mathbf{x}_0) = \frac{1}{2\pi}\int_{\partial\Omega}\left(\log\frac{1}{|\mathbf{x}_0 - \mathbf{y}|}\frac{\partial u}{\partial \nu}(\mathbf{y}) - u(\mathbf{y})\frac{\partial}{\partial \nu(\mathbf{y})}\log\frac{1}{|\mathbf{x}_0 - \mathbf{y}|}\right)ds(\mathbf{y}).$$

But on $\partial\Omega$ we have $1/|\mathbf{x}_0 - \mathbf{y}| = 1/\rho$ and $\dfrac{\partial}{\partial \nu(\mathbf{y})}\log(1/|\mathbf{x}_0 - \mathbf{y}|) = -(1/\rho)$, so

$$u(\mathbf{x}_0) = \frac{1}{2\pi}\log\frac{1}{\rho}\int_{\partial\Omega}\frac{\partial u}{\partial \nu}\,ds + \frac{1}{2\pi\rho}\int_{\partial\Omega}u\,ds.$$

The theorem now follows from (4.4). ■

We shall now prove the maximum principle for Laplace's equation. Physically, this states that in a state of thermal equilibrium there can be no local maximum or minimum in the interior of a body, because if there were, then heat would flow away from or into this point, contradicting the assumption of equilibrium.

THEOREM 28 Strong Maximum Principle

Let $u \in C^2(D) \cap C(\overline{D})$ be harmonic in D and not equal to a constant. Then $u(\mathbf{x})$ achieves its maximum and minimum only on ∂D.

■ ***Proof.*** Since $u(\mathbf{x})$ is a continuous function on a compact set, $u(\mathbf{x})$ achieves its maximum and minimum in $\overline{D}$. Suppose $u(\mathbf{x})$ achieves its maximum M at $\mathbf{x} = \mathbf{x}_0 \in D$. Let Ω_ρ be a disk of radius ρ centered at $\mathbf{x}_0$ such that $\Omega_\rho \subset D$. Then from the mean value theorem

$$u(\mathbf{x}_0) = \frac{1}{2\pi\rho}\int_{\partial\Omega_\rho}u\,ds \le M,$$

where equality can hold only if $u(\mathbf{x})$ is identically equal to M on $\partial\Omega_\rho$. But $u(\mathbf{x}_0) = M$, so this must be the case; and since ρ is arbitrary we conclude that $u(\mathbf{x})$ is equal to M in the closure of any disk centered at $\mathbf{x}_0$ and lying wholly inside D. We shall now show that this implies $u(\mathbf{x})$ equals M throughout D.

Let $\mathbf{y} \in D$. We shall show that $u(\mathbf{y}) = u(\mathbf{x}_0)$. Join $\mathbf{x}_0$ to $\mathbf{y}$ by a finite series of line segments l lying in D and let d be the shortest distance of l from ∂D. From the above analysis, $u(\mathbf{x}) = M$ for $\mathbf{x}$ in a disk centered at $\mathbf{x}_0$ and of radius $d/2$. Let $\mathbf{x}_1$ be the intersection of the boundary of this disk with l. Then $u(\mathbf{x}_1) = M$, and again using the above analysis we see that $u(\mathbf{x}) = M$ for $\mathbf{x}$ in a disk centered at $\mathbf{x}_1$ and of radius $d/2$. Repeating this procedure a finite number of times leads to the conclusion that $u(\mathbf{y}) = M$. Hence $u(\mathbf{x}) = M$ for all $\mathbf{x}$ in D, a contradiction to our hypothesis. Therefore $u(\mathbf{x})$ can achieve its maximum only at a point on ∂D.

An identical analysis shows that $u(\mathbf{x})$ can achieve its minimum only at a point on ∂D. ■

4.3 BOUNDARY VALUE PROBLEMS FOR LAPLACE'S EQUATION

In this section we formulate the classical boundary value problems for Laplace's equation. We do not, however, consider initial-value problems, since for Laplace's equation they are improperly posed. To see this, consider the Cauchy problem

$$\Delta_2 u = 0$$

$$u(x,0) = 0$$

$$\frac{\partial u}{\partial y}(x,0) = \frac{1}{n}\sin nx.$$

The (unique) solution of this problem is

$$u(x,y) = \frac{1}{n^2}\sinh ny \sin nx.$$

But as n tends to infinity the initial data tends to zero whereas the solution oscillates between arbitrarily large values, and the solution does not depend continuously on the initial data. Although the Cauchy problem for Laplace's equation is improperly posed, it does appear in certain situations of practical importance. For details of how to "solve" such a problem we refer the reader to Payne.

The problems which are well posed for Laplace's equation are **boundary-value problems** which we shall now formulate. In what follows, D is a bounded, simply connected domain in R^2 with C^2 boundary ∂D, and f is a continuous function defined on ∂D. The unit outward normal to ∂D will again be denoted by $\boldsymbol{\nu}$. The following boundary-value problems are the classical ones associated with Laplace's equation.

Interior Dirichlet Problem: Find a function $u \in C^2(D) \cap C(\bar{D})$ such that $u(\mathbf{x})$ is harmonic in D and $u(\mathbf{x}) = f(\mathbf{x})$ for $\mathbf{x}$ on ∂D.

Exterior Dirichlet Problem: Find a function $u \in C^2(R^2\backslash\bar{D}) \cap C(R^2\backslash D)$ such that $u(\mathbf{x})$ is harmonic in $R^2\backslash\bar{D}$, bounded in $R^2\backslash D$, and $u(\mathbf{x}) = f(\mathbf{x})$ for $\mathbf{x}$ on ∂D.

Interior Neumann Problem: Find a function $u \in C^2(D) \cap C^1(\bar{D})$ such that $u(\mathbf{x})$ is harmonic in D and $\partial u(\mathbf{x})/\partial \nu = f(\mathbf{x})$ for $\mathbf{x}$ on ∂D. Note that from (4.4) we must require $f(\mathbf{x})$ to satisfy

$$\int_{\partial D} f\,ds = 0.$$

Exterior Neumann Problem: Find a function $u \in C^2(R^2\backslash\bar{D}) \cap C^1(R^2\backslash D)$ such that $u(\mathbf{x})$ is harmonic in $R^2\backslash\bar{D}$, $u(\mathbf{x})$ is regular at infinity, i.e.,

$$\lim_{r\to\infty} |ru(\mathbf{x})| < \infty$$

$$\lim_{r\to\infty} \left| r^2 \frac{\partial u(\mathbf{x})}{\partial r} \right| < \infty$$

where $r = |\mathbf{x}|$, and $\partial u(\mathbf{x})/\partial \nu = f(\mathbf{x})$ for $\mathbf{x}$ on ∂D. We will see in Chapter 5 that the condition of regularity at infinity implies that

$$\int_{\partial D} f\, ds = 0.$$

Other boundary value problems also appear in applications, for example prescribing

$$\frac{\partial u}{\partial \nu} + a(\mathbf{x})u = f(\mathbf{x})$$

for $\mathbf{x}$ on ∂D where $a(\mathbf{x})$ is a given continuous function defined on ∂D. In this chapter we shall concentrate on the interior Dirichlet problem; we refer the reader to Section 5.3 for further discussion of both the interior and exterior Dirichlet and Neumann problems.

The following two theorems establish the uniqueness of the solution of the interior Dirichlet and Neumann problems. For uniqueness theorems for the exterior Dirichlet and Neumann problems see Section 5.3.

THEOREM 29

The solution of the interior Dirichlet problem, if it exists, is unique.

■ ***Proof.*** Suppose there exist two solutions $u_1(\mathbf{x})$ and $u_2(\mathbf{x})$ of the interior Dirichlet problem. Then $u(\mathbf{x}) = u_1(\mathbf{x}) - u_2(\mathbf{x})$ is harmonic in D and vanishes on ∂D. By the maximum principle, $u(\mathbf{x})$ is identically zero for $\mathbf{x}$ in D, i.e., $u_1(\mathbf{x}) = u_2(\mathbf{x})$ for $\mathbf{x}$ in D. ■

THEOREM 30

The solution of the interior Neumann problem, if it exists, is unique up to an additive constant.

■ ***Proof.*** Suppose there exist two solutions $u_1(\mathbf{x})$ and $u_2(\mathbf{x})$ of the interior Neumann problem. Then $u(\mathbf{x}) = u_1(\mathbf{x}) - u_2(\mathbf{x})$ is harmonic in D, continuously differentiable in $\overline{D}$, and satisfies

$$\frac{\partial u}{\partial \nu}(\mathbf{x}) = 0 \qquad (\mathbf{x} \in \partial D).$$

By Green's first identity ($u(\mathbf{x})$ is not known to be in class $C^2(\overline{D})$, so apply this identity to $D_n \subset D$ and let $\lim_{n\to\infty} D_n = D$),

$$0 = \int_{\partial D} u \frac{\partial u}{\partial \nu}\, ds = \iint_D (u\Delta_2 u + |\nabla u|^2)\, d\mathbf{x} = \iint_D |\nabla u|^2\, d\mathbf{x},$$

and hence $\nabla u(\mathbf{x})$ is identically zero for $\mathbf{x}$ in D, i.e., $u(\mathbf{x})$ is constant in D. ■

4.4 SEPARATION OF VARIABLES IN POLAR AND SPHERICAL COORDINATES

We now show how the interior Dirichlet problem for Laplace's equation in a disk or a ball can be formally solved by the method of separation of variables. A primary aim of this analysis is to present the basic theory of Legendre polynomials and spherical harmonics, two of the most important special functions of mathematical physics. We begin with Laplace's equation defined in a disk of radius a and look for a solution of the Dirichlet problem

$$\begin{aligned} \Delta_2 u &= 0 \quad \text{in} \quad D \\ u &= f \quad \text{on} \quad \partial D \end{aligned} \tag{4.6}$$

where $D = \{\mathbf{x}: |\mathbf{x}| < a\}$ and f is a continuous function defined on ∂D. In polar coordinates $x = r\cos\theta$, $y = r\sin\theta$, Laplace's equation becomes

$$\frac{\partial^2 u}{\partial r^2} + \frac{1}{r}\frac{\partial u}{\partial r} + \frac{1}{r^2}\frac{\partial^2 u}{\partial \theta^2} = 0. \tag{4.7}$$

Applying the method of separation of variables, we look for a solution of (4.7) in the form

$$u(r,\theta) = R(r)\Theta(\theta). \tag{4.8}$$

Substituting (4.8) into (4.7) gives

$$R''\Theta + \frac{1}{r}R'\Theta + \frac{1}{r^2}R\Theta'' = 0$$

or

$$r^2\left[\frac{R'' + \frac{1}{r}R'}{R}\right] = -\frac{\Theta''}{\Theta} = \lambda^2,$$

where λ is a constant. Hence

$$\Theta'' + \lambda^2\Theta = 0 \tag{4.9}$$

$$r^2R'' + rR' - \lambda^2 R = 0. \tag{4.10}$$

Since solutions of (4.8) should be periodic functions of θ of period 2π, we have from (4.9) that $\lambda = n$ where n is an integer, i.e., $\Theta(\theta) = c_1\cos n\theta + c_2\sin n\theta$, or, in complex notation

$$\Theta(\theta) = a_n e^{in\theta} \qquad (n = 0, \pm 1, \pm 2, \ldots) \tag{4.11}$$

where the a_n are constants. The solution of (4.10) (for $\lambda = n$) is

$$R(r) = \begin{cases} b_0 + c_0 \log r & (n = 0) \\ b_n r^n + c_n r^{-n} & (n \neq 0) \end{cases} \tag{4.12}$$

where the b_n and c_n are constants. Since solutions of (4.6) are required to be bounded at the origin, we have $c_n = 0$ for $n = 0,1,2,\ldots$, and $b_n = 0$ for $n = -1,-2,\ldots$. Relabeling constants, solutions of (4.7) of the form (4.8) that are periodic and bounded at the origin are of the form

$$u_n(r,\theta) = b_n r^{|n|} e^{in\theta} \qquad (n = 0,\pm 1,\pm 2,\ldots), \tag{4.13}$$

where the b_n are constants.

We now return to (4.6) and assume $f(\theta)$ has the Fourier expansion

$$f(\theta) = \sum_{n=-\infty}^{\infty} a_n e^{in\theta}. \tag{4.14}$$

We seek a formal solution of (4.6) in the form

$$u(r,\theta) = \sum_{n=-\infty}^{\infty} b_n r^{|n|} e^{in\theta} \qquad (r < a). \tag{4.15}$$

Then from (4.11), (4.15), and the boundary condition $u(a,\theta) = f(\theta)$ we see that

$$u(r,\theta) = \sum_{n=-\infty}^{\infty} a_n \left(\frac{r}{a}\right)^{|n|} e^{in\theta} \qquad (r < a), \tag{4.16}$$

i.e., $b_n = a_n a^{-|n|}$. Equation (4.16) is the formal solution to (4.6). We can rewrite it in the form of an integral representation by noting that from (4.14),

$$a_n = \frac{1}{2\pi}\int_{-\pi}^{\pi} f(\theta) e^{-in\theta} d\theta, \tag{4.17}$$

and from (4.16) and (4.17) we formally have

$$\begin{aligned}
u(r,\theta) &= \frac{1}{2\pi}\int_{-\pi}^{\pi} f(\phi)\left[\sum_{n=-\infty}^{\infty}\left(\frac{r}{a}\right)^{|n|} e^{in(\theta-\phi)}\right] d\phi \\
&= \frac{1}{2\pi}\int_{-\pi}^{\pi} f(\phi)\left[1 + \sum_{n=1}^{\infty}\left(\frac{r}{a}\right)^{n} e^{in(\theta-\phi)} + \sum_{n=1}^{\infty}\left(\frac{r}{a}\right)^{n} e^{-in(\theta-\phi)}\right] d\phi \\
&= \frac{1}{2\pi}\int_{-\pi}^{\pi} f(\phi)\, Re\left[1 + 2\sum_{n=1}^{\infty}\left(\frac{r}{a}\right)^{n} e^{in(\theta-\phi)}\right] d\phi \\
&= \frac{1}{2\pi}\int_{-\pi}^{\pi} f(\phi)\, Re\left[\frac{2}{1 - \frac{r}{a} e^{i(\theta-\phi)}} - 1\right] d\phi \\
&= \frac{1}{2\pi}\int_{-\pi}^{\pi} f(\phi)\frac{a^2 - r^2}{a^2 - 2ar\cos(\theta-\phi) + r^2}\, d\phi.
\end{aligned} \tag{4.18}$$

We shall rigorously show that (4.18) is the solution of (4.6) in the following section. Equation (4.18) is known as **Poisson's formula** for the solution of the Dirichlet problem for Laplace's equation in a disk.

We now imitate the above procedure for the three-dimensional Dirichlet problem for Laplace's equation in a ball,

$$\begin{aligned} \Delta_3 u &= 0 \quad \text{in} \quad D \\ u &= f \quad \text{on} \quad \partial D, \end{aligned} \tag{4.19}$$

where $D = \{\mathbf{x}: |\mathbf{x}| < a\}$ and f is a continuous function defined on ∂D. As we shall see, this will require us to introduce and examine Legendre polynomials and spherical harmonics. As usual in our treatment of separation of variable methods, we shall not attempt any rigorous justification (such as that given in Section 2.4). In spherical coordinates

$$x = r \sin\theta \cos\phi$$

$$y = r \sin\theta \sin\phi$$

$$z = r \cos\theta$$

Laplace's equation becomes

$$\frac{\partial^2 u}{\partial r^2} + \frac{2}{r}\frac{\partial u}{\partial r} + \frac{1}{r^2 \sin\theta}\frac{\partial}{\partial \theta}\left(\sin\theta \frac{\partial u}{\partial \theta}\right) + \frac{1}{r^2 \sin^2\theta}\frac{\partial^2 u}{\partial \theta^2} = 0. \tag{4.20}$$

Applying again the method of separation of variables we look for a solution of (4.20) in the form

$$u(r,\theta,\phi) = R(r)\Theta(\theta)\Phi(\phi). \tag{4.21}$$

Substituting (4.21) into (4.20) gives

$$\frac{r^2}{R(r)}\left[R'' + \frac{2}{r}R'\right] + \frac{1}{\sin\theta\Theta(\theta)}\frac{\partial}{\partial\theta}(\sin\theta\Theta') + \frac{1}{\sin^2\theta\Phi(\phi)}\Phi'' = 0$$

and hence

$$r^2\left[\frac{R'' + \frac{2}{r}R'}{R}\right] = \frac{1}{\sin\theta}\left[\frac{(\sin\theta\Theta')'}{\Theta} + \frac{\Phi''}{\sin\theta\Phi}\right] = -\mu \tag{4.22}$$

where μ is a constant. Proceeding further, we see that (4.22) implies that

$$\frac{\sin\theta(\sin\theta\Theta')'}{\Theta} + \mu\sin^2\theta = -\frac{\Phi''}{\Phi} = \nu^2 \tag{4.23}$$

where ν is a constant. Hence from (4.22) and (4.23),

$$\Phi'' + \nu^2\Phi = 0 \tag{4.24}$$

$$(\sin\theta\Theta')' + \left(\mu\sin\theta - \frac{\nu^2}{\sin\theta}\right)\Theta = 0 \tag{4.25}$$

$$R'' + \frac{2}{r}R' - \frac{\mu}{r^2}R = 0. \tag{4.26}$$

Since solutions of (4.19) should be periodic in ϕ of period 2π, we have as before that $\nu = m$ where m is an integer, i.e., in complex notation,

$$\Phi(\phi) = a_m e^{im\phi} \qquad (m = 0, \pm 1, \pm 2, \ldots). \tag{4.27}$$

Equation (4.25) is called the associated Legendre equation. To solve it we set $t = \cos\theta$ and arrive at

$$(1 - t^2)y'' - 2ty' + \left(\mu - \frac{m^2}{1 - t^2}\right) y = 0, \tag{4.28}$$

where $y(t) = \Theta(\theta)$.

We shall now proceed to examine solutions of this equation. We first consider the case when $m = 0$. An elementary power series analysis shows that at the regular singular point $t = 1$ the indicial equation is $\alpha^2 = 0$, and hence the only solution bounded at $t = 1$ is of the form

$$y(t) = \sum_{k=0}^{\infty} c_k (t - 1)^k. \tag{4.29}$$

The recursion formula for this solution is

$$c_{k+1} = -\frac{k(k + 1) - \mu}{2(k + 1)^2} c_k,$$

and hence

$$-\frac{c_{k+1}}{c_k} > \frac{1}{2}\frac{k - 1}{k + 1}$$

for $k + 1 > \mu$. This implies that in general (4.29) approaches infinity as t tends to -1, the only exception occurring if the series terminates, i.e.,

$$\mu = n(n + 1) \qquad (n = 0, 1, 2, \ldots).$$

Setting $c_0 = 1$ in (4.29) now gives us the **Legendre polynomial**

$$P_n(t) = \sum_{k=0}^{n} \frac{(n + k)!}{(n - k)!(k!)^2 2^k} (t - 1)^k \tag{4.30}$$

as the only bounded solution of (4.28) (where $m = 0$, $\mu = n(n + 1)$).

The Legendre polynomials are orthogonal. To see this, we write (4.28) for $m = 0$, $\mu = n(n + 1)$ in the form

$$((1 - t^2)y')' + n(n + 1)y = 0 \tag{4.31}$$

and set $y(t) = P_k(t)$ and $y(t) = P_l(t)$ respectively to get

$$((1 - t^2)P_k')' + k(k + 1)P_k = 0$$
$$((1 - t^2)P_l')' + l(l + 1)P_l = 0.$$

Multiplying the first equation by $P_l(t)$, the second by $P_k(t)$, subtracting, and integrating by parts gives

$$(1 - t^2)(P_k'(t)P_l(t) - P_l'(t)P_k(t))\Big|_{t=-1}^{t=1} = (k(k+1) - l(l+1)) \int_{-1}^{1} P_k(t)P_l(t)dt,$$

and hence if $k \neq l$,

$$\int_{-1}^{1} P_k(t)P_l(t)\, dt = 0. \tag{4.32}$$

Since $P_n(t)$ is of exact degree n, any polynomial of degree k can be expressed as a linear combination of Legendre polynomials of degree less than or equal to k. From this fact and the orthogonality property (4.32) it follows that

$$\int_{-1}^{1} t^k P_n(t)\, dt = 0 \qquad (k = 0,1,\ldots,n-1). \tag{4.33}$$

The n conditions (4.33), the fact that $P_n(t)$ is of degree n, and the normalization $P_n(1) = 1$ uniquely determine $P_n(t)$. This characterization now implies **Rodrigues' formula**

$$P_n(t) = \frac{1}{2^n n!}\frac{d^n}{dt^n}(t^2 - 1)^n$$

(see Exercise 13). Finally, we have the following theorem analogous to that for Fourier series (Lebedev):

THEOREM 31

Let $f(t)$ be piecewise differentiable. Then at each point $t \in (-1,1)$,

$$\frac{f(t+) + f(t-)}{2} = \sum_{n=0}^{\infty} a_n P_n(t), \tag{4.34}$$

where the coefficients a_n are given by

$$a_n = \frac{2n+1}{2}\int_{-1}^{1} f(t)P_n(t)\, dt \qquad (n = 0,1,\ldots).$$

We now return to (4.28) for $m \neq 0$. Assuming for now that m is positive, we differentiate (4.31) m times with respect to t:

$$(1 - t^2)u'' - 2(m+1)tu' + (n-m)(n-m+1)u = 0, \tag{4.35}$$

where

$$u(t) = \frac{d^m}{dt^m}P_n(t).$$

Setting

$$v(t) = (1 - t^2)^{m/2}u(t) = (1 - t^2)^{m/2} \frac{d^m}{dt^m} P_n(t)$$

now reduces (4.35) to

$$((1 - t^2)v')' + \left(n(n + 1) - \frac{m^2}{1 - t^2}\right) v = 0,$$

i.e., if $\mu = n(n + 1)$, then

$$P_n^m(t) = (1 - t^2)^{m/2} \frac{d^m}{dt^m} P_n(t) \tag{4.36}$$

is a solution of (4.28). The function $P_n^m(t)$ is called the **associated Legendre function.** As when $m = 0$, it can be shown that bounded solutions of (4.28) can occur only if $\mu = n(n + 1)$; in this case the bounded solution is given by (4.36). Note that $P_n^m(t)$ is identically zero for $m > n$. From the form of (4.36) we might expect that m must be non-negative. However, if $P_n(t)$ is expressed by Rodrigues' formula, this limitation on m can be relaxed and we may have $-n \leq m \leq n$, negative as well as positive values of m being permitted, i.e.,

$$P_n^m(t) = \frac{1}{2^n n!} (1 - t^2)^{m/2} \frac{d^{m+n}}{dt^{m+n}} (t^2 - 1) \qquad (-n \leq m \leq n). \tag{4.37}$$

Finally, as in the case with the Legendre polynomials, we have the orthogonality property

$$\int_{-1}^{1} P_k^m(t) P_l^m(t) dt = 0$$

if $k \neq l$.

Returning now to (4.26), we see that since $\mu = n(n + 1)$,

$$R(r) = a_n r^n + b_n r^{-n-1},$$

where a_n and b_n are constants. Hence the only solution of (4.26) that is bounded at the origin is

$$R(r) = a_n r^n.$$

Summarizing, the only solutions of Laplace's equation in R^3 of the form (4.21) that are of period 2π in ϕ and bounded for $0 \leq \theta \leq \pi$, $-\pi \leq \phi \leq \pi$, $r \geq 0$, are (in complex notation) the functions

$$a_{mn} r^n P_n^m(\cos\theta) e^{im\phi} \tag{4.38}$$

where $n = 0,1,2,\ldots,\ -n \le m \le n$, and the a_{mn} are constants. The functions (4.38) are known as **spherical harmonics** and correspond to the solutions (4.13) for Laplace's equation in R^2.

We now want to formally solve the Dirichlet problem (4.19). To do this we shall need the following formulas obtained from Rodrigues' formula and integration by parts. In the first formula we have used the residue theorem to evaluate

$$\int_{-1}^{1} (1 - t^2)^n dt = \int_0^{\pi} \sin^{2n+1}\theta \, d\theta = \frac{2^{2n+1}(n!)^2}{(2n+1)!}$$

whereas in the second formula we have used (4.33) and have denoted the coefficient of t^n in $P_n(t)$ by c_n.

$$\begin{aligned} \int_0^{\pi} (P_n(\cos\theta))^2 \sin\theta \, d\theta &= \int_{-1}^{1} (P_n(t))^2 dt \\ &= \frac{1}{2^{2n}(n!)^2} \int_{-1}^{1} \left(\frac{d^n}{dt^n}(1 - t^2)^n \right)^2 dt \\ &= \frac{(2n)!}{2^{2n}(n!)^2} \int_{-1}^{1} (1 - t^2) dt = \frac{2}{2n+1}, \end{aligned} \tag{4.39}$$

$$\begin{aligned} \int_0^{\pi} (P_n^m(\cos\theta))^2 \, sin\, \theta \, d\theta &= \int_{-1}^{1} (P_n^m(t))^2 dt \\ &= \int_{-1}^{1} \left((1 - t^2)^{m/2} \frac{d^m P_n(t)}{dt^m} \right)^2 dt \\ &= (-1)^m \int_{-1}^{1} P_n(t) \frac{d^m}{dt^m} \left((1 - t^2)^m \frac{d^m P_n(t)}{dt^m} \right) dt \\ &= \frac{(n+m)!}{(n-m)!} \int_{-1}^{1} P_n(t) c_n t^n dt \\ &= \frac{(n+m)!}{(n-m)!} \int_{-1}^{1} (P_n(t))^2 dt \\ &= \frac{2}{2n+1} \frac{(n+m)!}{(n-m)!}. \end{aligned} \tag{4.40}$$

Note that $(1 - t^2)^m (d^m P_n(t)/dt^m)$ is a polynomial of degree $(n + m)$ with leading coefficient $(-1)^m c_n t^{n+m}$, and in the last line of (4.40) we have used (4.39).

The formal solution of the Dirichlet problem (4.19) is now easy. We write

$$u(r,\theta,\phi) = \sum_{n=0}^{\infty} \sum_{m=-n}^{n} a_{mn} \left(\frac{r}{a} \right)^n P_n^m(\cos\theta) e^{im\phi} \qquad (r < a) \tag{4.41}$$

and determine the coefficients a_{mn} by the condition

$$u(a,\theta,\phi) = f(\theta,\phi),$$

i.e., from the orthogonality condition (4.40),

$$a_{mn} = \frac{(2n+1)(n-m)!}{4\pi(n+m)!}\int_{-\pi}^{\pi}\int_{0}^{\pi} f(\theta,\phi)P_n^m(\cos\theta)e^{-im\phi}\sin\theta\,d\theta d\phi. \tag{4.42}$$

For the use of (4.41) and (4.42) to derive Poisson's equation for a ball see Weinberger. (For an alternate derivation, see the following section.) Note that we have implicitly assumed that the expansion (4.41) is valid for $r = a$, and $f(\theta,\phi)$ can be expanded in the form

$$f(\theta,\phi) = \sum_{n=0}^{\infty}\sum_{m=-n}^{n} a_{mn}P_n^m(\cos\theta)e^{im\phi}.$$

4.5 GREEN'S FUNCTION AND POISSON'S FORMULA

Let D be a bounded, simply connected domain in R^2 with C^2 boundary ∂D, and $u \in C^2(\overline{D})$ be a harmonic function in D. Then for $\mathbf{x} \in D$ we can write

$$u(\mathbf{x}) = \frac{1}{2\pi}\int_{\partial D}\left(\log\frac{1}{|\mathbf{x}-\mathbf{y}|}\frac{\partial u}{\partial \nu}(\mathbf{y}) - u(\mathbf{y})\frac{\partial}{\partial \nu(\mathbf{y})}\log\frac{1}{|\mathbf{x}-\mathbf{y}|}\right)ds(\mathbf{y}). \tag{4.43}$$

Now suppose there exists a function $g(\mathbf{x},\mathbf{y})$ such that

1. As a function of $\mathbf{y}$, $g(\mathbf{x},\mathbf{y})$ is harmonic in $\overline{D}$ for each fixed $\mathbf{x} \in D$.
2. For $\mathbf{y} \in \partial D$, $g(\mathbf{x},\mathbf{y}) = -\log(1/|\mathbf{x}-\mathbf{y}|)$ for each fixed $\mathbf{x} \in D$.

It can be shown (see Vekua) that since ∂D is in class C^2, for $\mathbf{x} \in D$, $g(\mathbf{x},\mathbf{y})$ is a continuously differentiable function of $\mathbf{y}$ in $\overline{D}$. Then from Green's second identity (applied to $D_n \subset D$ and letting D_n tend to D) we have for fixed $\mathbf{x} \in D$,

$$\begin{aligned} 0 &= \frac{1}{2\pi}\int_{\partial D}\left(g(\mathbf{x},\mathbf{y})\frac{\partial u}{\partial \nu}(\mathbf{y}) - u(\mathbf{y})\frac{\partial}{\partial \nu(\mathbf{y})}g(\mathbf{x},\mathbf{y})\right)ds(\mathbf{y}) \\ &= -\frac{1}{2\pi}\int_{\partial D}\left(\log\frac{1}{|\mathbf{x}-\mathbf{y}|}\frac{\partial u}{\partial \nu}(\mathbf{y}) + u(\mathbf{y})\frac{\partial}{\partial \nu(\mathbf{y})}g(\mathbf{x},\mathbf{y})\right)ds(\mathbf{y}) \end{aligned}$$

and hence from (4.43) we have

$$u(\mathbf{x}) = -\frac{1}{2\pi}\int_{\partial D}u(\mathbf{y})\frac{\partial}{\partial \nu(\mathbf{y})}\left[\log\frac{1}{|\mathbf{x}-\mathbf{y}|} + g(\mathbf{x},\mathbf{y})\right]ds(\mathbf{y}). \tag{4.44}$$

The function

$$G(\mathbf{x},\mathbf{y}) = \log\frac{1}{|\mathbf{x}-\mathbf{y}|} + g(\mathbf{x},\mathbf{y})$$

is known as **Green's function,** and from (4.44) we see that the solution of the interior Dirichlet problem

$$\Delta_2 u = 0 \quad \text{in} \quad D$$
$$u = f \quad \text{on} \quad \partial D$$

is given by

$$u(\mathbf{x}) = -\frac{1}{2\pi}\int_{\partial D} f(\mathbf{y})\,\frac{\partial G(\mathbf{x},\mathbf{y})}{\partial \nu(\mathbf{y})}\,ds(\mathbf{y}) \qquad (\mathbf{x} \in D). \tag{4.45}$$

This assumes that the solution $u(\mathbf{x})$ of the interior Dirichlet problem exists and that this solution is twice continuously differentiable in $\overline{D}$, and thus (4.45) is an integral representation for sufficiently smooth solutions of the Dirichlet problem (assuming both $u(\mathbf{x})$ and $G(\mathbf{x},\mathbf{y})$ exist).

In an analogous manner we can define the **Neumann function**

$$N(\mathbf{x},\mathbf{y}) = \log\frac{1}{|\mathbf{x}-\mathbf{y}|} + n(\mathbf{x},\mathbf{y})$$

where, as a function of $\mathbf{y}$ for $\mathbf{x} \in D$, $n(\mathbf{x},\mathbf{y})$ is harmonic in $\overline{D}$ and on ∂D

$$\frac{\partial n(\mathbf{x},\mathbf{y})}{\partial \nu(\mathbf{y})} = -\frac{2\pi}{L} - \frac{\partial}{\partial \nu(\mathbf{y})}\log\frac{1}{|\mathbf{x}-\mathbf{y}|} \qquad (\mathbf{y} \in \partial D,\ \mathbf{x} \in D),$$

where L is the arc length of ∂D. Note that the boundary condition allows $n(\mathbf{x},\mathbf{y})$ to satisfy (4.4), i.e., using (4.5) with $u(\mathbf{y}) = 1$ we have

$$\int_{\partial D}\frac{\partial n(\mathbf{x},\mathbf{y})}{\partial \nu(\mathbf{y})}\,ds(\mathbf{y}) = -\frac{2\pi}{L}\int_{\partial D} ds(\mathbf{y}) - \int_{\partial D}\frac{\partial}{\partial \nu(\mathbf{y})}\log\frac{1}{|\mathbf{x}-\mathbf{y}|}\,ds(\mathbf{y})$$
$$= -2\pi + 2\pi = 0.$$

These requirements determine the Neumann function only to within an additive constant, and to complete our definition we require

$$\int_{\partial D} N(\mathbf{x},\mathbf{y})\,ds(\mathbf{y}) = 0,$$

which has the advantage of ensuring the symmetry condition (see the following theorem) $N(\mathbf{x},\mathbf{y}) = N(\mathbf{y},\mathbf{x})$. If the Neumann function is known, then the solution of the interior Neumann problem

$$\Delta_2 u = 0 \quad \text{in} \quad D$$
$$\frac{\partial u}{\partial \nu} = f \quad \text{on} \quad \partial D$$

can be represented in the form

$$u(\mathbf{x}) = \frac{1}{2\pi}\int_{\partial D} f(\mathbf{y}) N(\mathbf{x},\mathbf{y})\,ds(\mathbf{y}) \qquad (\mathbf{x} \in D),$$

where, since the solution of the interior Neumann problem is determined only up to an additive constant, we have imposed the normalization condition

$$\int_{\partial D} u\, ds = 0.$$

Aside from the above definitions, we shall not pursue a study of the Neumann function, but instead concentrate on Green's function.

THEOREM 32

Assume Green's function $G(\mathbf{x},\mathbf{y})$ exists. Then for $\mathbf{x},\mathbf{y} \in D$, $G(\mathbf{x},\mathbf{y}) = G(\mathbf{y},\mathbf{x})$.

■ ***Proof.*** Let $\mathbf{x}_1,\mathbf{x}_2 \in D$ and apply Green's second identity to $G(\mathbf{x}_1,\mathbf{y})$ and $G(\mathbf{x}_2,\mathbf{y})$ over the region obtained from D by excluding two disks Ω_1 and Ω_2 of radius ϵ centered at $\mathbf{x}_1$ and $\mathbf{x}_2$ respectively and contained in D. Since $G(\mathbf{x}_1,\mathbf{y})$ and $G(\mathbf{x}_2,\mathbf{y})$ vanish for $\mathbf{y}$ on ∂D,

$$\int_{\partial\Omega_1} \left(G(\mathbf{x}_1,\mathbf{y}) \frac{\partial G(\mathbf{x}_2,\mathbf{y})}{\partial \nu(\mathbf{y})} - G(\mathbf{x}_2,\mathbf{y}) \frac{\partial G(\mathbf{x}_1,\mathbf{y})}{\partial \nu(\mathbf{y})} \right) ds(\mathbf{y})$$
$$+ \int_{\partial\Omega_2} \left(G(\mathbf{x}_1,\mathbf{y}) \frac{\partial G(\mathbf{x}_2,\mathbf{y})}{\partial \nu(\mathbf{y})} - G(\mathbf{x}_2,\mathbf{y}) \frac{\partial G(\mathbf{x}_1,\mathbf{y})}{\partial \nu(\mathbf{y})} \right) ds(\mathbf{y}) = 0.$$

Letting ϵ tend to zero shows that

$$-2\pi G(\mathbf{x}_2,\mathbf{x}_1) + 2\pi G(\mathbf{x}_1,\mathbf{x}_2) = 0$$

and this proves the theorem. ■

We shall now construct Green's function by the **method of images** for the special case when D is a disk Ω with center at the origin and radius a. Let $\mathbf{x} \in \Omega$ and $r = |\mathbf{x}|$. Define the image of $\mathbf{x}$ with respect to $\partial\Omega$ by

$$\mathbf{x}_1 = \frac{a^2}{r^2}\mathbf{x}, \tag{4.46}$$

and for $\mathbf{y} \in \overline{\Omega}$ let

$$R = |\mathbf{x} - \mathbf{y}|$$
$$R_1 = |\mathbf{x}_1 - \mathbf{y}|.$$

The geometry of this distribution of points is pictured in Figure 4.1 for $\mathbf{y} \in \partial\Omega$. For $\mathbf{y} \in \partial\Omega$, the triangles $\mathbf{Oxy}$ and $\mathbf{Ox}_1\mathbf{y}$ are similiar since they have a common angle at $\mathbf{O}$ and, in view of (4.46), the sides forming this angle are proportional, i.e.,

$$\frac{a}{|\mathbf{x}_1|} = \frac{r}{a}.$$

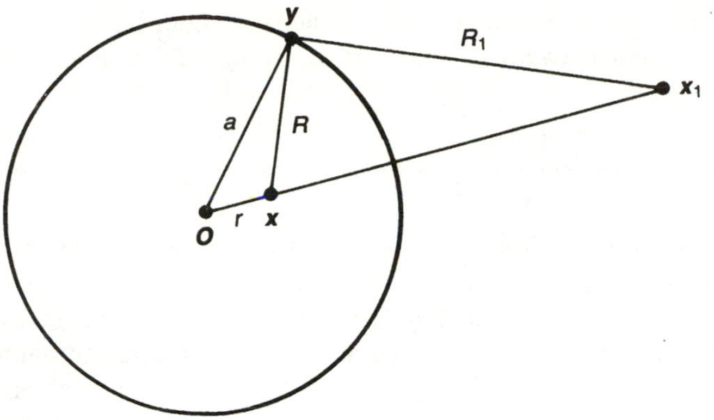

Figure 4.1

Therefore, for $\mathbf{y} \in \partial\Omega$, $R/R_1 = r/a$, so

$$\frac{1}{R} - \frac{a}{r}\frac{1}{R_1} = 0. \tag{4.47}$$

We now claim that the Green function for Ω is given by

$$G(\mathbf{x},\mathbf{y}) = \log\frac{1}{R} - \log\frac{a}{rR_1} = \log\frac{rR_1}{aR}\,.$$

Indeed, if

$$g(\mathbf{x},\mathbf{y}) = -\log\frac{a}{rR_1}\,,$$

then $g(\mathbf{x},\mathbf{y})$ is a harmonic function of $\mathbf{y}$ and for $\mathbf{y} \in \partial\Omega$,

$$g(\mathbf{x},\mathbf{y}) = -\log\frac{1}{R} = -\log\frac{1}{|\mathbf{x}-\mathbf{y}|}\,.$$

From (4.45) the solution of the interior Dirichlet problem for Ω can be represented in the form (for $u(\mathbf{x})$ sufficiently regular)

$$u(\mathbf{x}) = -\frac{1}{2\pi}\int_{\partial\Omega} f(\mathbf{y})\frac{\partial}{\partial\nu(\mathbf{y})}\log\frac{rR_1}{aR}\,ds(\mathbf{y}). \tag{4.48}$$

We now want to simplify this expression. We have

$$\begin{aligned}\log\frac{rR_1}{aR} &= \frac{1}{2}\log\frac{r^2(|\mathbf{x}_1|^2 - 2\mathbf{x}_1\cdot\mathbf{y} + |\mathbf{y}|^2)}{a^2(|\mathbf{x}|^2 - 2\mathbf{x}\cdot\mathbf{y} + |\mathbf{y}|^2)}\\ &= \frac{1}{2}\log\frac{a^2 - 2r\rho\cos(\theta-\phi) + r^2(\rho^2/a^2)}{r^2 - 2r\rho\cos(\theta-\phi) + \rho^2}\end{aligned}$$

where (ρ,ϕ) are the polar coordinates of $\mathbf{y}$ and (r,θ) are the polar coordinates of $\mathbf{x}$. From (4.48), the fact that the unit outward normal $\boldsymbol{\nu}$ is in the radial direction, and $\rho = a$ for $\mathbf{y} \in \partial\Omega$, we now have

$$u(r,\theta) = \frac{1}{2\pi}\int_{-\pi}^{\pi} f(\phi)\frac{a^2 - r^2}{a^2 - 2ar\cos(\theta - \phi) + r^2}\,d\phi \qquad (r < a), \tag{4.49}$$

where $u(a,\theta) = f(\theta)$. Equation (4.49) is Poisson's formula (it was previously derived in Section 4.4 by the method of separation of variables). If a solution $u(\mathbf{x})$ of the interior Dirichlet problem for a disk exists, and $u \in C^2(\overline{D})$, then $u(\mathbf{x})$ is given by (4.49) for $\mathbf{x} \in D$. We note also that the method of images can be carried through to obtain an integral representation for the solution of the interior Dirichlet problem for Laplace's equation in a ball of radius a:

$$u(r,\theta,\phi) = \frac{a}{4\pi}\int_{-\pi}^{\pi}\int_{0}^{\pi} f(\theta_0,\phi_0)\frac{(a^2 - r^2)\sin\theta_0}{(a^2 - 2ar\cos\gamma + r^2)^{3/2}}\,d\theta_0 d\phi_0 \qquad (r < a),$$

where

$$\cos\gamma = \cos\theta\cos\theta_0 + \sin\theta\sin\theta_0\cos(\phi - \phi_0)$$

and $u(a,\theta,\phi) = f(\theta,\phi)$.

We now want to show that if $f(\theta)$ is continuous, then (4.49) is the solution of the interior Dirichlet problem for a disk. For existence theorems for boundary value problems for Laplace's equation defined in more general domains, the reader is referred to Section 5.3 of the following chapter.

THEOREM 33

Let $f(\theta)$ be continuous and of period 2π. Then $u(r,\theta)$ as defined by (4.49) is harmonic for $r < a$ and continuously assumes the boundary data $f(\theta)$ as r tends to a.

■ ***Proof.*** One can easily verify directly that for $r < a$, $u(r,\theta)$ is harmonic. In order to show that $u(r,\theta)$ continuously assumes the boundary data $f(\theta)$, it suffices by symmetry to prove that

$$\lim_{(r,\theta)\to(a,0)} u(r,\theta) = f(0). \tag{4.50}$$

For $r < 1$ we have (since $u(r,\theta) = 1$ satisfies all the assumptions under which (4.49) was derived) that

$$1 = \frac{1}{2\pi}\int_{-\pi}^{\pi}\frac{a^2 - r^2}{a^2 - 2ar\cos(\theta - \phi) + r^2}\,d\phi, \tag{4.51}$$

and hence for $r < a$,

$$u(r,\theta) - f(0) = \frac{1}{2\pi}\int_{-\pi}^{\pi}(f(\phi) - f(0))\frac{a^2 - r^2}{a^2 - 2ar\cos(\theta - \phi) + r^2}\,d\phi.$$

Since $f(\phi)$ is continuous, for every $\epsilon > 0$ there exists a $\delta > 0$ such that $|f(\phi) - f(0)| < \epsilon$ for $-\delta \le \phi \le \delta$. Furthermore, since for $r < a$ the **Poisson kernel**

$$\frac{a^2 - r^2}{a^2 - 2ar\cos(\theta - \phi) + r^2}$$

is positive, we have

$$\begin{aligned}|u(r,\theta) - f(0)| \le \frac{\epsilon}{2\pi}\int_{-\delta}^{\delta} \frac{a^2 - r^2}{a^2 - 2ar\cos(\theta - \phi) + r^2}\,d\phi \\ + \frac{M}{\pi}\int_{-\pi}^{-\delta} \frac{a^2 - r^2}{a^2 - 2ar\cos(\theta - \phi) + r^2}\,d\phi \\ + \frac{M}{\pi}\int_{\delta}^{\pi} \frac{a^2 - r^2}{a^2 - 2ar\cos(\theta - \phi) + r^2}\,d\phi,\end{aligned} \tag{4.52}$$

where

$$M = \max_{-\pi \le \phi \le \pi} |f(\phi)|.$$

By (4.51) and the positivity of the Poisson kernel, the first integral on the right side of (4.52) is less than ϵ; (4.50) will be established if we can show that the second and third integrals on the right side of (4.52) tend to zero as (ρ,ϕ) tends to $(a,0)$.

It suffices to show that the second integral in (4.52) tends to zero, since the proof that the third integral tends to zero is exactly the same. To this end, let $-\delta/2 \le \theta \le \delta/2$. Then for $-\pi \le \phi \le -\delta$,

$$\begin{aligned}a^2 - 2ar\cos(\theta - \phi) + r^2 &\ge a^2 - 2ar\cos\frac{\delta}{2} + r^2 \\ &> \left(a - r\cos\frac{\delta}{2}\right)^2 > a^2\left(1 - \cos\frac{\delta}{2}\right)^2,\end{aligned}$$

so

$$0 < \frac{a^2 - r^2}{a^2 - 2ar\cos(\theta - \phi) + r^2} < \frac{a^2 - r^2}{a^2\left(1 - \cos\frac{\delta}{2}\right)^2}.$$

Therefore,

$$0 < \int_{-\pi}^{-\delta} \frac{a^2 - r^2}{a^2 - 2ar\cos(\theta - \phi) + r^2}\,d\phi < \frac{(a^2 - r^2)(\pi - \delta)}{a^2\left(1 - \cos\frac{\delta}{2}\right)^2} \tag{4.53}$$

and letting r tend to a shows that the left side of (4.53) tends to zero. The theorem is now proved. ■

The Poisson formula (4.49) has many applications. One of the more important of these is the following:

THEOREM 34 Harnack's Inequality

Let $u \in C^2(\overline{\Omega})$ be harmonic and non-negative in the disk $\Omega = \{\mathbf{x}: |\mathbf{x} - \mathbf{x}_0| < a\}$. Then for $\mathbf{x} \in \Omega$ we have

$$\frac{a-r}{a+r}\, u(\mathbf{x}_0) \le u(\mathbf{x}) \le \frac{a+r}{a-r}\, u(\mathbf{x}_0)$$

where $r = |\mathbf{x} - \mathbf{x}_0|$.

■ ***Proof.*** For $r < a$ and arbitrary θ, ϕ, we have

$$\frac{1}{2\pi a}\frac{a-r}{a+r} \le \frac{1}{2\pi a}\frac{a^2 - r^2}{a^2 - 2ar\cos(\theta - \phi) + r^2} \le \frac{1}{2\pi a}\frac{a+r}{a-r},$$

because

$$\begin{aligned}(a - r)^2 = a^2 - 2ar + r^2 &\le a^2 - 2ar\cos(\theta - \phi) + r^2 \\ &\le a^2 + 2ar + r^2 = (a + r)^2.\end{aligned}$$

Hence from Poisson's formula (applied to $|\mathbf{x} - \mathbf{x}_0| < a$ instead of $|\mathbf{x}| < a$ and recalling that $ds = a\, d\theta$) we have

$$\frac{a-r}{a+r}\frac{1}{2\pi a}\int_{\partial\Omega} u\, ds \le u(\mathbf{x}) \le \frac{a+r}{a-r}\frac{1}{2\pi a}\int_{\partial\Omega} u\, ds,$$

and from the mean value theorem,

$$\frac{a-r}{a+r}\, u(\mathbf{x}_0) \le u(\mathbf{x}) \le \frac{a+r}{a-r}\, u(\mathbf{x}_0).$$

■

COROLLARY Liouville's Theorem

A function harmonic in the whole plane and bounded either above or below is a constant.

■ ***Proof.*** Suppose $u(\mathbf{x}) < M$ for all $\mathbf{x} \in R^2$ where M is a constant. Consider the non-negative harmonic function $M - u(\mathbf{x})$. Then from Harnack's inequality with $\mathbf{x}_0 = 0$, we have for any $\mathbf{x} \in R^2$, $a > r = |\mathbf{x}|$,

$$\frac{a-r}{a+r}(M - u(0)) \le M - u(\mathbf{x}) \le \frac{a+r}{a-r}(M - u(0)).$$

Letting a tend to infinity now implies that

$$M - u(0) \le M - u(\mathbf{x}) \le M - u(0),$$

i.e., $u(\mathbf{x}) = u(0) =$ constant. If $u(\mathbf{x}) > m$ for all $\mathbf{x} \in R^2$ where m is a constant, the proof proceeds in the same manner by considering the non-negative harmonic function $u(\mathbf{x}) - m$. ■

4.6 FINITE DIFFERENCE METHODS FOR LAPLACE'S EQUATION

Except for extremely simple shapes, such as disks and half planes (cf. Exercises 18–20) no analytic expression for the solution of the Dirichlet problem for Laplace's equation is available, so one must resort to some approximation scheme. One of the most popular methods is that of finite differences, a method we shall outline here. (An alternative approach is the method of integral equations, cf. Sections 5.3 and 5.5.) One of the attractions of the finite difference method is its applicability to a wide variety of partial differential equations of hyperbolic, parabolic, and elliptic types. We can only touch the surface of this approach; for more details the reader is referred to Forsythe and Wasow.

The basic idea of the finite difference method is extremely simple: replace the second order derivatives in Laplace's equation by finite differences. If we define $\Delta_x^+ u$ and $\Delta_x^- u$ by

$$\Delta_x^+ u = u(x+h,y) - u(x,y)$$
$$\Delta_x^- u = u(x,y) - u(x-h,y),$$

where h is a fixed positive number called the **step size**, and define $\Delta_{xx}^2 u$ by

$$\Delta_{xx}^2 u = \Delta_x^+ u - \Delta_x^- u = u(x+h,y) - u(x-h,y) - 2u(x,y)$$

with corresponding expressions for $\Delta_y^+ u$, $\Delta_y^- u$, and $\Delta_{yy}^2 u$, then the finite difference approximation for Laplace's equation

$$\Delta_2 u = \frac{\partial^2 u}{\partial x^2} + \frac{\partial^2 u}{\partial y^2} = 0$$

has the form

$$\frac{\Delta_h u}{h^2} = \frac{1}{h^2}[\Delta_{xx}^2 u + \Delta_{yy}^2 u] = \frac{1}{h}\left[\frac{\Delta_x^+ u}{h} - \frac{\Delta_x^- u}{h}\right] + \frac{1}{h}\left[\frac{\Delta_y^+ u}{h} - \frac{\Delta_y^- u}{h}\right]$$
$$= \frac{1}{h^2}[u(x+h,y) + u(x,y+h) + u(x-h,y) + u(x,y-h) - 4u(x,y)] = 0.$$

Now consider the Dirichlet problem

$$\begin{aligned} \Delta_2 u &= 0 \quad \text{in} \quad D \\ u &= f \quad \text{on} \quad \partial D, \end{aligned} \tag{4.54}$$

where D is a bounded, simply connected domain in R^2 with C^2 boundary ∂D, and f is a continuous function prescribed on ∂D. For arbitrary positive step size h construct a grid on the x-y plane consisting of squares having sides of length h parallel to the coordinate axes, and let D_h denote the union of the squares contained in D with boundary ∂D_h (see Figure 4.2). At the corner points of ∂D_h, define the boundary function f_h equal to $f(x)$ for x the point on ∂D nearest to the corner point (chosen arbitrarily if there is more than one nearest point). Denote the corners of the squares of D_h (called the **nodes**) by $M_{i,k} = (x_i, y_k)$. Then the finite difference method for approximating the solution of (4.54) finds a function $u_{i,k}$ defined on M_{ik} such that at each interior node

$$u_{i+1,k} + u_{i,k+1} + u_{i-1,k} + u_{i,k-1} - 4u_{i,k} = 0 \qquad \text{in} \qquad D_h, \tag{4.55}$$

and for corner points of ∂D_h

$$u_{i,k} = f_h \qquad \text{on} \qquad \partial D_h. \tag{4.56}$$

If there are n nodes in D_h, then this is a system of n linear equations in n unknowns.

We shall show that the system (4.55), (4.56) has a unique solution. To this end, it suffices to show that the homogeneous equation ($f_h = 0$) has only the trivial solution $u_{i,k} = 0$. Suppose for the homogeneous system $u_{i_1,k_1} \neq 0$ for some integers i_1, k_1. Without loss of generality, assume that u_{i_1,k_1} is positive. Let u_{i_0,k_0} be the maximum value of the solution $u_{i,k}$ of the homogeneous equation, i.e.,

$$u_{i,k} \leq u_{i_0,k_0}$$

for all points $M_{i,k}$. Note that M_{i_0,k_0} is in D_h and not on ∂D_h, because we are considering the case $f_h = 0$. Then from (4.55), (4.56),

$$4u_{i_0,k_0} = u_{i_0+1,k_0} + u_{i_0,k_0+1} + u_{i_0-1,k_0} + u_{i_0,k_0-1},$$

which implies that

$$u_{i_0,k_0} = u_{i_0+1,k_0} = u_{i_0,k_0+1} = u_{i_0-1,k_0} = u_{i_0,k_0-1}.$$

Proceeding in this way for $u_{i_0+1,k_0}, u_{i_0+2,k_0}, \ldots$, we finally arrive at a boundary point M_{i_2,k_2} where

$$u_{i_2,k_2} = u_{i_0,k_0} > 0.$$

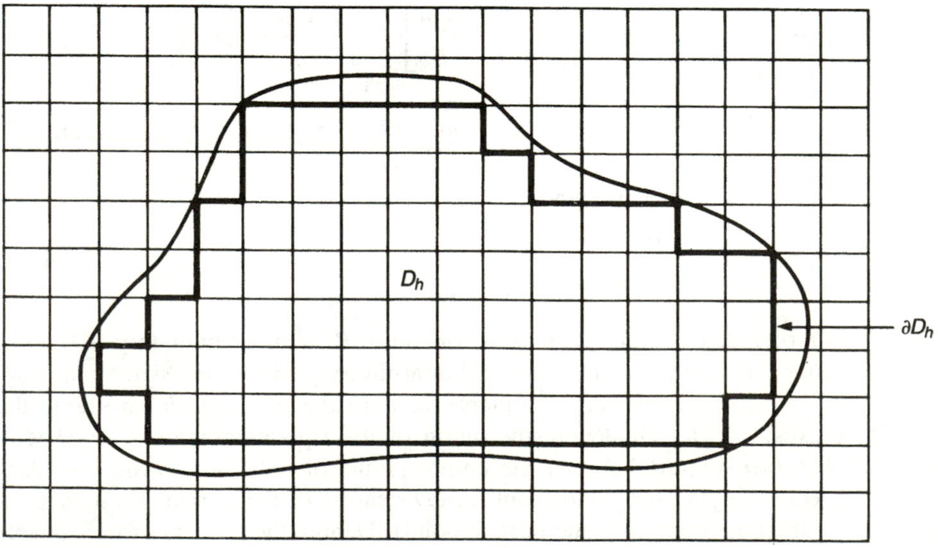

Figure 4.2

This is a contradiction, since $u_{i_2,k_2} = 0$. Therefore, $u_{i,k} = 0$ for all points $M_{i,k}$, and the system has a unique solution.

For a proof of the fact that the solution of (4.55), (4.56) approximates the solution of (4.54) for small h, we refer the reader to Forsythe and Wasow.

4.7 POISSON'S EQUATION

Let D be a bounded domain in R^2. Poisson's equation is

$$\Delta_2 u = \rho(\mathbf{x}) \qquad (\mathbf{x} \in D), \tag{4.57}$$

where $\rho(\mathbf{x})$ is a given function defined in D.

In this section we construct a solution of (4.57); note that to this particular solution, we can always add a function harmonic in D and still have a solution of (4.57). In what follows, $\mathbf{x} = (x_1,x_2)$ and $\mathbf{y} = (y_1,y_2)$.

We first recall some properties of improper integrals depending on a parameter. Consider the improper integral

$$\phi(\mathbf{x}) = \iint_D f(\mathbf{x},\mathbf{y})\, d\mathbf{y} \qquad (\mathbf{x} \in D) \tag{4.58}$$

where $f(\mathbf{x},\mathbf{y})$ is continuous for $\mathbf{x},\mathbf{y} \in \overline{D}$ except for $\mathbf{x} = \mathbf{y}$ and

$$|f(\mathbf{x},\mathbf{y})| \leq \frac{M}{|\mathbf{x} - \mathbf{y}|^\alpha},$$

where M is a positive constant and $0 < \alpha < 2$. In this case it can be easily shown that (4.58) converges absolutely.

DEFINITION 10

The integral (4.58) is said to be **uniformly convergent** at a point $\mathbf{x}_0 \in D$ if for every $\epsilon > 0$ there exists a $\delta > 0$ such that

$$\left| \iint_{D_\delta} f(\mathbf{x},\mathbf{y})\, d\mathbf{y} \right| < \epsilon$$

for $|\mathbf{x} - \mathbf{x}_0| < \delta$ where $D_\delta \subset D$ is any domain containing $\mathbf{x}_0$ and having diameter less than or equal to δ.

The proof of the following theorem can be found in Apostol:

THEOREM 35

If (4.58) is uniformly convergent at $\mathbf{x}_0 \in D$, then $\phi(\mathbf{x})$ is continuous at $\mathbf{x} = \mathbf{x}_0$.

We now consider the **volume potential**

$$\phi(\mathbf{x}) = \iint_D \rho(\mathbf{y}) \log \frac{1}{|\mathbf{x}-\mathbf{y}|}\, d\mathbf{y} \tag{4.59}$$

where D is a bounded domain and $\rho(\mathbf{y})$ is bounded and integrable in D. If $\mathbf{x}$ lies outside $\overline{D}$, then (4.59) is a proper integral and in this case $\phi(\mathbf{x})$ is infinitely differentiable and the derivatives of $\phi(\mathbf{x})$ can be obtained by differentiating under the integral sign, i.e.,

$$\Delta_2 \phi = 0 \qquad (\mathbf{x} \in R^2 \backslash \overline{D}).$$

Now suppose that $\mathbf{x} \in D$. Then (4.59) is improper, but, since $\rho(\mathbf{y})$ is bounded, the integral converges, because for any $\delta > 0$ there exists a positive constant M such that for $\mathbf{x}$ and $\mathbf{y}$ in compact subsets of R^2,

$$\left| \rho(\mathbf{y}) \log \frac{1}{|\mathbf{x}-\mathbf{y}|} \right| \leq \frac{M}{|\mathbf{x}-\mathbf{y}|^{\delta}}.$$

THEOREM 36

If $\rho(\mathbf{y})$ is bounded and integrable in D, then $\phi(\mathbf{x})$ is continuously differentiable in D, and the derivatives of $\phi(\mathbf{x})$ can be obtained by differentiating under the integral sign.

■ ***Proof.*** We begin by showing that (4.59) and the integrals

$$v_i(\mathbf{x}) = -\iint_D \rho(\mathbf{y}) \frac{x_i - y_i}{|\mathbf{x}-\mathbf{y}|^2}\, d\mathbf{y},$$

$i = 1,2$ (obtained by differentiating (4.59) under the integral sign with respect to x_i, where $\mathbf{x} = (x_1,x_2)$, $\mathbf{y} = (y_1,y_2)$), converge uniformly at any point $\mathbf{x}_0 \in D$. Let $\mathbf{x}_0 \in D_\delta \subset D$, where the diameter of D_δ is less than δ. Then for $|\mathbf{x} - \mathbf{x}_0| < \delta$ there exists a positive constant M such that

$$\left| \iint_{D_\delta} \rho(\mathbf{y}) \log \frac{1}{|\mathbf{x}-\mathbf{y}|}\, d\mathbf{y} \right| < M \iint_{\Omega_{2\delta}} \frac{1}{|\mathbf{x}-\mathbf{y}|}\, d\mathbf{y},$$

where $\Omega_{2\delta} \supset D_\delta$ is a disk of radius 2δ centered at $\mathbf{x}$. Then

$$M \iint_{\Omega_{2\delta}} \frac{1}{|\mathbf{x}-\mathbf{y}|}\, d\mathbf{y} = M \int_{-\pi}^{\pi} \int_0^{2\delta} dr\, d\theta = M 4\pi\delta,$$

where we have changed variables to polar coordinates (r,θ) centered at $\mathbf{x}$. Then

$$\left| \iint_{D_\delta} \rho(\mathbf{y}) \log \frac{1}{|\mathbf{x}-\mathbf{y}|}\, d\mathbf{y} \right| < 4\pi M\delta,$$

and the right side tends to zero as δ tends to zero independently of $\mathbf{x}$ for $|\mathbf{x} - \mathbf{x}_0| < \delta$, i.e., (4.59) converges uniformly at any point $\mathbf{x}_0 \in D$. Repeating this argument for $v_i(\mathbf{x})$ shows that there exists a positive constant M such that

$$\left| \iint_{D_\delta} \rho(\mathbf{y}) \frac{x_i - y_i}{|\mathbf{x} - \mathbf{y}|^2} d\mathbf{y} \right| < M \iint_{\Omega_{2\delta}} \frac{1}{|\mathbf{x} - \mathbf{y}|} d\mathbf{y} = 4\pi M \delta,$$

so $v_i(\mathbf{x})$ also converges uniformly at any point $\mathbf{x}_0 \in D$.

We now must show that $v_i(\mathbf{x}) = \partial\phi/\partial x_i(\mathbf{x})$. We first consider $v_1(\mathbf{x})$. Let

$$\begin{aligned} I &= \frac{\phi(\mathbf{x}_h) - \phi(\mathbf{x})}{h} - v_1(\mathbf{x}) \\ &= \frac{1}{h} \iint_D \rho(\mathbf{y}) \left(\log \frac{1}{|\mathbf{x}_h - \mathbf{y}|} - \log \frac{1}{|\mathbf{x} - \mathbf{y}|} \right) d\mathbf{y} - \iint_D \rho(\mathbf{y}) \frac{y_1 - x_1}{|\mathbf{x} - \mathbf{y}|^2} d\mathbf{y}, \end{aligned}$$

where $\mathbf{x}_h = (x_1 + h, x_2)$. We shall show that I tends to zero as h tends to zero. Let Ω_{δ_1} be a disk of radius δ_1 centered at $\mathbf{x}$ and lying inside D and let $D_1 = D \backslash \Omega_{\delta_1}$. Then writing

$$\begin{aligned} \phi(\mathbf{x}) &= \iint_{\Omega_{\delta_1}} \rho(\mathbf{y}) \log \frac{1}{|\mathbf{x} - \mathbf{y}|} d\mathbf{y} + \iint_{D_1} \rho(\mathbf{y}) \log \frac{1}{|\mathbf{x} - \mathbf{y}|} d\mathbf{y} \\ &= \phi^{(1)}(\mathbf{x}) + \phi^{(2)}(\mathbf{x}) \end{aligned}$$

$$v_1(\mathbf{x}) = -\iint_{\Omega_{\delta_1}} \rho(\mathbf{y}) \frac{x_1 - y_1}{|\mathbf{x} - \mathbf{y}|^2} d\mathbf{y} - \iint_{D_1} \rho(\mathbf{y}) \frac{x_1 - y_1}{|\mathbf{x} - \mathbf{y}|^2} d\mathbf{y} = v^{(1)}(\mathbf{x}) + v^{(2)}(\mathbf{x}),$$

we have

$$I = \left(\frac{\phi^{(1)}(\mathbf{x}_h) - \phi^{(1)}(\mathbf{x})}{h} \right) - v^{(1)}(\mathbf{x}) + \left(\frac{\phi^{(2)}(\mathbf{x}_h) - \phi^{(2)}(\mathbf{x})}{h} \right) - v^{(2)}(\mathbf{x}).$$

Assume h is small enough such that $\mathbf{x}_h \in \Omega_{\delta_1}$. Then using the elementary inequalities

$$\frac{|\mathbf{x} - \mathbf{y}|}{|\mathbf{x}_h - \mathbf{y}|} \leq 1 + \frac{|h|}{|\mathbf{x}_h - \mathbf{y}|}$$

$$\frac{|\mathbf{x}_h - \mathbf{y}|}{|\mathbf{x} - \mathbf{y}|} \leq 1 + \frac{|h|}{|\mathbf{x} - \mathbf{y}|}$$

$$|\log(1 + t)| \leq t \qquad (t \geq 0),$$

we have

$$\begin{aligned}\left|\frac{\phi^{(1)}(\mathbf{x}_h) - \phi^{(1)}(\mathbf{x})}{h}\right| &= \left|\frac{1}{h}\iint_{\Omega_{\delta_1}} \rho(\mathbf{y})\left(\log\frac{1}{|\mathbf{x}_h - \mathbf{y}|} - \log\frac{1}{|\mathbf{x} - \mathbf{y}|}\right) d\mathbf{y}\right| \\ &\le \frac{1}{|h|}\iint_{\Omega_{\delta_1}} |\rho(\mathbf{y})| \left|\log\frac{|\mathbf{x} - \mathbf{y}|}{|\mathbf{x}_h - \mathbf{y}|}\right| d\mathbf{y} \\ &\le M \iint_{\Omega_{\delta_1}} \left(\frac{1}{|\mathbf{x} - \mathbf{y}|} + \frac{1}{|\mathbf{x}_h - \mathbf{y}|}\right) d\mathbf{y} \\ &\le M \iint_{\Omega_{\delta_1}} \frac{1}{|\mathbf{x} - \mathbf{y}|} d\mathbf{y} + M \iint_{\Omega_{2\delta_1}} \frac{1}{|\mathbf{x}_h - \mathbf{y}|} d\mathbf{y} \\ &= 2\pi M\delta_1 + 4\pi M\delta_1 = 6\pi M\delta_1,\end{aligned}$$

where M is a positive constant and $\Omega_{2\delta_1}$ is a disk of radius $2\delta_1$ centered at $\mathbf{x}_h$. Furthermore,

$$\begin{aligned}|v^{(1)}(\mathbf{x})| &= \left|\iint_{\Omega_{\delta_1}} \rho(\mathbf{y}) \frac{x_1 - y_1}{|\mathbf{x} - \mathbf{y}|^2} d\mathbf{y}\right| \le M \iint_{\Omega_{\delta_1}} \frac{1}{|\mathbf{x} - \mathbf{y}|} d\mathbf{y} \\ &= 2\pi M\delta_1 < 6\pi M\delta_1,\end{aligned}$$

where M is a positive constant (chosen, without loss of generality, to be the same as in the previous estimate). Now let $\epsilon > 0$ and choose δ_1 such that $6\pi M\delta_1 < \epsilon/3$. Then we have

$$\left|\frac{\phi^{(1)}(\mathbf{x}_h) - \phi^{(1)}(\mathbf{x})}{h} - v^{(1)}(\mathbf{x})\right| < \frac{2\epsilon}{3}$$

for all $\mathbf{x}_h \in \Omega_{\delta_1}$. The remaining terms in I tend to zero as h tends to zero, since

$$\frac{\partial \phi^{(2)}(\mathbf{x})}{\partial x_1} = v^{(2)}(\mathbf{x}),$$

so for $|h| < \delta_2$ we can make these terms less than $\epsilon/3$. Therefore, if $|h| < \delta = \min(\delta_1, \delta_2)$ we have $|I| < \epsilon$, and

$$\frac{\partial \phi(\mathbf{x})}{\partial x_1} = v_1(\mathbf{x})$$

as desired. In an identical manner we can show that

$$\frac{\partial \phi(\mathbf{x})}{\partial x_2} = v_2(\mathbf{x}).$$

■

THEOREM 37

Let $\rho \in C^1(D)$ be bounded and integrable in D and let $\phi(\mathbf{x})$ be defined by (4.59). Then $\phi \in C^2(D)$ and $\Delta_2\phi = -2\pi\rho(\mathbf{x})$ for $\mathbf{x} \in D$.

■ ***Proof.*** Let $\mathbf{x}_0 \in D$ and $\Omega_\delta \subset D$ be a disk of radius δ centered at $\mathbf{x}_0$. Let $D_1 = D\backslash\Omega_\delta$ and write

$$\phi(\mathbf{x}) = \iint_{\Omega_\delta} \rho(\mathbf{y}) \log \frac{1}{|\mathbf{x}-\mathbf{y}|}\, d\mathbf{y} + \iint_{D_1} \rho(\mathbf{y}) \log \frac{1}{|\mathbf{x}-\mathbf{y}|}\, d\mathbf{y}$$

$$= \phi^{(1)}(\mathbf{x}) + \phi^{(2)}(\mathbf{x}).$$

Then from the previous theorem,

$$\frac{\partial\phi}{\partial x_1} = \iint_{\Omega_\delta} \rho(\mathbf{y}) \frac{\partial}{\partial x_1} \log \frac{1}{|\mathbf{x}-\mathbf{y}|}\, d\mathbf{y} + \iint_{D_1} \rho(\mathbf{y}) \frac{\partial}{\partial x_1} \log \frac{1}{|\mathbf{x}-\mathbf{y}|}\, d\mathbf{y},$$

and using the divergence theorem,

$$\frac{\partial\phi}{\partial x_1} = -\iint_{\Omega_\delta} \rho(\mathbf{y}) \frac{\partial}{\partial y_1} \log \frac{1}{|\mathbf{x}-\mathbf{y}|}\, d\mathbf{y} + \iint_{D_1} \rho(\mathbf{y}) \frac{\partial}{\partial x_1} \log \frac{1}{|\mathbf{x}-\mathbf{y}|}\, d\mathbf{y}$$

$$= \iint_{\Omega_\delta} \frac{\partial\rho(\mathbf{y})}{\partial y_1} \log \frac{1}{|\mathbf{x}-\mathbf{y}|}\, d\mathbf{y} - \int_{\partial\Omega_\delta} \rho(\mathbf{y}) \log \frac{1}{|\mathbf{x}-\mathbf{y}|}\, \boldsymbol{\nu}\cdot\mathbf{e}_1\, ds(\mathbf{y})$$

$$+ \iint_{D_1} \rho(\mathbf{y}) \frac{\partial}{\partial x_1} \log \frac{1}{|\mathbf{x}-\mathbf{y}|}\, d\mathbf{y},$$

where $\boldsymbol{\nu}$ is the unit outward normal to $\partial\Omega_\delta$, and $\mathbf{e}_1$ is the unit vector in the y_1 direction. The second and third integrals on the right side of the expression are infinitely differentiable for $\mathbf{x} \in \Omega_\delta$, whereas from the previous theorem the first integral is continuously differentiable for $\mathbf{x} \in \Omega_\delta$. Hence, $\partial\phi/\partial x_1 \in C^1(\Omega_\delta)$, and since $\mathbf{x}_0$ was arbitrary, $\partial\phi/\partial x_1 \in C^1(D)$. Similarly, $\partial\phi/\partial x_2 \in C^1(D)$, and hence we can conclude that $\phi \in C^2(D)$.

We shall now show that

$$\Delta_2\phi = -2\pi\rho(\mathbf{x})$$

for $\mathbf{x} \in D$. Note first that

$$\Delta_2\phi^{(2)} = 0$$

for $\mathbf{x} \in \Omega_\delta$, and hence

$$\Delta_2\phi = \Delta_2\phi^{(1)}$$

for $\mathbf{x} \in \Omega_\delta$. Therefore, for $\mathbf{x} \in \Omega_\delta$

$$\Delta_2\phi(\mathbf{x}) = \Delta_2 \iint_{\Omega_\delta} \rho(\mathbf{y}) \log \frac{1}{|\mathbf{x} - \mathbf{y}|} \, d\mathbf{y}.$$

The above analysis now implies that

$$\Delta_2\phi(\mathbf{x}_0) = \iint_{\Omega_\delta} \nabla\rho(\mathbf{y}) \cdot \frac{(\mathbf{y} - \mathbf{x}_0)}{|\mathbf{x}_0 - \mathbf{y}|^2} \, d\mathbf{y} - \int_{\partial\Omega_\delta} \rho(\mathbf{y}) \frac{(\mathbf{y} - \mathbf{x}_0) \cdot \boldsymbol{\nu}}{|\mathbf{x}_0 - \mathbf{y}|^2} \, ds(\mathbf{y}). \tag{4.60}$$

Let

$$M = \max_{y \in \Omega_\delta} |\nabla\rho(\mathbf{y})|.$$

Then

$$\left| \iint_{\Omega_\delta} \nabla\rho(\mathbf{y}) \cdot \frac{(\mathbf{y} - \mathbf{x}_0)}{|\mathbf{x}_0 - \mathbf{y}|^2} \, d\mathbf{y} \right| \le M \iint_{\Omega_\delta} \frac{1}{|\mathbf{x}_0 - \mathbf{y}|} \, d\mathbf{y} = 2\pi M \delta$$

and since $\boldsymbol{\nu} = (\mathbf{y} - \mathbf{x}_0)/|\mathbf{y} - \mathbf{x}_0|$,

$$\int_{\partial\Omega_\delta} \rho(\mathbf{y}) \frac{(\mathbf{y} - \mathbf{x}_0) \cdot \boldsymbol{\nu}}{|\mathbf{x}_0 - \mathbf{y}|^2} \, ds(\mathbf{y}) = \frac{1}{\delta} \int_{\partial\Omega_\delta} \rho(\mathbf{y}) \, ds(\mathbf{y}).$$

Letting δ tend to zero in (4.60) now gives

$$\Delta_2\phi(\mathbf{x}_0) = -2\pi\rho(\mathbf{x}_0),$$

and since $\mathbf{x}_0$ was an arbitrary point of D the theorem is proved. ■

We conclude this section by making a few remarks. The hypothesis of the above theorem can be weakened to the requirement that $\rho(\mathbf{x})$ is bounded, integrable, and **Hölder continuous** in D, i.e., there exist positive constants M and α, $0 < \alpha \le 1$, such that for $\mathbf{x}, \mathbf{y} \in D$,

$$|\rho(\mathbf{x}) - \rho(\mathbf{y})| \le M|\mathbf{x} - \mathbf{y}|^\alpha.$$

However, assuming only that $\rho(\mathbf{x})$ is bounded, integrable, and continuous is not enough (cf. Epstein). The above analysis can be extended in a straightforward fashion to show that if $\rho(\mathbf{x})$ is bounded, integrable, and continuously differentiable in a domain in R^3, then

$$\phi(\mathbf{x}) = \iiint_D \rho(\mathbf{y}) \frac{1}{|\mathbf{x} - \mathbf{y}|} \, d\mathbf{y}$$

satisfies

$$\Delta_3\phi = -4\pi\rho(\mathbf{x})$$

for $\mathbf{x} \in D$. This result will be used in the following section where we consider a scattering problem in R^3. (We have avoided treating the analogous problem in R^2 in order not to invoke the theory of Hankel functions which we prefer to delay until Chapter 6.)

4.8 TIME HARMONIC WAVE PROPAGATION IN A NONHOMOGENEOUS MEDIUM

Consider a time harmonic acoustic wave passing through a pocket of rarefied or condensed air. Then, assuming that the density $\rho_0 = \rho_0(\mathbf{x})$ is slowly varying (i.e., terms involving $\nabla\rho_0$ can be neglected), we see from Section 1.1 that the velocity potential $U(\mathbf{x},t)$ satisfies the wave equation

$$\Delta_3 U = \frac{1}{c^2(\mathbf{x})}\frac{\partial^2 U}{\partial t^2},$$

where $c \in C^1(R^3)$ denotes the speed of sound and $c(\mathbf{x}) = c_0$ for $|\mathbf{x}| \geq a$ where c_0 and a are positive constants. Since the wave propagation is time harmonic, in complex notation we have

$$U(\mathbf{x},t) = u(\mathbf{x})e^{-i\omega t},$$

where ω is the frequency. The (complex valued) function $u(\mathbf{x})$ thus satisfies the reduced wave equation

$$\Delta_3 u + k^2(1 - m(\mathbf{x}))u = 0,$$

where $k^2 = \omega^2/c_0^2$ and

$$m(\mathbf{x}) = 1 - \frac{c_0^2}{c^2(\mathbf{x})}.$$

Note that $m(\mathbf{x})$ is identically zero for $|\mathbf{x}| \geq a$, i.e., $m(\mathbf{x})$ has compact support in R^3.

We now decompose $U(\mathbf{x},t)$ into the sum of two quantities, the incident wave $U^i(\mathbf{x},t)$ and the "scattered wave" $U^s(\mathbf{x},t) = U(\mathbf{x},t) - U^i(\mathbf{x},t)$. Note that if there is no pocket of rarefied or condensed air, then $U(\mathbf{x},t) = U^i(\mathbf{x},t)$, so $U^s(\mathbf{x},t)$ is identically zero. We shall assume that $U^i(\mathbf{x},t)$ is the **plane wave**

$$U^i(\mathbf{x},t) = \exp i(k\boldsymbol{\alpha}\cdot\mathbf{x} - \omega t)$$

moving in the direction $\boldsymbol{\alpha}$ where $|\boldsymbol{\alpha}| = 1$. Turning now to the scattered wave, we make the assumption that for large distances away from the air pocket, $U^s(\mathbf{x},t)$ behaves like an outgoing time harmonic spherical wave, i.e., $U^s(\mathbf{x},t)$ is a solution of the wave equation

$$\Delta_3 U^s = \frac{1}{c_0^2}\frac{\partial^2 U^s}{\partial t^2}$$

having the form

$$U^s(\mathbf{x},t) \sim \frac{F(\theta,\phi)}{r} \exp i(kr - \omega t)$$

for large r, where (r,θ,ϕ) are the spherical coordinates of $\mathbf{x}$. This condition is guaranteed by the **Sommerfeld radiation condition**

$$\lim_{r\to\infty} r\left(\frac{\partial U^s}{\partial r} - ikU^s\right) = 0$$

where the limit is assumed to hold uniformly in all directions. In particular, the Sommerfeld radiation condition guarantees that $U^s(\mathbf{x},t)$ is a wave going *out* from the origin instead of *in* to the origin, i.e., no energy is radiated in from infinity.

These considerations lead us to the problem of finding a (complex valued) function $u \in C^2(R^3)$ such that

$$\Delta_3 u + k^2(1 - m(\mathbf{x}))u = 0 \qquad \text{in} \qquad R^3 \tag{4.61}$$

$$u(\mathbf{x}) = \exp[ik\boldsymbol{\alpha}\cdot\mathbf{x}] + u^s(\mathbf{x}) \tag{4.62}$$

$$\lim_{r\to\infty} r\left(\frac{\partial u^s}{\partial r} - iku^s\right) = 0. \tag{4.63}$$

We shall now proceed to construct a solution to (4.61)–(4.63) by using the results of the previous section.

Let Ω_δ be a ball of radius δ centered at $\mathbf{x}$ and contained in the ball $\Omega = \{\mathbf{x}\colon |\mathbf{x}| \le a\}$ and consider the integral

$$I(\mathbf{x}) = \iiint_\Omega \frac{\exp[ik|\mathbf{x}-\mathbf{y}|]}{|\mathbf{x}-\mathbf{y}|} m(\mathbf{y})u(\mathbf{y})\, d\mathbf{y}$$

for $\mathbf{x} \in D$. We first rewrite $I(\mathbf{x})$ as the sum of two integrals,

$$I(\mathbf{x}) = \iiint_{\Omega_\delta} \frac{\exp[ik|\mathbf{x}-\mathbf{y}|]}{|\mathbf{x}-\mathbf{y}|} m(\mathbf{y})u(\mathbf{y})\, d\mathbf{y} + \iiint_{\Omega\setminus\Omega_\delta} \frac{\exp[ik|\mathbf{x}-\mathbf{y}|]}{|\mathbf{x}-\mathbf{y}|} m(\mathbf{y})u(\mathbf{y})\, d\mathbf{y}$$

$$= I^{(1)}(\mathbf{x}) + I^{(2)}(\mathbf{x}).$$

For $\mathbf{x} \neq \mathbf{y}$, $\exp[ik|\mathbf{x}-\mathbf{y}|]/|\mathbf{x}-\mathbf{y}|$ is a solution of

$$\Delta_3 u + k^2 u = 0,$$

so

$$\Delta_3 I + k^2 I = \Delta_3 I^{(1)} + k^2 I^{(1)}.$$

But

$$I^{(1)}(\mathbf{x}) = \iiint\limits_{\Omega_\delta} \left(\frac{\exp[ik|\mathbf{x}-\mathbf{y}|] - 1}{|\mathbf{x}-\mathbf{y}|}\right) m(\mathbf{y})u(\mathbf{y})\,d\mathbf{y} + \iiint\limits_{\Omega_\delta} \frac{1}{|\mathbf{x}-\mathbf{y}|} m(\mathbf{y})u(\mathbf{y})\,d\mathbf{y},$$

and using the analysis of the previous section, we see that

$$\begin{aligned}
\Delta_3 I^{(1)} + k^2 I^{(1)} &= \iiint\limits_{\Omega_\delta} (\Delta_3 + k^2)\left(\frac{\exp[ik|\mathbf{x}-\mathbf{y}|] - 1}{|\mathbf{x}-\mathbf{y}|}\right) m(\mathbf{y})u(\mathbf{y})\,d\mathbf{y} \\
&\quad + \Delta_3 \iiint\limits_{\Omega_\delta} \frac{1}{|\mathbf{x}-\mathbf{y}|} m(\mathbf{y})u(\mathbf{y})\,d\mathbf{y} \\
&\quad + k^2 \iiint\limits_{\Omega_\delta} \frac{1}{|\mathbf{x}-\mathbf{y}|} m(\mathbf{y})u(\mathbf{y})\,d\mathbf{y} \\
&= \iiint\limits_{\Omega_\delta} (\Delta_3 + k^2)\left(\frac{\exp[ik|\mathbf{x}-\mathbf{y}|] - 1}{|\mathbf{x}-\mathbf{y}|}\right) m(\mathbf{y})u(\mathbf{y})\,d\mathbf{y} \\
&\quad - 4\pi m(\mathbf{y})u(\mathbf{y}) + k^2 \iiint\limits_{\Omega_\delta} \frac{1}{|\mathbf{x}-\mathbf{y}|} m(\mathbf{y})u(\mathbf{y})\,d\mathbf{y}.
\end{aligned}$$

Hence, letting δ tend to zero, we have

$$\Delta_3 I + k^2 I = -4\pi m(\mathbf{x})u(\mathbf{x}).$$

Noting that $\exp[ik\boldsymbol{\alpha}\cdot\mathbf{x}]$ is a solution of $\Delta_3 u + k^2 u = 0$, we see now that if $u(\mathbf{x})$ is a solution of the integral equation

$$u(\mathbf{x}) = \exp[ik\boldsymbol{\alpha}\cdot\mathbf{x}] - \frac{k^2}{4\pi}\iiint\limits_{\Omega} \frac{\exp[ik|\mathbf{x}-\mathbf{y}|]}{|\mathbf{x}-\mathbf{y}|} m(\mathbf{y})u(\mathbf{y})\,d\mathbf{y}, \tag{4.64}$$

then, for $\mathbf{x} \in \Omega$, $u(\mathbf{x})$ is a solution of (4.61). In deriving (4.64) we have assumed that $\mathbf{x} \in \Omega$. However, if $\mathbf{x} \in R^3\backslash\Omega$ then, since $m(\mathbf{x}) = 0$ for $|\mathbf{x}| \geq a$ and $\exp[ik|\mathbf{x}-\mathbf{y}|]/|\mathbf{x}-\mathbf{y}|$ has no singularity in Ω, it is clear that $u(\mathbf{x})$ is a solution of (4.61). Hence, if $u(\mathbf{x})$ is a solution of (4.64) for $\mathbf{x} \in \Omega$, then (4.64) defines a solution $u(\mathbf{x})$ of (4.61) for all $\mathbf{x} \in R^3$.

Suppose we can find a solution $u(\mathbf{x})$ of (4.64) that is continuous in Ω. Then we see that $u(\mathbf{x}) \in C^2(R^3)$ and is a solution of (4.61) in R^3. Define $u^s(\mathbf{x})$ by

$$u^s(\mathbf{x}) = -\frac{k^2}{4\pi} \iiint_\Omega \frac{\exp[ik|\mathbf{x}-\mathbf{y}|]}{|\mathbf{x}-\mathbf{y}|} m(\mathbf{y})u(\mathbf{y})\, d\mathbf{y}. \tag{4.65}$$

Then for $r > a$ we compute

$$r\left(\frac{\partial u^s}{\partial r} - iku^s\right) = -\frac{k^2}{4\pi} \iiint_\Omega r\left(\frac{\partial}{\partial r} - ik\right) \frac{\exp[ik|\mathbf{x}-\mathbf{y}|]}{|\mathbf{x}-\mathbf{y}|} m(\mathbf{y})u(\mathbf{y})\, d\mathbf{y}.$$

Since for $r = |\mathbf{x}|$, $\mathbf{y} \in \Omega$,

$$|\mathbf{x}-\mathbf{y}| = \sqrt{r^2 - 2r|\mathbf{y}|\cos(\mathbf{x},\mathbf{y}) + |\mathbf{y}|^2}$$

$$= r\sqrt{1 - 2\frac{|\mathbf{y}|}{r}\cos(\mathbf{x},\mathbf{y}) + \frac{|\mathbf{y}|^2}{r^2}} = r\left(1 + 0\left(\frac{1}{r}\right)\right)$$

and similarly

$$\frac{\partial|\mathbf{x}-\mathbf{y}|}{\partial r} = 1 + 0\left(\frac{1}{r}\right),$$

we have

$$r\left(\frac{\partial}{\partial r} - ik\right)\frac{\exp[ik|\mathbf{x}-\mathbf{y}|]}{|\mathbf{x}-\mathbf{y}|} = 0\left(\frac{1}{r}\right),$$

where $f(r) = 0(1/r)$ means $|f(r)| \leq M/r$ for $r \geq r_0$ for some positive constants M and r_0. Hence if $u(\mathbf{x})$ is a solution of (4.64), then (4.62) and (4.63) are satisfied witih $u^s(\mathbf{x})$ defined by (4.65). Therefore to find a solution of (4.61)–(4.63) it suffices to find a (continuous) solution of the integral equation (4.64). We now proceed to do this for k sufficiently small, using the method of successive approximations (see Section 2.2).

For the sake of notational convenience, we define the linear operator $\mathbf{T}: C(\Omega) \to C(\Omega)$ by

$$(\mathbf{T}u)(\mathbf{x}) = -\frac{k^2}{4\pi} \iiint_\Omega \frac{\exp[ik|\mathbf{x}-\mathbf{y}|]}{|\mathbf{x}-\mathbf{y}|} m(\mathbf{y})u(\mathbf{y})\, d\mathbf{y}.$$

Then our integral equation (4.64) can be written as

$$u(\mathbf{x}) = \exp[ik\boldsymbol{\alpha}\cdot\mathbf{x}] + (\mathbf{T}u)(\mathbf{x}).$$

We define the sequence of successive approximations by

$$u_{n+1}(\mathbf{x}) = \exp[ik\boldsymbol{\alpha}\cdot\mathbf{x}] + (\mathbf{T}u_n)(\mathbf{x}) \qquad (n \geq 0)$$

$$u_0(\mathbf{x}) = 0.$$

The sequence $\{u_n(\mathbf{x})\}$ is convergent if and only if the series

$$u(\mathbf{x}) = \sum_{n=0}^{\infty} (u_{n+1}(\mathbf{x}) - u_n(\mathbf{x})) \tag{4.66}$$

is convergent. This series will be (uniformly) convergent for $\mathbf{x} \in \Omega$ if

$$\|u_{n+1} - u_n\| \le M\gamma^n, \tag{4.67}$$

where M and γ are positive constants such that $0 < \gamma < 1$ and

$$\|u\| = \max_{\mathbf{x}\in\Omega} |u(\mathbf{x})|.$$

We shall now show that (4.67) is valid if $\mathbf{T}$ is a **contraction mapping**;

$$\|\mathbf{T}u - \mathbf{T}v\| \le \gamma\|u - v\| \tag{4.68}$$

for some γ, $0 < \gamma < 1$, and all $u,v \in C(\Omega)$. Note also that if (4.68) is true, then $\|u_n - u\|$ tending to zero as n tends to infinity implies that $\|\mathbf{T}u_n - \mathbf{T}u\|$ tends to zero as n tends to infinity, so $u(\mathbf{x})$ is a solution of the integral equation (4.64). Since $\mathbf{T}$ is linear, (4.68) is equivalent to

$$\|\mathbf{T}u\| \le \gamma\|u\| \tag{4.69}$$

for $u \in C(\Omega)$.

Suppose that (4.69) is valid. Then

$$\|u_{n+1} - u_n\| = \|\mathbf{T}(u_n - u_{n-1})\| = \|\mathbf{T}^n(u_1 - u_0)\| \le \gamma^n\|u_1 - u_0\|,$$

since $\|\mathbf{T}^n u\| = \|\mathbf{T}(\mathbf{T}^{n-1}u)\| \le \gamma\|\mathbf{T}^{n-1}u\|$ and hence $\|\mathbf{T}^n u\| \le \gamma^n\|u\|$ by induction. Therefore

$$\|u_{n+1} - u_n\| \le \gamma^n\|u_1 - u_0\| = \gamma^n\|e^{ik\boldsymbol{\alpha}\cdot\mathbf{x}}\| = \gamma^n,$$

so (4.67) is valid for $M = 1$. Thus, if (4.69) is valid, the series (4.66) is uniformly convergent, and there exists a function $u \in C(\Omega)$ such that $\|u_n - u\|$ tends to zero as n tends to infinity. We now know that $u(\mathbf{x})$ is a solution of (4.64). Our problem is solved provided we can establish the inequality (4.69) for some γ, $0 < \gamma < 1$.

To do so, we first note that for $u \in C(\Omega)$,

$$\begin{aligned}\|\mathbf{T}u\| &= \max_{\mathbf{x}\in\Omega} \frac{k^2}{4\pi} \left| \iiint_{\Omega} \frac{\exp[ik|\mathbf{x} - \mathbf{y}|]}{|\mathbf{x} - \mathbf{y}|} m(\mathbf{y})u(\mathbf{y})\, d\mathbf{y} \right| \\ &\le \frac{k^2\mu\|u\|}{4\pi} \left\| \iiint_{\Omega} \frac{1}{|\mathbf{x} - \mathbf{y}|} d\mathbf{y} \right\|,\end{aligned} \tag{4.70}$$

where

$$\mu = \max_{\mathbf{x}\in\Omega} |m(\mathbf{x})|.$$

But

$$h(\mathbf{x}) = \iiint_{\Omega} \frac{1}{|\mathbf{x} - \mathbf{y}|}\, d\mathbf{y}$$

is a solution of $\Delta_3 h = -4\pi$ for $\mathbf{x} \in \Omega$, and by symmetry $h(\mathbf{x})$ depends only on $r = |\mathbf{x}|$. Hence

$$\frac{\partial^2 h}{\partial r^2} + \frac{2}{r}\frac{\partial h}{\partial r} = -4\pi,$$

and using the fact that $h(\mathbf{x})$ is bounded at $r = 0$ we have

$$h(\mathbf{x}) = -\frac{2\pi}{3} r^2 + \text{constant}.$$

At $r = 0$,

$$h(0) = \iiint_{\Omega} \frac{1}{|\mathbf{y}|}\, d\mathbf{y} = \int_0^a \int_{-\pi}^{\pi} \int_0^{\pi} \rho \sin\theta \, d\theta d\phi d\rho = 2\pi a^2,$$

so

$$\|h\| = \max_{\mathbf{x}\in\Omega} \left| -\frac{2\pi}{3} r^2 + 2\pi a^2 \right| = 2\pi a^2,$$

and from (4.70),

$$\|\mathbf{T}u\| \le \frac{1}{2} k^2 \mu a^2 \|u\|. \tag{4.71}$$

Therefore, if $k^2\mu a^2/2 < 1$, i.e.,

$$k^2 < \frac{2}{\mu a^2}, \tag{4.72}$$

then we can solve (4.64) by successive approximations. In particular, from (4.66) we have

$$u(\mathbf{x}) = \sum_{n=0}^{\infty} (u_{n+1}(\mathbf{x}) - u_n(\mathbf{x})) = \sum_{n=0}^{\infty} (\mathbf{T}^n \exp[ik\boldsymbol{\alpha}\cdot\mathbf{y}])(\mathbf{x}),$$

where from (4.71)

$$\|\mathbf{T}^n \exp[ik\boldsymbol{\alpha}\cdot\mathbf{y}]\| \le \frac{k^n \mu^n a^{2n}}{2^n}.$$

For small k, we have the **Born approximation**

$$u(\mathbf{x}) \approx \exp[ik\boldsymbol{\alpha}\cdot\mathbf{x}] + (\mathbf{T}\exp[ik\boldsymbol{\alpha}\cdot\mathbf{y}])(\mathbf{x})$$
$$= \exp[ik\boldsymbol{\alpha}\cdot\mathbf{x}] - \frac{k^2}{4\pi}\iiint_\Omega \frac{1}{|\mathbf{x}-\mathbf{y}|}\exp[ik(|\mathbf{x}-\mathbf{y}| + \boldsymbol{\alpha}\cdot\mathbf{y})]m(\mathbf{y})\,d\mathbf{y},$$

which gives a first order correction to the incident plane wave and, for small k, an approximate solution to (4.61)–(4.63).

Now we must show uniqueness. The solution of (4.64) is unique if k satisfies (4.72). To see this, assume two solutions $u_1(\mathbf{x})$, $u_2(\mathbf{x})$. Then $v(\mathbf{x}) = u_1(\mathbf{x}) - u_2(\mathbf{x})$ satisfies $v = \mathbf{T}v$, and $v = \mathbf{T}^n v$ for every integer n. But this implies that

$$\|v\| = \|\mathbf{T}^n v\| \le \frac{k^n \mu^n a^{2n}}{2^n}\|v\|$$

for every integer n. Letting n tend to infinity and using (4.72) shows that $v(\mathbf{x})$ is identically zero.

It remains to show that the solution of (4.61)–(4.63) is unique if k satisfies (4.72), i.e., if $u \in C^2(R^3)$ satisfies

$$\Delta_3 u + k^2(1 - m(\mathbf{x}))u = 0 \qquad \text{in} \qquad R^3 \tag{4.73}$$

$$\lim_{r\to\infty} r\left(\frac{\partial u}{\partial r} - iku\right) = 0, \tag{4.74}$$

where (4.72) is valid and (4.74) holds uniformly with respect to the spherical angles θ and ϕ, then $u(\mathbf{x})$ is identically zero.

In order to show that the only solution of (4.73), (4.74) is $u(\mathbf{x})$ identically zero, we note that (4.74) implies that

$$\lim_{a\to\infty}\iint_{|\mathbf{y}|=a}\left|\frac{\partial u}{\partial r} - iku\right|^2 ds = 0, \tag{4.75}$$

and from Green's second identity (in R^3),

$$\iint_{|\mathbf{y}|=a}\left(\bar{u}\frac{\partial u}{\partial r} - u\frac{\partial \bar{u}}{\partial r}\right) ds = 0 \tag{4.76}$$

where $\overline{u(\mathbf{y})}$ is the complex conjugate of $u(\mathbf{y})$ and ds denotes an element of surface area. Equations (4.75) and (4.76) imply that

$$0 = \lim_{a\to\infty}\iint_{|\mathbf{y}|=a}\left|\frac{\partial u}{\partial r} - iku\right|^2 ds$$
$$= \lim_{a\to\infty}\iint_{|\mathbf{y}|=a}\left(\left|\frac{\partial u}{\partial r}\right|^2 + k^2|u|^2 + ik\left(\bar{u}\frac{\partial u}{\partial r} - u\frac{\partial \bar{u}}{\partial r}\right)\right) ds \tag{4.77}$$
$$= \lim_{a\to\infty}\iint_{|\mathbf{y}|=a}\left(\left|\frac{\partial u}{\partial r}\right|^2 + k^2|u|^2\right) ds.$$

From Green's second identity (in R^3), we have, in a manner exactly analogous to the analysis of Section 4.1, that for arbitrary $a > 0$

$$u(\mathbf{x}) = \frac{1}{4\pi} \iint\limits_{|\mathbf{y}|=a} \left(\frac{\exp[ik|\mathbf{x}-\mathbf{y}|]}{|\mathbf{x}-\mathbf{y}|} \frac{\partial u}{\partial r}(\mathbf{y}) - u(\mathbf{y}) \frac{\partial}{\partial r} \frac{\exp[ik|\mathbf{x}-\mathbf{y}|]}{|\mathbf{x}-\mathbf{y}|} \right) ds(\mathbf{y})$$
$$- \frac{k^2}{4\pi} \iiint\limits_{|\mathbf{y}|\le a} \frac{\exp[ik|\mathbf{x}-\mathbf{y}|]}{|\mathbf{x}-\mathbf{y}|} m(\mathbf{y})u(\mathbf{y})\, d\mathbf{y} \tag{4.78}$$

for $|\mathbf{x}| < a$, $r = |\mathbf{y}|$, and by Schwarz's inequality (cf. Section 1.4) we have

$$\frac{1}{4\pi} \left| \iint\limits_{|\mathbf{y}|=a} \left(\frac{\exp[ik|\mathbf{x}-\mathbf{y}|]}{|\mathbf{x}-\mathbf{y}|} \frac{\partial u}{\partial r}(\mathbf{y}) - u(\mathbf{y}) \frac{\partial}{\partial r} \frac{\exp[ik|\mathbf{x}-\mathbf{y}|]}{|\mathbf{x}-\mathbf{y}|} \right) ds(\mathbf{y}) \right|$$
$$\le \frac{1}{4\pi} \left[\iint\limits_{|\mathbf{y}|=a} \left(\left| \frac{\partial u}{\partial r} \right|^2 + |u|^2 \right) ds \right]^{1/2} \tag{4.79}$$
$$\cdot \left[\iint\limits_{|\mathbf{y}|=a} \left(\left| \frac{\partial}{\partial r} \frac{\exp[ik|\mathbf{x}-\mathbf{y}|]}{|\mathbf{x}-\mathbf{y}|} \right|^2 + \left| \frac{\exp[ik|\mathbf{x}-\mathbf{y}|]}{|\mathbf{x}-\mathbf{y}|} \right|^2 \right) ds(\mathbf{y}) \right]^{1/2}$$

Since from (4.77) we see that

$$\lim_{a\to\infty} \iint\limits_{|\mathbf{y}|=a} \left| \frac{\partial u}{\partial r} \right|^2 ds = \lim_{a\to\infty} \iint\limits_{|\mathbf{y}|=a} |u|^2\, ds = 0, \tag{4.80}$$

and for $r = |\mathbf{x}|$,

$$\frac{\exp[ik|\mathbf{x}-\mathbf{y}|]}{|\mathbf{x}-\mathbf{y}|} = 0\left(\frac{1}{r}\right)$$
$$\frac{\partial}{\partial r} \frac{\exp[ik|\mathbf{x}-\mathbf{y}|]}{|\mathbf{x}-\mathbf{y}|} = 0\left(\frac{1}{r}\right), \tag{4.81}$$

it follows that the left side of (4.79) tends to zero as a tends to infinity. Since $m(\mathbf{x})$ is identically zero for $|\mathbf{x}| \ge a$, we can now conclude from (4.78) (by letting a tend to infinity) that

$$u(\mathbf{x}) = -\frac{k^2}{4\pi} \iint\limits_{|\mathbf{y}|\le a} \frac{\exp[ik|\mathbf{x}-\mathbf{y}|]}{|\mathbf{x}-\mathbf{y}|} m(\mathbf{y})u(\mathbf{y})\, d\mathbf{y} \tag{4.82}$$

for $|\mathbf{x}| < a$. But for k satisfying (4.72), we have already established that the only solution of (4.82) is $u(\mathbf{x})$ identically equal to zero. Since a was arbitrary and $\mathbf{x}$ is any point such that $|\mathbf{x}| < a$, we are done.

In closing, we want to point out several aspects of our method for solving the scattering problem (4.61)–(4.63). We used a fundamental solution of the Helmholtz equation to rewrite (4.61)–(4.63) as an integral equation. In this context, the fundamental solution is often referred to as a **parametrix**. To solve the resulting integral equation, it was necessary only to show that the operator **T** was a contraction mapping; this was accomplished by making a few relatively simple estimates. The same method is applicable to any other integral equation of the form (4.64), provided the desired estimates are available. On the other hand, if (4.72) is not valid, **T** is no longer a contraction mapping, so our analysis is no longer valid. Then we must resort to more sophisticated methods for treating the integral equation. One such method is the Fredholm alternative, to be discussed in the next chapter.

Exercises

1. Let $q \in C^1(\overline{D})$, where D is a bounded, simply connected domain in the plane with C^2 boundary ∂D. Let $u \in C^2(D) \cap C^1(\overline{D})$ be the solution of

$$\Delta_2 u + q(\mathbf{x})u = 0 \qquad \text{in} \qquad D$$

$$\frac{\partial u}{\partial \nu} = f \qquad \text{on} \qquad \partial D,$$

where $\boldsymbol{\nu}$ is the unit outward normal to ∂D, and f is a prescribed continuous function on ∂D. If $q(\mathbf{x}) < 0$ for $\mathbf{x} \in \overline{D}$, show that the solution to this problem, if it exists, is unique.

2. Let D be a bounded, simply connected domain in the plane with C^2 boundary ∂D. By using the divergence theorem, show that the only solution $u \in C^2(\overline{D})$ of

$$\frac{\partial}{\partial x}\left(e^x \frac{\partial u}{\partial x}\right) + \frac{\partial}{\partial y}\left(e^y \frac{\partial u}{\partial y}\right) = 0 \qquad \text{in} \qquad D$$

$$u = 0 \qquad \text{on} \qquad \partial D$$

is $u(\mathbf{x})$ identically equal to zero.

3. Show that

$$\Delta_2 u = u^3 \qquad \text{in} \qquad x^2 + y^2 < 1$$

$$u = 0 \qquad \text{on} \qquad x^2 + y^2 = 1$$

has no twice continuously differentiable solution other than $u(x,y)$ identically zero.

4. Show that if $u(\mathbf{x})$, $\mathbf{x} \in R^n$, is a harmonic function, then so is

$$v(\mathbf{x}) = \frac{1}{|\mathbf{x}|^{n-2}}\, u\!\left(\frac{\mathbf{x}}{|\mathbf{x}|^2}\right)$$

at any point where it is defined. This is known as **Kelvin's inversion formula.**

5. Let D be a domain in R^2 and $u \in C^1(D)$ such that the integral of the normal derivative $\partial u/\partial \nu$ over any circle lying in D is eqis not zero. Show that $u(\mathbf{x})$ is harmonic in D.

6. Let $u(x,y)$ be the solution of

$$\Delta_2 u = -1 \qquad \text{in} \qquad |x| < 1,\ |y| < 1$$

$$u = 0 \qquad \text{on} \qquad |x| = 1,\ |y| = 1.$$

Find upper and lower bounds for $u(0,0)$. (Hint: Consider the function $v(x,y) = u(x,y) + \frac{1}{4}(x^2 + y^2)$.)

7. Let D be a bounded, simply connected domain in the plane with C^2 boundary ∂D. Let $u(\mathbf{x})$ be the harmonic function (assumed to exist) satisfying

$$\Delta_2 u = 0 \qquad \text{in} \qquad D$$

$$u = f \qquad \text{on} \qquad \partial D,$$

where f is a prescribed continuous function on ∂D. Show that for all functions $v \in C^1(D) \cap C(\overline{D})$ such that $v(\mathbf{x}) = f(\mathbf{x})$ for $\mathbf{x} \in \partial D$, the harmonic function $u(\mathbf{x})$ minimizes the integral

$$\iint_D |\nabla v|^2 \, d\mathbf{x}.$$

This is known as **Dirichlet's principle**.

8. Let k be a positive constant.

(a) Find all solutions $u(\mathbf{x})$ of $\Delta_2 u + k^2 u = 0$ that depend only on $r = \sqrt{x^2 + y^2}$ and are bounded at $r = 0$.

(b) Show that any solution $u(\mathbf{x})$ of $\Delta_2 u + k^2 u = 0$ in the disk $|\mathbf{x} - \mathbf{x}_0| < \rho$ such that $u(\mathbf{x}) \in C^2(D) \cap C(\overline{D})$ and $J_0(k\rho)$ is not zero satisfies the mean value property

$$u(\mathbf{x}_0) = \frac{1}{2\pi\rho J_0(k\rho)} \int_{|\mathbf{x}-\mathbf{x}_0|=\rho} u \, ds.$$

(Hint: Show that $\frac{1}{2\pi\rho} \int_{|\mathbf{x}-\mathbf{x}_0|=\rho} u \, ds$ is a solution of Bessel's equation.)

9. Let D be a domain in R^2 and define $u \in C(D)$ such that

$$u(\mathbf{x}_0) = \frac{1}{2\pi\rho} \int_{\partial\Omega} u \, ds$$

for every disk $\Omega = \{\mathbf{x}: |\mathbf{x} - \mathbf{x}_0| < \rho\}$ such that $\overline{\Omega} \subset D$. Show that $u(\mathbf{x})$ is harmonic in D. (Hint: Let $v(\mathbf{x})$ be the harmonic function defined in Ω such that $v(\mathbf{x}) = u(\mathbf{x})$ for $\mathbf{x} \in \partial\Omega$. Consider $u(\mathbf{x}) - v(\mathbf{x})$ and apply the arguments used at the beginning of the proof of the strong maximum principle for harmonic functions.)

10. Let $D = \{(x,y): x^2 + y^2 < \rho^2, y > 0\}$ and $u \in C^2(D) \cap C(\overline{D})$ a harmonic function. Show that if $u(x,0) = 0$ and for $(x,y) \in \Omega = \{(x,y): x^2 + y^2 < \rho^2\}$, we define $u(x,y)$ by

$$u(x,y) = \begin{cases} u(x,y) & (y \geq 0) \\ -u(x,-y) & (y \leq 0), \end{cases}$$

then $u(x,y)$ is harmonic in Ω. This result is known as the **Schwarz reflection principle.** (Hint: Use Exercise 9.)

11. Show that for $n \geq 1$ the Legendre polynomial $P_n(t)$ satisfies the recurrence relation

$$(2n + 1)tP_n(t) = (n + 1)P_{n+1}(t) + nP_{n-1}(t).$$

12. Derive the recurrence relation

$$P_n(t) = \frac{1}{2n + 1}(P'_{n+1}(t) - P'_{n-1}(t)) \qquad (n \geq 1)$$

for Legendre polynomials.

13. Prove Rodrigues' formula

$$P_n(t) = \frac{1}{2^n n!}\frac{d^n}{dt^n}(t^2 - 1)^n$$

for the Legendre polynomial $P_n(t)$.

14. Show that if $P_n(t)$ is a Legendre polynomial, then

$$P_n(\cos\theta) = \frac{1}{\pi}\int_0^{\pi}(\cos\theta + i\sin\theta\cos t)^n\,dt.$$

Deduce that $|P_n(t)| \leq 1$ for $-1 < t < 1$.

15. Use Rolle's theorem from calculus to show that all the roots of the equation $P_n(t) = 0$ where $P_n(t)$ is the Legendre polynomial lie in the interval $(-1,1)$.

16. Deduce the generating function for Legendre polynomials

$$(1 - 2xt + t^2)^{-1/2} = \sum_{n=0}^{\infty} t^n P_n(x) \qquad (-1 < t < 1,\ -1 \leq x \leq 1).$$

17. **(a)** Show that for the Legendre polynomial $P_n(t)$

$$\int_0^{\pi/2} P_n(\cos\theta)\sin\theta\,d\theta = \frac{P'_n(0)}{n(n + 1)}.$$

(b) Use the method of separation of variables and part (a) to find a solution $u(r,\theta,\phi)$ of Laplace's equation in the sphere $r < a$ such that

$$u(a,\theta,\phi) = \begin{cases} 1 & \left(0 < \theta < \dfrac{\pi}{2}\right) \\ 0 & \left(\dfrac{\pi}{2} < \theta < \pi\right). \end{cases}$$

(Answer: $u(r,\theta,\phi) = \dfrac{1}{2} + \dfrac{1}{2}\sum_{n=1}^{\infty}\left(\dfrac{r}{a}\right)^n P_n'(0)\dfrac{2n+1}{n(n+1)}P_n(\cos\theta)$.**)**

18. By considering $\log (R^*/R)$ where $R = |\mathbf{x} - \mathbf{y}|$, $R^* = |\mathbf{x}^* - \mathbf{y}|$, with $\mathbf{x}^*$ the mirror image of $\mathbf{x} = (x,y)$ with respect to $y = 0$, construct Green's function for Laplace's equation in the half plane $y > 0$ and derive the Poisson formula

$$u(x,y) = \frac{y}{\pi}\int_{-\infty}^{\infty} \frac{u(\xi,0)\,d\xi}{(\xi - x)^2 + y^2}.$$

19. Use the method of images to construct Green's function for Laplace's equation in $D = \{(x,y)\colon 0 < y < 1\}$. (Hint: You will need an infinite number of image points.)

20. Find an integral representation analogous to Poisson's formula in Exercise 18 for the Dirichlet problem

$$\Delta_2 u = 0 \qquad (x > 0,\ y > 0)$$

$$u(x,0) = f(x) \qquad (x \geq 0)$$

$$u(0,y) = g(y) \qquad (y \geq 0),$$

where $f(x)$ and $g(y)$ are continuous and bounded.

21. By reducing the problem to a Dirichlet problem, find Poisson's formula for the solution of the Neumann problem

$$\Delta_2 u = 0 \qquad (-\infty < x < \infty,\ y > 0)$$

$$\frac{\partial u}{\partial y}(x,0) = f(x) \qquad (-\infty < x < \infty),$$

where $f(x)$ is continuous and $|x^k f(x)|$ is bounded for some $k > 1$.

(Answer: $u(x,y) = \dfrac{1}{2\pi}\displaystyle\int_{-\infty}^{\infty} f(\xi)\log\,[(x - \xi)^2 + y^2]\,d\xi$.**)**

22. Construct the Neumann function for Laplace's equation in the disk $|\mathbf{x}| < a$.

(Answer: $N(\mathbf{x},\mathbf{y}) = \log\,(RR_1 r/a^3)$ where R, R_1, and r are defined as in Section 4.5.**)**

23. State why the Neumann problem

$$\Delta_2 u = 0 \qquad (r < 1, -\pi \le \theta \le \pi)$$

$$\frac{\partial u}{\partial r} = \sin^2\theta \qquad (-\pi \le \theta \le \pi)$$

has no solution. However, show that

$$\Delta_2 u = 1 \qquad (r < 1, -\pi \le \theta \le \pi)$$

$$\frac{\partial u}{\partial r} = \sin^2\theta \qquad (-\pi \le \theta \le \pi)$$

has a solution.

24. Use Fourier's method to solve formally the following boundary value problems where $R = \{(x,y)\colon 0 < x < a,\ 0 < y < b\}$ and $f \in C[0,a]$.

(a) $\Delta_2 u = 0 \quad \text{in} \quad R$

$$u(0,y) = u(a,y) = 0 \qquad (0 \le y \le b)$$

$$u(x,0) = 0 \qquad (0 \le x \le a)$$

$$u(x,b) = f(x) \qquad (0 \le x \le a).$$

(b) $\Delta_2 u = 0 \quad \text{in} \quad R$

$$u(0,y) = \frac{\partial u}{\partial x}(0,y) = 0 \qquad (0 \le y \le b)$$

$$u(x,0) = 0 \qquad (0 \le x \le a)$$

$$u(x,b) = f(x) \qquad (0 \le x \le a).$$

25. Use Fourier's method to solve the boundary value problem

$$\Delta_2 u = 0 \qquad (0 < x < \infty, 0 < y < b)$$

$$u(x,0) = u(x,b) = 0$$

$$u(0,y) = f(y),$$

u bounded for $0 \le x < \infty,\ 0 \le y \le b$,

where $f \in C[0,b]$.

26. By considering an appropriate boundary value problem in a square, find the harmonic function, defined in the right isosceles triangle bounded by the lines $y = 0$, $x = \pi$, $x = y$, that satisfies the boundary conditions $u(x,0) = 0$, $u(\pi,y) = f(y)$, $u(x,x) = 0$ where $f \in C[0,\pi]$.

27. Let $u(\mathbf{x})$ be harmonic in a bounded domain D in R^2. Show that for $\mathbf{x} \in D$

$$|\nabla u(\mathbf{x})| \leq \frac{2\sqrt{2}M}{d},$$

where d is the distance from $\mathbf{x}$ to ∂D and $|u(\mathbf{x})| \leq M$ for $\mathbf{x} \in \partial D$. (Hint: Show that $u(\mathbf{x}) = \dfrac{1}{\pi\rho^2} \displaystyle\iint_{|\mathbf{x}-\mathbf{y}|\leq\rho} u(\mathbf{y})\, d\mathbf{y}$ and apply this to the derivative of $u(\mathbf{x})$.)

28. Let $u_n(x,y)$ ($n = 1,2,\ldots$) be a nonincreasing sequence of harmonic functions in a domain D. Show that if $\{u_n(x,y)\}$ converges at a single point (x_0,y_0) of D, it converges for every point in D.

29. Let $u(\mathbf{x})$ be harmonic in the entire plane such that $|u(\mathbf{x})| \leq M|\mathbf{x}|$ for some positive constant M and all $\mathbf{x} \in R^2$. Show that $u(\mathbf{x}) = \mathbf{c}\cdot\mathbf{x}$ for some constant vector $\mathbf{c}$. (Hint: Use Exercise 27.)

30. Let $u(\mathbf{x})$ be the solution of

$$\Delta_2 u = \rho(\mathbf{x}) \quad \text{in} \quad \Omega$$

$$u = 0 \quad \text{on} \quad \partial\Omega,$$

where $\Omega = \{\mathbf{x}\colon |\mathbf{x}| < a\}$ and $\rho \in C^1(\overline{\Omega})$. Determine a constant M such that

$$\max_{\mathbf{x}\in\overline{\Omega}} |u(\mathbf{x})| \leq M \max_{\mathbf{x}\in\overline{\Omega}} |\rho(\mathbf{x})|.$$

(Hint: Use Green's function to represent $u(\mathbf{x})$ as an integral over Ω and show that $|G(\mathbf{x},\mathbf{y})| \leq \log(2a/|\mathbf{x}-\mathbf{y}|)$. Then deduce that $M = a^2$ suffices.)

31. Use the method of successive approximations to show that if $|\lambda| < 3$ then the solution of

$$u(x) = f(x) + \lambda \int_0^1 xtu(t)\, dt,$$

where $f \in C[0,1]$ is given by

$$u(x) = f(x) + \frac{3\lambda x}{3-\lambda}\int_0^1 tf(t)\, dt.$$

32. Consider the scattering problem (4.61)–(4.63) and define the **far field pattern** $F(\hat{x},k)$ by

$$F(\hat{\mathbf{x}};k) = \lim_{r\to\infty} re^{-ikr}u^s(\mathbf{x}),$$

where $\mathbf{x} = r\hat{\mathbf{x}}$, $|\hat{\mathbf{x}}| = 1$. Show that for small $k > 0$,

$$F(\hat{\mathbf{x}};k) = -\frac{k^2}{4\pi}\iiint_{\Omega} \exp\,[ik(\boldsymbol{\alpha} - \hat{\mathbf{x}})\cdot\mathbf{y}]\, m(\mathbf{y})\, d\mathbf{y} + 0(k^4).$$

Chapter 5

Potential Theory and Fredholm Integral Equations

In the previous chapter we used volume potentials and integral equation methods to solve the problem of the scattering of time harmonic acoustic waves by a nonhomogeneous medium. We now use *surface* potentials and integral equation methods to solve the Dirichlet and Neumann problems for Laplace's equation in an arbitrary domain and *volume* potentials and integral equation methods to solve the Dirichlet eigenvalue problem for Laplace's equation, i.e., to find constants λ such that there exist nontrivial solutions of

$$\Delta_2 u + \lambda u = 0 \qquad \text{in} \qquad D$$

$$u = 0 \qquad \text{on} \qquad \partial D,$$

where D is an arbitrary domain in the plane. For alternate methods for solving the Dirichlet and Neumann problems for Laplace's equation, see Garabedian 1964 and Epstein.

In order to accomplish these tasks, we need to establish the Fredholm alternative for integral equations with continuous kernels and the Hilbert-Schmidt theory of integral equations with singular symmetric kernels. We shall be dealing with integral equations of the *second kind,* i.e., integral equations of the form

$$\phi(s) - \lambda \int_a^b K(s,t)\phi(t)\, dt = f(s) \qquad (a \le s \le b).$$

In many applications, one also encounters integral equations of the *first kind,* i.e., of the form

$$\int_a^b K(s,t)\phi(t)\, dt = f(s) \qquad (a \le s \le b);$$

however, we shall not discuss such equations in this book, noting merely that they are improperly posed: in general, no solution exists, and if one does, it does not depend continuously on the data $f(s)$ (cf. Groetsch, Hsiao, and Nashed).

5.1 POTENTIAL THEORY

In this section we introduce the concept of a surface potential and use it to reformulate the Dirichlet and Neumann problems for Laplace's equation as an integral equation. In addition, we shall reformulate the above Dirichlet eigenvalue problem as an integral equation. In all of what follows, we assume that D is a bounded, simply connected domain in R^2 such that ∂D has continuous curvature, i.e., ∂D is in class C^2.

We first establish the **Gauss formula**

$$\int_{\partial D} \frac{\partial}{\partial \nu(\mathbf{y})} \log \frac{1}{|\mathbf{x} - \mathbf{y}|}\, ds(\mathbf{y}) = \begin{cases} -2\pi & (\mathbf{x} \in D) \\ -\pi & (\mathbf{x} \in \partial D) \\ 0 & (\mathbf{x} \in R^2 \backslash \overline{D}), \end{cases} \tag{5.1}$$

where $\boldsymbol{\nu}$ is the unit outward normal to ∂D. The equality for $\mathbf{x} \in R^2\backslash\overline{D}$ follows from the fact that here $\log 1/|\mathbf{x} - \mathbf{y}|$ is harmonic for $\mathbf{y} \in \overline{D}$. Now let $\mathbf{x} \in D$, and for

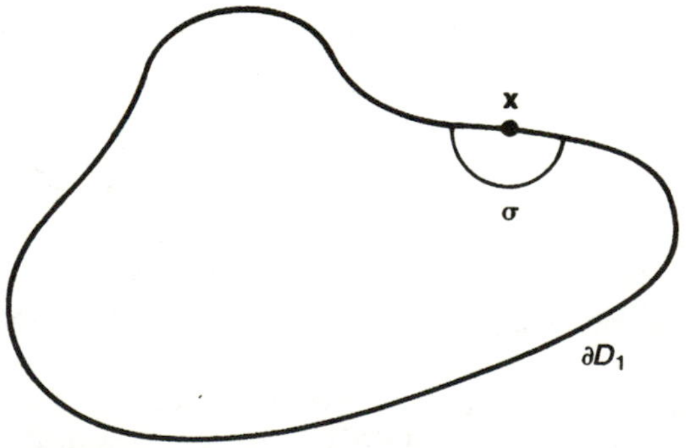

Figure 5.1

$\epsilon > 0$ sufficiently small let $\{\mathbf{y}: |\mathbf{x} - \mathbf{y}| \le \epsilon\}$ be a disk contained in D. Then

$$\int_{\partial D} \frac{\partial}{\partial \nu(\mathbf{y})} \log \frac{1}{|\mathbf{x} - \mathbf{y}|}\, ds(\mathbf{y}) = -\int_{|\mathbf{x}-\mathbf{y}|=\epsilon} \frac{\partial}{\partial \nu(\mathbf{y})} \log \frac{1}{|\mathbf{x} - \mathbf{y}|}\, ds(\mathbf{y})$$
$$= -\int_{-\pi}^{\pi} \frac{1}{\epsilon}\, \epsilon\, d\theta$$
$$= -2\pi.$$

For $\mathbf{x} \in \partial D$, let σ be the portion of the circle $\{\mathbf{y}: |\mathbf{x} - \mathbf{y}| = \epsilon\}$ lying in D, and let $\partial D_1 = \partial D \backslash (\partial D \cap \{\mathbf{y}: |\mathbf{x} - \mathbf{y}| \le \epsilon\})$.

Then

$$\int_{\partial D_1} \frac{\partial}{\partial \nu(\mathbf{y})} \log \frac{1}{|\mathbf{x} - \mathbf{y}|}\, ds(\mathbf{y}) = -\int_{\sigma} \frac{\partial}{\partial \nu(\mathbf{y})} \log \frac{1}{|\mathbf{x} - \mathbf{y}|}\, ds(\mathbf{y}),$$

and letting ϵ tend to zero gives

$$\int_{\partial D} \frac{\partial}{\partial \nu(\mathbf{y})} \log \frac{1}{|\mathbf{x} - \mathbf{y}|}\, ds(\mathbf{y}) = \int_{0}^{-\pi} \frac{1}{\epsilon}\, \epsilon\, d\theta = -\pi.$$

This proves the Gauss formula.

DEFINITION 11

Let $\psi(\mathbf{y})$ be a continuous function defined for $\mathbf{y} \in \partial D$, and let $\boldsymbol{\nu}$ be the unit outward normal to ∂D. Then

$$u(\mathbf{x}) = \frac{1}{2\pi} \int_{\partial D} \psi(\mathbf{y}) \frac{\partial}{\partial \nu(\mathbf{y})} \log \frac{1}{|\mathbf{x} - \mathbf{y}|}\, ds(\mathbf{y}) \qquad (\mathbf{x} \in R^2 \backslash \partial D)$$

is called a **double layer potential** with density $\psi(\mathbf{y})$.

Note that the double layer potential is a solution of Laplace's equation for $\mathbf{x} \in R^2 \backslash \partial D$.

We first want to show that the double layer potential is well defined for $\mathbf{x} = \mathbf{x}_0 \in \partial D$. Let s and t denote the arc length along ∂D to $\mathbf{x}_0$ and $\mathbf{y} \in \partial D$ respectively, measured from a fixed point on ∂D, and define

$$K(s,t) = \frac{\partial}{\partial \nu(\mathbf{y})} \log \frac{1}{|\mathbf{x}_0 - \mathbf{y}|} \qquad (\mathbf{x}_0 = \mathbf{x}_0(s),\ \mathbf{y} = \mathbf{y}(t)).$$

Then

$$K(s,t) = \frac{1}{|\mathbf{x}_0 - \mathbf{y}|^2} \sum_{i=1}^{2} (x_i^0 - y_i) \frac{\partial y_i}{\partial \nu(\mathbf{y})} \tag{5.2}$$
$$= \frac{\cos \phi}{|\mathbf{x}_0 - \mathbf{y}|}$$

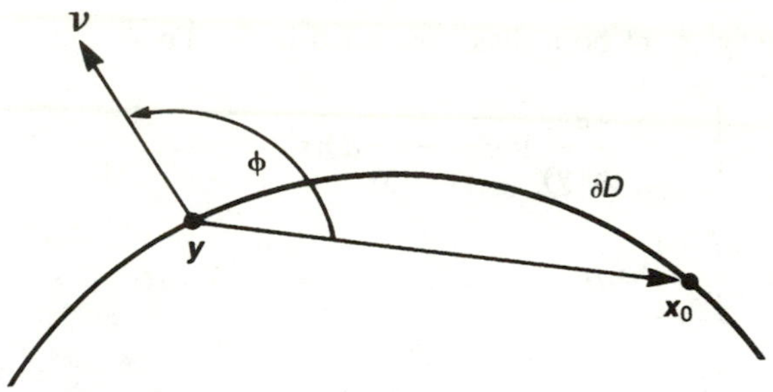

Figure 5.2

where $\mathbf{x}_0 = (x_1^0, x_2^0)$, $\mathbf{y} = (y_1, y_2)$, and

$$\cos\phi = \frac{(\mathbf{x}_0 - \mathbf{y}) \cdot \boldsymbol{\nu}(\mathbf{y})}{|\mathbf{x}_0 - \mathbf{y}|}$$

with ϕ the angle between $\boldsymbol{\nu}$ and $\mathbf{x}_0 - \mathbf{y}$. We claim that $K(s,t)$ is continuous. To see this, assume without loss of generality that $\mathbf{y} = (0,0)$ and the tangent to ∂D at $\mathbf{y}$ coincides with the x_1 axis, where (x_1,x_2) are Cartesian coordinates in R^2. Then $\boldsymbol{\nu}$ lies along the x_2 axis (see Figure 5.3). Let ∂D be described locally by $x_2 = x_2(x_1)$ for x_1 in a neighborhood of the origin. Then

$$\begin{aligned} x_2(x_1) &= x_2(0) + x_1 x_2'(0) + \frac{x_1^2}{2} x_2''(\theta x_1) \\ &= \frac{1}{2} x_1^2 x_2''(\theta x_1), \end{aligned}$$

where $0 < \theta < 1$ and

$$\begin{aligned} |\mathbf{x}_0 - \mathbf{y}| &= \sqrt{(x_1^0)^2 + (x_2^0)^2} \\ &= \sqrt{(x_1^0)^2 + \frac{1}{4}(x_1^0)^4 (x_2''(\theta x_1^0))^2} \\ &= x_1^0 \sqrt{1 + \frac{1}{4}(x_1^0)^2 (x_2''(\theta x_1^0))^2}\,. \end{aligned}$$

Hence,

$$\begin{aligned} \frac{\cos\phi}{|\mathbf{x}_0 - \mathbf{y}|} &= \frac{x_2^0}{|\mathbf{x}_0 - \mathbf{y}|^2} \\ &= \frac{x_2''(\theta x_1^0)}{2\left[1 + \frac{1}{4}(x_1^0)^2 (x_2''(\theta x_1^0))^2\right]}, \end{aligned}$$

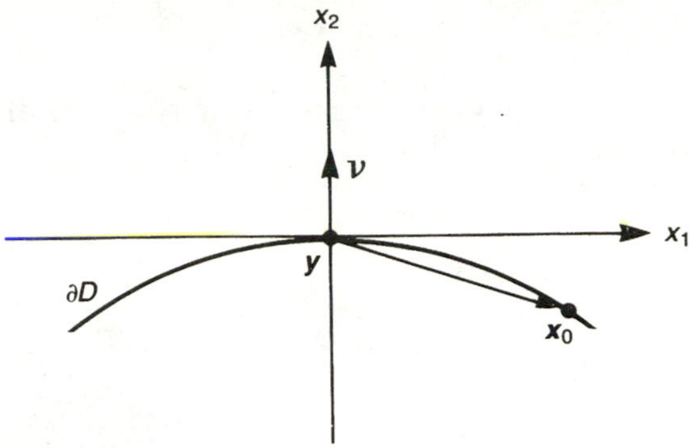

Figure 5.3

and therefore,

$$\lim_{\mathbf{x}_0 \to \mathbf{y}} \frac{\cos \phi}{|\mathbf{x}_0 - \mathbf{y}|} = \frac{1}{2} x_2''(0).$$

Recalling that the curvature of ∂D at $\mathbf{y} = \mathbf{y}(t)$ is given by

$$\begin{aligned} \kappa(t) &= \frac{x_2''(0)}{(1 + (x_2'(0))^2)^{3/2}} \\ &= x_2''(0), \end{aligned}$$

we now see that

$$\lim_{\mathbf{x}_0 \to \mathbf{y}} \frac{\cos \phi}{|\mathbf{x}_0 - \mathbf{y}|} = \frac{1}{2} \text{ curvature at } \mathbf{y}.$$

Since by assumption ∂D has continuous curvature, and $K(s,t)$ is continuous for $s \neq t$, we can now conclude that if we define $K(t,t)$ by

$$K(t,t) = \frac{1}{2} \kappa(t),$$

then $K(s,t)$ is continuous for $s,t \in [0,l]$ where l is the arc length of ∂D. Hence, the double layer potential

$$u(\mathbf{x}) = \frac{1}{2\pi} \int_{\partial D} \psi(\mathbf{y}) \frac{\partial}{\partial \nu(\mathbf{y})} \log \frac{1}{|\mathbf{x} - \mathbf{y}|} \, ds(\mathbf{y}) \tag{5.3}$$

is defined and continuous for $\mathbf{x} \in \partial D$.

We now want to deduce the continuity properties of the double layer potential (5.3) as $\mathbf{x} \in R^2 \backslash \partial D$ tends to a point $\mathbf{x}_0 \in \partial D$. Let $v \in C^2(\overline{D})$, where v will be defined more precisely in the sequel. Then from Green's first identity and the analysis used in the derivation of Gauss' formula, we can deduce in a manner which we hope is by now familiar that

$$\iint_D \nabla v(\mathbf{y})\nabla_{\mathbf{y}} \log \frac{1}{|\mathbf{x}-\mathbf{y}|}\, d\mathbf{y} + pv(\mathbf{x}) = \int_{\partial D} v(\mathbf{y})\frac{\partial}{\partial \nu(\mathbf{y})} \log \frac{1}{|\mathbf{x}-\mathbf{y}|}\, ds(\mathbf{y}), \tag{5.4}$$

where

$$p = \begin{cases} -2\pi & (\mathbf{x} \in D) \\ -\pi & (\mathbf{x} \in \partial D) \\ 0 & (\mathbf{x} \in R^2 \backslash \overline{D}). \end{cases}$$

Similarly,

$$\begin{aligned} pv(\mathbf{x}) = & \iint_D \Delta_2 v(\mathbf{y}) \log \frac{1}{|\mathbf{x}-\mathbf{y}|}\, d\mathbf{y} \\ & + \int_{\partial D} \left(v(\mathbf{y}) \frac{\partial}{\partial \nu(\mathbf{y})} \log \frac{1}{|\mathbf{x}-\mathbf{y}|} - \log \frac{1}{|\mathbf{x}-\mathbf{y}|} \frac{\partial v}{\partial \nu}(\mathbf{y}) \right) ds(\mathbf{y}). \end{aligned} \tag{5.5}$$

Now consider the double layer potential (5.3), and assume for the moment that $\psi \in C^2(\partial D)$. Then ψ can be continued into $\overline{D}$ as a twice continuously differentiable function. Define

$$u^+(\mathbf{x}_0) = \lim_{\substack{\mathbf{x}\to\mathbf{x}^0 \\ \mathbf{x}\in D^+}} u(\mathbf{x})$$

$$u^-(\mathbf{x}_0) = \lim_{\substack{\mathbf{x}\to\mathbf{x}^0 \\ \mathbf{x}\in D^-}} u(\mathbf{x}),$$

where $D^+ = D$, $D^- = R^2\backslash\overline{D}$, $\mathbf{x}_0 \in \partial D$, and set $v(\mathbf{x}) = \psi(\mathbf{x})$ in (5.4). From our study of Poisson's equation, we know that the integral over D in (5.4) is continuous, so

$$\begin{aligned} u^+(\mathbf{x}_0) - u(\mathbf{x}_0) &= -\frac{1}{2}\psi(\mathbf{x}_0) \\ u^-(\mathbf{x}_0) - u(\mathbf{x}_0) &= \frac{1}{2}\psi(\mathbf{x}_0). \end{aligned} \tag{5.6}$$

Returning to (5.5), from our study of Poisson's equation, we again see that the integral over D in (5.5) is continuously differentiable as $\mathbf{x}$ passes from D^+ to D^- through $\mathbf{x}_0$. Assume that $v(\mathbf{x})$ has been chosen such that $v \in C^2(\overline{D})$, $v(\mathbf{x}) = \psi(\mathbf{x})$ for $\mathbf{x} \in \partial D$, and $\partial v(\mathbf{x})/\partial \nu = 0$ for $\mathbf{x} \in \partial D$. Then the derivatives of the double layer potential have the same discontinuity properties as $pv(\mathbf{x})/2\pi$, i.e.,

$$\frac{\partial u^+}{\partial \nu}(\mathbf{x}_0) = \frac{\partial u^-}{\partial \nu}(\mathbf{x}_0), \tag{5.7}$$

where

$$\frac{\partial u^+}{\partial \nu}(\mathbf{x}_0) = \lim_{\substack{\mathbf{x}\to\mathbf{x}^0 \\ \mathbf{x}\in D^+}} \frac{\partial u}{\partial \nu}(\mathbf{x})$$

$$\frac{\partial u^-}{\partial \nu}(\mathbf{x}_0) = \lim_{\substack{\mathbf{x}\to\mathbf{x}^0 \\ \mathbf{x}\in D^-}} \frac{\partial u}{\partial \nu}(\mathbf{x}).$$

Therefore, the derivative of the double layer potential in the direction normal to ∂D varies continuously when $\mathbf{x}$ passes through the boundary along the normal at $\mathbf{x}_0$.

These continuity properties remain valid when $\psi(\mathbf{x})$ is assumed only to be continuous for $\mathbf{x} \in \partial D$ (cf. Colton and Kress). Indeed, (5.6) and (5.7) are also valid when $\psi(\mathbf{x})$ is assumed to be square integrable on ∂D, and in this case (5.6) and (5.7) hold for all $\mathbf{x}_0 \in \partial D$, except possibly for a set of points that can be covered by a sequence of open segments of ∂D whose total length is arbitrarily small (Kersten). In this case (5.6) and (5.7) are said to hold **almost everywhere**. In particular, it can be shown that

$$\int_{\partial D} |f(s)|^2\, ds = 0$$

if and only if $f(s) = 0$ almost everywhere on ∂D.

We can now reformulate the Dirichlet problem for Laplace's equation as an integral equation. Consider the interior Dirichlet problem

$$\begin{aligned} \Delta_2 u &= 0 \quad \text{in} \quad D \\ u &= f \quad \text{on} \quad \partial D, \end{aligned} \tag{5.8}$$

where $u \in C^2(D) \cap C(\overline{D})$, $f \in C(\partial D)$, and ∂D has continuous curvature. We look for a solution of (5.8) in the form of a double layer potential (5.3). Letting $\mathbf{x}$ tend to $\mathbf{x}_0 \in \partial D$ and using (5.6) leads to the integral equation

$$\psi(\mathbf{x}_0) - \frac{1}{\pi}\int_{\partial D} \psi(\mathbf{y}) \frac{\partial}{\partial \nu(\mathbf{y})} \log \frac{1}{|\mathbf{x}_0 - \mathbf{y}|}\, ds(\mathbf{y}) = -2f(\mathbf{x}_0) \qquad (\mathbf{x}_0 \in \partial D) \tag{5.9}$$

for the unknown density $\psi(\mathbf{x})$. As previously shown, the kernel

$$\frac{\partial}{\partial \nu(\mathbf{y})} \log \frac{1}{|\mathbf{x}_0 - \mathbf{y}|}$$

is continuous for $\mathbf{x}_0, \mathbf{y} \in \partial D$. Thus, the Dirichlet problem (5.8) reduces to the problem of solving the integral equation (5.9) with continuous kernel. We shall study this problem in the next section.

The above considerations apply to the *interior* Dirichlet problem. For the *exterior* Dirichlet problem

$$\begin{aligned} \Delta_2 u &= 0 \quad \text{in} \quad R^2\backslash\overline{D} \\ u &= f \quad \text{on} \quad \partial D, \end{aligned}$$

we are led by the same considerations to the integral equation

$$\psi(\mathbf{x}_0) + \frac{1}{\pi}\int_{\partial D} \psi(\mathbf{y}) \frac{\partial}{\partial \nu(\mathbf{y})} \log \frac{1}{|\mathbf{x}_0 - \mathbf{y}|} \, ds(\mathbf{y}) = 2f(\mathbf{x}_0) \qquad (\mathbf{x}_0 \in \partial D). \tag{5.10}$$

We are assuming that the solution to the interor or exterior Dirichlet problem can be represented in the form of a double layer potential; but there is no guarantee that this is possible, and it may turn out that the above integral equations are not soluble for an arbitrary continuous function $f(\mathbf{x})$. Of course, if the above integral equations are soluble, we have constructed a solution to the corresponding Dirichlet problem. We postpone until Section 5.3 the answer to whether the integral equations derived above are soluble.

We now turn our attention to the interior Neumann problem

$$\begin{aligned} \Delta_2 u &= 0 \qquad \text{in} \qquad D \\ \frac{\partial u}{\partial \nu} &= g \qquad \text{on} \qquad \partial D, \end{aligned} \tag{5.11}$$

where $u \in C^2(D) \cap C^1(\overline{D})$, $g \in C(\partial D)$, and as usual $\boldsymbol{\nu}$ denotes the unit outward normal to ∂D. Our aim again is to reformulate this problem as an integral equation.

DEFINITION 12

Let $\psi(\mathbf{y})$ be a continuous function defined for $\mathbf{y} \in \partial D$. Then

$$u(\mathbf{x}) = \frac{1}{2\pi}\int_{\partial D} \psi(\mathbf{y}) \log \frac{1}{|\mathbf{x} - \mathbf{y}|} \, ds(\mathbf{y}) \qquad (\mathbf{x} \in R^2 \backslash \partial D)$$

is called a **single layer potential** with density $\psi(\mathbf{y})$.

Note that the single layer potential is a solution of Laplace's equation for $\mathbf{x} \in R^2\backslash\partial D$.

In order to derive the continuity properties of the single layer potential, we temporarily assume that $\psi \in C^1(\partial D)$. Then in (5.5) we choose $v(\mathbf{x})$ such that $v \in C^2(\overline{D})$, $v(\mathbf{x}) = 0$ for $\mathbf{x} \in \partial D$, $\partial v(\mathbf{x})/\partial \nu = \psi(\mathbf{x})$ for $\mathbf{x} \in \partial D$, to conclude that the single layer potential is continuous for $\mathbf{x} \in R^2$. On the other hand, (5.5) also shows that the normal derivative of the single layer potential has the same discontinuity properties as $(-p/2\pi)(\partial v(\mathbf{x})/\partial \nu)$, i.e.,

$$\begin{aligned} \frac{\partial u^+}{\partial \nu}(\mathbf{x}_0) - \frac{\partial u}{\partial \nu}(\mathbf{x}_0) &= \frac{1}{2}\psi(\mathbf{x}_0) \\ \frac{\partial u^-}{\partial \nu}(\mathbf{x}_0) - \frac{\partial u}{\partial \nu}(\mathbf{x}_0) &= -\frac{1}{2}\psi(\mathbf{x}_0), \end{aligned} \tag{5.12}$$

where $\mathbf{x}_0 \in \partial D$ and

$$\frac{\partial u}{\partial \nu}(\mathbf{x}_0) = \frac{1}{2\pi} \int_{\partial D} \psi(\mathbf{y}) \frac{\partial}{\partial \nu(\mathbf{x}_0)} \log \frac{1}{|\mathbf{x}_0 - \mathbf{y}|} \, ds(\mathbf{y}).$$

We now look for a solution of the interior Neumann problem (5.11) in the form of a single layer potential. Letting $\mathbf{x} \in D$ tend to $\mathbf{x}_0 \in \partial D$, we see from (5.12) that $\psi(\mathbf{x})$ must satisfy the integral equation

$$\psi(\mathbf{x}_0) + \frac{1}{\pi} \int_{\partial D} \psi(\mathbf{y}) \frac{\partial}{\partial \nu(\mathbf{x}_0)} \log \frac{1}{|\mathbf{x}_0 - \mathbf{y}|} \, ds(\mathbf{y}) = 2g(\mathbf{x}_0) \qquad (\mathbf{x}_0 \in \partial D). \tag{5.13}$$

Note that since the kernel

$$K^*(s,t) = \frac{\partial}{\partial \nu(\mathbf{x}_0)} \log \frac{1}{|\mathbf{x}_0 - \mathbf{y}|} \qquad (\mathbf{x}_0 = \mathbf{x}_0(s),\ \mathbf{y} = \mathbf{y}(t))$$

satisfies $K^*(s,t) = K(t,s)$, where $K(s,t)$ is the kernel of the double layer potential evaluated on ∂D, we can conclude that $K^*(s,t)$ is continuous for $s,t \in [0,l]$ where l is the arc length of ∂D.

We again note that (5.12) and (5.13) remain valid if $\psi(\mathbf{x})$ is assumed only to be continuous for $\mathbf{x} \in \partial D$, and if $\psi(\mathbf{x})$ is assumed only to be square integrable for $\mathbf{x} \in \partial D$ then (5.12) and (5.13) hold almost everywhere in the sense previously defined.

From (5.12) we see that the integral equation corresponding to (5.13) for the *exterior* Neumann problem

$$\Delta_2 u = 0 \quad \text{in} \quad R^2 \setminus \overline{D}$$

$$\frac{\partial u}{\partial \nu} = g \quad \text{on} \quad \partial D$$

is

$$\psi(\mathbf{x}_0) - \frac{1}{\pi} \int_{\partial D} \psi(\mathbf{y}) \frac{\partial}{\partial \nu(\mathbf{x}_0)} \log \frac{1}{|\mathbf{x}_0 - \mathbf{y}|} \, ds(\mathbf{y}) = -2g(\mathbf{x}_0) \qquad (\mathbf{x}_0 \in \partial D). \tag{5.14}$$

We again postpone until Section 5.3 the question of whether the solution of the interior or exterior Neumann problem can be represented as a single layer potential.

In all of these integral equations we have assumed that $f \in C(\partial D)$ or $g \in C(\partial D)$. Since the kernels of these integral equations are continuous, we know that if $\psi(\mathbf{x})$ is a square integrable solution, then in fact $\psi(\mathbf{x})$ must be continuous. Therefore, it suffices to seek a solution of the integral equations in the class of square integrable functions defined on ∂D.

We now consider the Dirichlet eigenvalue problem

$$\begin{aligned} \Delta_2 u + \lambda u &= 0 \quad \text{in} \quad D \\ u &= 0 \quad \text{on} \quad \partial D, \end{aligned} \tag{5.15}$$

where we seek an eigenvalue λ and eigenfunction $u \in C^2(D) \cap C(\overline{D})$ such that (5.15) is satisfied and $u(\mathbf{x})$ is not identically zero. By considering the Neumann function instead of Green's function as below, we can also treat the corresponding Neumann eigenvalue problem; however, in this book we shall only consider the Dirichlet eigenvalue problem.

We begin by noting that if we can solve (5.9) for

$$f(\mathbf{x}_0) = -\log \frac{1}{|\mathbf{x}_0 - \mathbf{y}|} \qquad (\mathbf{x}_0 \in \partial D, \mathbf{y} \in D),$$

then we can construct Green's function $G(\mathbf{x},\mathbf{y})$ for Laplace's equation in D. That this can indeed be done will be shown in Section 5.3. From our study of Poisson's equation, we see that if $u \in C^1(D) \cap C(\overline{D})$ is a solution of the integral equation

$$u(\mathbf{x}) = \frac{\lambda}{2\pi} \iint_D G(\mathbf{x},\mathbf{y})u(\mathbf{y})\, d\mathbf{y} \qquad (\mathbf{x} \in D), \tag{5.16}$$

then $u(\mathbf{x})$ is a solution of $\Delta_2 u + \lambda u = 0$ in D. More generally, we shall show that if $u(\mathbf{x})$ is a square integrable solution of (5.16), then $u \in C(\overline{D})$, and from our analysis of Poisson's equation $u \in C^1(D)$. Hence, a square integrable solution of (5.16) will be a solution of (5.15) provided we can show that $u \in C(\overline{D})$ and

$$\lim_{\substack{\mathbf{x}\to\mathbf{x}^0 \\ \mathbf{x}\in D}} u(\mathbf{x}) = 0 \qquad (\mathbf{x}_0 \in \partial D). \tag{5.17}$$

(Although $G(\mathbf{x},\mathbf{y}) = 0$ for $\mathbf{x} \in \partial D$, $\mathbf{y} \in D$, it is not clear how $G(\mathbf{x},\mathbf{y})$ behaves as both $\mathbf{x}$ and $\mathbf{y}$ approach ∂D, so (5.17) is not immediately obvious.) Conversely, it is possible to show (cf. Vekua) that since ∂D has continuous curvature, every solution of (5.15) is in fact continuously differentiable in $\overline{D}$, and for $\mathbf{x} \in D$, $G(\mathbf{x},\mathbf{y}) \in C^1(\overline{D}) \cap C^2(D)$ as a function of $\mathbf{y}$. This is sufficient to apply Green's second identity (applied to $D_n \subset D$ and then letting D_n tend to D) to conclude that for $\mathbf{x} \in D$, a solution of (5.15) is

$$\begin{aligned} u(\mathbf{x}) &= \frac{1}{2\pi} \int_{\partial D} \left(\frac{\partial u}{\partial \nu}(\mathbf{y}) G(\mathbf{x},\mathbf{y}) - u(\mathbf{y}) \frac{\partial}{\partial \nu} G(\mathbf{x},\mathbf{y}) \right) ds(\mathbf{y}) \\ &\quad - \frac{1}{2\pi} \iint_D G(\mathbf{x},\mathbf{y}) \Delta_2 u(\mathbf{y})\, d\mathbf{y} \\ &= \frac{\lambda}{2\pi} \iint_D G(\mathbf{x},\mathbf{y}) u(\mathbf{y})\, d\mathbf{y}. \end{aligned}$$

Therefore, $u(\mathbf{x})$ satisfies the integral equation (5.16). Hence the integral equation (5.16) defined in the space of square integrable functions on D is equivalent to (5.15) provided we can show that square integrable solutions of (5.16) are continuous in $\overline{D}$ and (5.17) is valid. We shall now show that these facts are true.

Recall that for $\mathbf{x} \in D$,

$$G(\mathbf{x},\mathbf{y}) = \log \frac{1}{|\mathbf{x} - \mathbf{y}|} + g(\mathbf{x},\mathbf{y}),$$

where $g(\mathbf{x},\mathbf{y})$ is harmonic in D and continuous in $\overline{D}$. Hence, for $\mathbf{x}_1,\mathbf{x}_2 \in D$ and $u(\mathbf{x})$ a square integrable solution of (5.16), we have

$$\begin{aligned} |u(\mathbf{x}_1) - u(\mathbf{x}_2)|^2 &= \left| \frac{\lambda}{2\pi} \iint_D (G(\mathbf{x}_1,\mathbf{y}) - G(\mathbf{x}_2,\mathbf{y}))u(\mathbf{y})\, d\mathbf{y} \right|^2 \\ &\leq \frac{|\lambda|}{2\pi} \iint_D |G(\mathbf{x}_1,\mathbf{y}) - G(\mathbf{x}_2,\mathbf{y})|^2\, d\mathbf{y} \iint_D |u(\mathbf{y})|^2\, d\mathbf{y} \end{aligned} \tag{5.18}$$

by Schwarz's inequality (cf. Section 1.4). Since the integral of $(\log |\mathbf{x} - \mathbf{y}|)^2$ is uniformly convergent, we can conclude from (5.18) that $u(\mathbf{x})$ is continuous in D. In order to show that $u(\mathbf{x})$ is continuous in $\overline{D}$ it now suffices to show (5.17).

To show that (5.17) is valid, we first need to give upper and lower bounds to Green's function $G(\mathbf{x},\mathbf{y})$. Let $\mathbf{y} \in D$ and h be the diameter of D. Then for $\mathbf{x} \in \partial D$, the regular part of $G(\mathbf{x},\mathbf{y})$ satisfies

$$g(\mathbf{x},\mathbf{y}) = -\log \frac{1}{|\mathbf{x} - \mathbf{y}|} \leq \log h, \tag{5.19}$$

and hence by the maximum principle (5.19) is valid for $\mathbf{x} \in \overline{D}$. Now let $\epsilon > 0$ be such that $\Omega = \{\mathbf{x}: |\mathbf{x} - \mathbf{y}| < \epsilon\}$ is contained in D, and consider $G(\mathbf{x},\mathbf{y})$ for $\mathbf{x} \in D\backslash\Omega$. Since $G(\mathbf{x},\mathbf{y}) = 0$ for $\mathbf{x} \in \partial D$ and for ϵ sufficiently small $G(\mathbf{x},\mathbf{y}) > 0$ for $\mathbf{x} \in \partial\Omega$, we can conclude by the maximum principle that $G(\mathbf{x},\mathbf{y}) \geq 0$ for $\mathbf{x} \in \overline{D}\backslash\Omega$. Since ϵ can be arbitrarily small,

$$G(\mathbf{x},\mathbf{y}) \geq 0 \qquad (\mathbf{x} \in \overline{D},\ \mathbf{y} \in D). \tag{5.20}$$

Putting (5.19) and (5.20) together gives

$$0 \leq G(\mathbf{x},\mathbf{y}) = \log \frac{1}{|\mathbf{x} - \mathbf{y}|} + g(\mathbf{x},\mathbf{y}) \leq \log \frac{h}{|\mathbf{x} - \mathbf{y}|} \qquad (\mathbf{x} \in \overline{D},\ \mathbf{y} \in D). \tag{5.21}$$

Now let D_δ be a domain whose boundary is a distance δ from ∂D measured along the inner normal. Then since the logarithm is uniformly square integrable, we have that for $\mathbf{x} \in \overline{D}$,

$$\begin{aligned} \left| \iint_D G(\mathbf{x},\mathbf{y})u(\mathbf{y})\, d\mathbf{y} \right|^2 &\leq \iint_D |G(\mathbf{x},\mathbf{y})|^2\, d\mathbf{y} \iint_D |u(\mathbf{y})|^2\, d\mathbf{y} \\ &\leq \iint_{D_\delta} |G(\mathbf{x},\mathbf{y})|^2\, d\mathbf{y} \iint_D |u(\mathbf{y})|^2\, d\mathbf{y} + \epsilon(\delta,\mathbf{x}), \end{aligned}$$

where

$$\epsilon(\delta,\mathbf{x}) = \iint_{D\backslash D_\delta} \left|\log\frac{h}{|\mathbf{x}-\mathbf{y}|}\right|^2 d\mathbf{y} \iint_D |u(\mathbf{y})|^2\, d\mathbf{y}.$$

Note that $\epsilon(\delta,\mathbf{x})$ tends to zero as δ tends to zero uniformly for $\mathbf{x} \in \overline{D}$ and is continuous for $\mathbf{x} \in \overline{D}$. Hence, for $\mathbf{x}_0 \in \partial D$,

$$\lim_{\mathbf{x}\to\mathbf{x}_0} \left|\iint_D G(\mathbf{x},\mathbf{y})u(\mathbf{y})\, d\mathbf{y}\right|^2 \tag{5.22}$$
$$\leq \lim_{\mathbf{x}\to\mathbf{x}_0} \iint_{D_\delta} |G(\mathbf{x},\mathbf{y})|^2\, d\mathbf{y} \iint_D |u(\mathbf{y})|^2\, d\mathbf{y} + \epsilon(\delta,\mathbf{x}_0) = \epsilon(\delta,\mathbf{x}_0),$$

and since $\epsilon(\delta,\mathbf{x}_0)$ can be made arbitrarily small for δ sufficiently small we can conclude that (5.17) is valid.

We have established that the eigenvalue problem (5.15) is equivalent to finding a nontrivial square integrable solution of (5.16). This problem will be studied in Section 5.4, where we will use the fact that Green's function is symmetric (cf. Section 4.5), i.e., $G(\mathbf{x},\mathbf{y}) = G(\mathbf{y},\mathbf{x})$, and satisfies the inequalities (5.21).

5.2 The Fredholm Alternative

In the previous section we showed how the Dirichlet and Neumann problems for Laplace's equation can be reformulated as a Fredholm integral equation of the second kind, i.e., an integral equation of the form

$$\phi(s) - \int_0^l K(s,t)\phi(t)\, dt = f(s) \qquad (0 \leq s \leq l) \tag{5.23}$$

for an unknown function $\phi(s)$, where l is a real number. (In particular, l is the arc length of ∂D and $\phi(s) = \psi(\mathbf{x}(s))$, $\mathbf{x} \in \partial D$, where s denotes the arc length measured along ∂D.) Motivated by the analysis of the previous section, we assume that the kernel $K(s,t)$ is continuous for s, t in the square $0 \leq s \leq l$, $0 \leq t \leq l$. We further assume that $f(s)$ is square integrable over $0 \leq s \leq l$ and look for a square integrable solution $\phi(s)$ of (5.23). For the purpose of future applications in Chapter 6, we assume that $K(s,t)$ and $f(s)$ may be complex valued. Our aim is to prove the **Fredholm alternative** for (5.23), which roughly states that if the homogeneous equation $f(s) = 0$ has only the trivial solution $\phi(s) = 0$, then (5.23) has a solution for any square integrable right side.

We begin by introducing some notation and definitions. We define the operator $\mathbf{K}$ by

$$\mathbf{K}\phi = \int_0^l K(s,t)\phi(t)\, dt,$$

and so (5.23) becomes

$$(\mathbf{I} - \mathbf{K})\phi = f. \tag{5.24}$$

(Henceforth, when considering operator equations of the form (5.24), we shall, in conformance with the standard notation of functional analysis, indicate square integrable functions by $\phi, f, \ldots$ instead of $\phi(s), f(s), \ldots$.) The **adjoint** of the integral equation (5.23) is

$$\phi(s) - \int_0^l \overline{K(t,s)}\psi(t)\,dt = g(s),$$

where $g(s)$ is a square integrable function or, in operator notation,

$$(\mathbf{I} - \mathbf{K}^*)\psi = g,$$

where

$$\mathbf{K}^*\psi = \int_0^l \overline{K(t,s)}\psi(t)\,dt. \tag{5.25}$$

Finally, we remind the reader that the inner product is defined as

$$(\phi,\psi) = \int_0^l \phi(t)\overline{\psi(t)}\,dt$$

and the norm as

$$\|\phi\| = \sqrt{(\phi,\phi)}.$$

We now investigate the method of successive approximations (cf. Sections 2.2 and 4.8) for solving (5.23) or (5.24). If we define

$$\phi_{n+1} = f + \mathbf{K}\phi_n \qquad (n = 0,1,2,\ldots)$$
$$\phi_0 = 0,$$

then, when the sequence $\{\phi_n\}$ converges (i.e., there exists a square integrable function ϕ such that $\lim_{n\to\infty} \|\phi_n - \phi\| = 0$), the **Neumann series**

$$\begin{aligned} \phi &= \sum_{k=0}^{\infty} (\phi_{k+1} - \phi_k) \\ &= f + \mathbf{K}f + \mathbf{K}^2 f + \mathbf{K}^3 f + \ldots \end{aligned} \tag{5.26}$$

provides a solution of (5.24). In order to establish when this series is convergent, we define the constant $B > 0$ by

$$B^2 = \int_0^l \int_0^l |K(s,t)|^2\,dt\,ds. \tag{5.27}$$

Then Schwarz's inequality (cf. Section 1.4) shows that

$$|(\mathbf{K}f)(s)|^2 \leq \|f\|^2 \int_0^l |K(s,t)|^2\, dt$$

and integrating with respect to s gives

$$\|\mathbf{K}f\| \leq \|f\|B.$$

By induction, we have

$$\|\mathbf{K}^n f\| \leq \|f\|B^n. \tag{5.28}$$

By the triangle inequality (cf. Section 1.4) and (5.28), (5.26) is majorized by the geometric series

$$\|f\|(1 + B + B^2 + \ldots).$$

Hence, (5.26) is convergent to a square integrable function ϕ, provided this geometric series converges. We have made use of the fact from Lebesgue integration theory that the space integrable functions is complete, that is, every Cauchy sequence of square integrable functions converges to a square integrable function. For details, consult any book on Lebesgue integration. An easily accessible text (claimed by the author to be understandable by a senior in high school!) which we recommend is Weir. We now present the following definition and theorem:

DEFINITION 13

A sequence $\{\phi_n\}$ **converges in the mean** to a function ϕ provided

$$\lim_{n\to\infty} \|\phi_n - \phi\| = 0.$$

THEOREM 38

If f is square integrable and $B < 1$, the Neumann series (5.26) converges in the mean to a square integrable function ϕ that satisfies (5.24) almost everywhere (recall the definition in the previous section).

This theorem says only that (5.23) or (5.24) is satisfied almost everywhere, since we are considering mean convergence, i.e.,

$$\left\| \phi - \sum_{n=0}^{\infty} \mathbf{K}^n f \right\| = 0.$$

If $f(s)$ is continuous, then from (5.23), $\phi(s)$ is also; in this case (5.23) is satisfied pointwise. We have previously shown in a slightly different context (Section 4.8) that the solution obtained by the method of successive approximations is unique. Hence, we can introduce the **resolvent operator** defined as

$$(\mathbf{I} - \mathbf{K})^{-1} = \mathbf{I} + \mathbf{K} + \mathbf{K}^2 + \ldots \tag{5.29}$$

and represent the solution of (5.24) in the form

$$\phi = (\mathbf{I} - \mathbf{K})^{-1} f,$$

provided $B < 1$.

We now drop the assumption that $B < 1$. We shall first derive *necessary* conditions for a solution of (5.24) to exist. Suppose there exists a nontrivial solution, or eigenfunction, of the homogeneous adjoint equation

$$(\mathbf{I} - \mathbf{K}^*)\psi = 0. \tag{5.30}$$

Then multiplying (5.24) by $\bar{\psi}$ and integrating gives

$$\begin{aligned}(f,\psi) &= (\phi - \mathbf{K}\phi,\psi) \\ &= \int_0^l \overline{\psi(s)}\left[\phi(s) - \int_0^l K(s,t)\phi(t)\,dt\right] ds \\ &= \int_0^l \phi(s)\left[\overline{\psi(s)} - \int_0^l K(t,s)\overline{\psi(t)}\,dt\right] ds \\ &= (\phi,\psi - \mathbf{K}^*\psi)\end{aligned}$$

and from (5.30) we have

$$(f,\psi) = 0. \tag{5.31}$$

Therefore, a necessary condition for (5.24) to be soluble is that $(f,\psi) = 0$ for every nontrivial solution of the adjoint equation (5.30), i.e., f is orthogonal to all of the eigenfunctions of the adjoint equation. The Fredholm alternative, which we are about to prove, states that this necessary condition is also sufficient!

THEOREM 39 Fredholm Alternative

Either

$$\phi - \mathbf{K}\phi = 0 \tag{5.32}$$

and

$$\psi - \mathbf{K}^*\psi = 0 \tag{5.33}$$

have only the trivial solutions $\phi = \psi = 0$, or they have the same number of linearly independent solutions $\phi_1,\phi_2,\ldots,\phi_m$ and $\psi_1,\psi_2,\ldots,\psi_m$, where m is an integer. In the first case,

$$\phi - \mathbf{K}\phi = f \tag{5.34}$$

and

$$\psi - \mathbf{K}^*\psi = g \tag{5.35}$$

have unique square integrable solutions ϕ and ψ for every square integrable function f and g. In the second case, (5.34) has a solution if and only if f is orthogonal to the m solutions of (5.33), and (5.35) has a solution if and only if g is orthogonal to the m solutions of (5.32).

■ ***Proof.*** We shall first prove the theorem for **degenerate kernels,** i.e., kernels of the form

$$K_n(s,t) = \sum_{j=1}^{n} \alpha_j(s)\overline{\beta_j(t)},$$

where $\alpha_j, \beta_j \in C[0,l]$ and the $\beta_j(t)$ are linearly independent. Without loss of generality, we can assume that the $\beta_j(t)$ are orthonormal, that is, $(\beta_i,\beta_j) = 0$ for $i \neq j$ and $\|\beta_i\| = 1$, since if this were not true we could apply the **Gram-Schmidt method** to construct the orthogonal basis $\tilde{\beta}_j(t)$ defined by

$$\tilde{\beta}_1(t) = \beta_1(t)$$

$$\tilde{\beta}_m(t) = \beta_m(t) - \sum_{k=1}^{m-1} \frac{(\beta_m,\tilde{\beta}_k)}{\|\tilde{\beta}_k\|^2}\,\tilde{\beta}_k(t)$$

and then orthonormalize by dividing each element $\tilde{\beta}_j(t)$ by its norm. Having proved the theorem for degenerate kernels, we shall then prove the general result by approximating the kernel by a degenerate kernel and applying the method of successive approximations to the difference of these two kernels.

To treat the case of degenerate kernels, we define the operator $\mathbf{K}_n$ by

$$\mathbf{K}_n\phi = \int_0^l K_n(s,t)\phi(t)\,dt$$

and consider the Fredholm integral equation (written in operator notation)

$$\phi - \mathbf{K}_n\phi = f. \tag{5.36}$$

The homogeneous adjoint equation

$$\psi - \mathbf{K}_n^*\psi = 0 \tag{5.37}$$

can be written in the form

$$\psi - \sum_{i=1}^{n} (\psi,\alpha_i)\beta_i = 0, \tag{5.38}$$

so all solutions of (5.37) are linear combinations of the β_i. We now write f as the sum

$$f = f_1 + f_2,$$

where f_1 is a linear combination of the β_i and $(f_2,\beta_i) = 0$ for every integer $i = 1,2,\ldots,n$. In particular,

$$f_1 = \sum_{i=1}^{n} (f,\beta_i)\beta_i$$

$$f_2 = f - f_1.$$

Since

$$\mathbf{K}_n f_2 = \sum_{j=1}^{n} (f_2,\beta_j)\alpha_j = 0,$$

setting $\phi = \phi_1 + f_2$ in (5.36) gives

$$\phi_1 - \sum_{j=1}^{n} (\phi_1,\beta_j)\alpha_j = f_1. \tag{5.39}$$

Therefore, it suffices to consider (5.36) in the special case when $f = f_1$ and $\phi = \phi_1$, i.e., (5.39). Multiplying (5.39) by $\bar{\beta}_i$ and integrating gives

$$x_i - \sum_{j=1}^{n} a_{ij}x_j = y_i \qquad (i = 1,2,\ldots,n), \tag{5.40}$$

where

$$\begin{aligned} x_i &= (\phi_1,\beta_i) = (\phi - f_2,\beta_i) = (\phi,\beta_i) - (f_2,\beta_i) = (\phi,\beta_i) \\ y_i &= (f_1,\beta_i) = (f - f_2,\beta_i) = (f,\beta_i) - (f_2,\beta_i) = (f,\beta_i) \\ a_{ij} &= (\alpha_j,\beta_i). \end{aligned} \tag{5.41}$$

If we can solve the linear algebraic system (5.40), then a solution of (5.39) is

$$\phi_1 = f_1 + \sum_{j=1}^{n} x_j\alpha_j,$$

so a solution of (5.36) is $\phi = \phi_1 + f_2$, or

$$\phi = f + \sum_{j=1}^{n} x_j\alpha_j.$$

Thus, solving the integral equation (5.36) is equivalent to solving the algebraic system (5.40). Similarly, the integral equation (5.37) is equivalent to the algebraic system

$$\begin{aligned} z_j - \sum_{i=1}^{n} z_i\overline{a_{ij}} &= 0 \\ z_j &= (\psi,\alpha_j). \end{aligned} \tag{5.42}$$

Finally, if ψ is a solution of (5.37), then from (5.38),

$$\psi = \sum_{i=1}^{n} z_i\beta_i,$$

so from (5.40),

$$\sum_{i=1}^{n} \bar{z}_i y_i = \sum_{i=1}^{n} \bar{z}_i x_i - \sum_{i,j=1}^{n} \bar{z}_i a_{ij} x_j = \sum_{j=1}^{n} \left[\bar{z}_j - \sum_{i=1}^{n} \bar{z}_i a_{ij} \right] x_j = 0. \tag{5.43}$$

But

$$\sum_{i=1}^{n} \bar{z}_i y_i = \sum_{i=1}^{n} \bar{z}_i (f, \beta_i) = \left(f, \sum_{i=1}^{n} z_i \beta_i \right) = (f, \psi),$$

so (5.43) expresses the orthogonality requirement appearing in the statement of the Fredholm alternative.

From linear algebra, we know that (5.40) will have a solution for every vector (y_i) if the homogeneous equation has only the trivial solution, i.e., $\det |\delta_{ij} - a_{ij}| \neq 0$ where δ_{ij} is the **Kronecker delta function,** defined by

$$\delta_{ij} = \begin{cases} 0 & (i \neq j) \\ 1 & (i = j). \end{cases}$$

But $\det |\delta_{ij} - a_{ij}| = \det |\delta_{ij} - a_{ji}|$, so this statement is equivalent to the case of the Fredholm alternative for degenerate kernels where the homogeneous equation has only the trivial solution. On the other hand, if $\det |\delta_{ij} - a_{ij}| = 0$ and the matrix $[\delta_{ij} - a_{ij}]$ has rank $r < n$, the vectors (y_i) for which (5.40) has solutions span a subspace of dimension r. Since $[\delta_{ij} - \overline{a_{ji}}]$ also has rank r, the set of solutions of (5.42) span a subspace of dimension $n - r$. But by (5.43) these two subspaces are orthogonal, so together they span the whole space. Hence, the vectors (y_i) for which (5.40) is soluble are exactly those that are orthogonal to the solutions (z_i) of (5.42). Translating this statement over to the integral equations (5.36), (5.37) gives us the Fredholm alternative in the case of degenerate kernels.

Before proving the Fredholm alternative for general continuous kernels, we present an example illustrating the above analysis.

□ EXAMPLE 20

Consider the integral equation

$$\phi(s) - \int_0^1 st\phi(t)\,dt = f(s).$$

The kernel $K_n(s,t) = st = ((1/\sqrt{3})s)(\sqrt{3}t)$ is clearly degenerate and

$$\int_0^1 (\sqrt{3}t)^2\,dt = 1.$$

Following the above analysis, we have that the solution of the integral equation is given by

$$\phi(s) = f(s) + x\left(\frac{1}{\sqrt{3}}s\right),$$

where x is the solution of

$$x\left(1 - \int_0^1 s^2\,ds\right) = \sqrt{3}\int_0^1 tf(t)\,dt,$$

i.e.,

$$x = \frac{3\sqrt{3}}{2}\int_0^1 tf(t)\,dt.$$

Hence

$$\phi(s) = \frac{3s}{2}\int_0^1 tf(t)\,dt + f(s). \qquad \square$$

We now turn to the case of a general continuous kernel. From the Weierstrass approximation theorem (cf. Section 3.3) generalized to complex valued functions of two independent variables defined in a square, we can construct a degenerate kernel K_n such that $K_\epsilon(s,t) = K(s,t) - K_n(s,t)$ satisfies

$$\int_0^l\int_0^l |K_\epsilon(s,t)|^2\,dtds < 1. \tag{5.44}$$

Then we can write the solution of

$$\phi_\epsilon - \mathbf{K}_\epsilon\phi_\epsilon = f_\epsilon\,,$$

where

$$\mathbf{K}_\epsilon\phi_\epsilon = \int_0^l K_\epsilon(s,t)\phi_\epsilon(t)\,dt,$$

in the form of a Neumann series, or more concisely (cf. (5.29))

$$\phi_\epsilon = (\mathbf{I} - \mathbf{K}_\epsilon)^{-1}f_\epsilon\,.$$

We can now rewrite the unrestricted Fredholm integral equation

$$\phi - \mathbf{K}\phi = f \tag{5.45}$$

in the form

$$\phi - \mathbf{K}_\epsilon\phi = f + \mathbf{K}_n\phi,$$

so

$$\phi - (\mathbf{I} - \mathbf{K}_\epsilon)^{-1}\mathbf{K}_n\phi = (\mathbf{I} - \mathbf{K}_\epsilon)^{-1}f. \tag{5.46}$$

But

$$\mathbf{K}_n\phi = \int_0^l\left(\sum_{j=1}^n \alpha_j(s)\overline{\beta_j(t)}\right)\phi(t)\,dt,$$

and

$$(\mathbf{I}-\mathbf{K}_\epsilon)^{-1}\mathbf{K}_n\phi = \int_0^l \left(\sum_{j=1}^n ((\mathbf{I}-\mathbf{K}_\epsilon)^{-1}\alpha_j)(s)\overline{\beta_j(t)}\right)\phi(t)\,dt\,; \tag{5.47}$$

thus (5.46) is an integral equation with degenerate kernel whose solutions ϕ are identical with those of (5.45)! From our previous analysis, (5.46) (and hence (5.45)) is soluble if and only if

$$((\mathbf{I}-\mathbf{K}_\epsilon)^{-1}f,\tilde{\psi}) = 0 \tag{5.48}$$

for every solution $\tilde{\psi}$ of the adjoint equation

$$\tilde{\psi} - [(\mathbf{I}-\mathbf{K}_\epsilon)^{-1}\mathbf{K}_n]^*\tilde{\psi} = 0. \tag{5.49}$$

It remains to show that (5.48) can be translated into the orthogonality relation

$$(f,\psi) = 0 \tag{5.50}$$

for every solution ψ of

$$\psi - \mathbf{K}^*\psi = 0. \tag{5.51}$$

We first note that

$$[(\mathbf{I}-\mathbf{K}_\epsilon)^{-1}\mathbf{K}_n]^*\tilde{\psi} = \mathbf{K}_n^*(\mathbf{I}-\mathbf{K}_\epsilon^*)^{-1}\tilde{\psi}. \tag{5.52}$$

To see this we observe that for any operator $\mathbf{K} = \mathbf{AB}$ we have (see the calculation preceding (5.31))

$$(\mathbf{K}^*\phi,\psi) = (\phi,\mathbf{K}\psi) = (\phi,\mathbf{AB}\psi) = (\mathbf{A}^*\phi,\mathbf{B}\psi) = (\mathbf{B}^*\mathbf{A}^*\phi,\psi)$$

for all functions ϕ, ψ, and hence

$$([\mathbf{K}^* - \mathbf{B}^*\mathbf{A}^*]\phi,\psi) = 0.$$

Setting $\psi = (\mathbf{K}^* - \mathbf{B}^*\mathbf{A}^*)\phi$ now implies that

$$\|(\mathbf{K}^* - \mathbf{B}^*\mathbf{A}^*)\phi\| = 0$$

for all ϕ, i.e., $(\mathbf{K}^* - \mathbf{B}^*\mathbf{A}^*)\phi = 0$ for all ϕ or $\mathbf{K}^* = \mathbf{B}^*\mathbf{A}^*$. Equation (5.52) now follows from the fact that

$$[(\mathbf{I}-\mathbf{K}_\epsilon)^{-1}]^* = (\mathbf{I}-\mathbf{K}_\epsilon^*)^{-1}.$$

Returning to (5.48)–(5.51), we note that if ψ is a solution of (5.51), then

$$\tilde{\psi} = (\mathbf{I}-\mathbf{K}_\epsilon^*)\psi$$

is a solution of (5.49), since

$$\begin{aligned}(\mathbf{I}-\mathbf{K}_\epsilon^*)\psi - [(\mathbf{I}-\mathbf{K}_\epsilon)^{-1}\mathbf{K}_n]^*(\mathbf{I}-\mathbf{K}_\epsilon^*)\psi \\ &= (\mathbf{I}-\mathbf{K}_\epsilon^*)\psi - \mathbf{K}_n^*(\mathbf{I}-\mathbf{K}_\epsilon^*)^{-1}(\mathbf{I}-\mathbf{K}_\epsilon^*)\psi \\ &= (\mathbf{I}-\mathbf{K}_\epsilon^*)\psi - \mathbf{K}_n^*\psi = (\mathbf{I}-\mathbf{K}^*+\mathbf{K}_n^*)\psi - \mathbf{K}_n^*\psi \\ &= (\mathbf{I}-\mathbf{K}^*)\psi = 0.\end{aligned}$$

Conversely, if $\tilde{\psi}$ is a solution of (5.49) then $\psi = (\mathbf{I} - \mathbf{K}_\epsilon^*)^{-1}\tilde{\psi}$ is a solution of (5.51). Hence, $\tilde{\psi} = 0$ if and only if $\psi = 0$. Furthermore, if (5.48) is valid for $\tilde{\psi}$ not zero, then

$$
\begin{aligned}
0 &= ((\mathbf{I} - \mathbf{K}_\epsilon)^{-1}f,\tilde{\psi}) = ((\mathbf{I} - \mathbf{K}_\epsilon)^{-1}f,(\mathbf{I} - \mathbf{K}_\epsilon^*)\psi) \\
&= (f,(\mathbf{I} - \mathbf{K}_\epsilon^*)^{-1}(\mathbf{I} - \mathbf{K}_\epsilon^*)\psi) = (f,\psi)
\end{aligned}
$$

for a solution ψ of (5.51), and conversely if (5.50) is valid then so is (5.48) for ψ a solution of (5.49). Hence (5.48) and (5.50) are equivalent, and this completes the proof of the Fredholm alternative. ■

The Fredholm alternative remains valid for integral equations in R^n where the kernel is only assumed to have a weak singularity of the form

$$K(\mathbf{s},\mathbf{t}) = \frac{\kappa(\mathbf{s},\mathbf{t})}{|\mathbf{s} - \mathbf{t}|^\alpha} \qquad (0 \leq \alpha < n;\ \mathbf{s},\mathbf{t} \in R^n)$$

where $\kappa(\mathbf{s},\mathbf{t})$ is a bounded function. This is particularly relevant for potential theory in the physically important case of Laplace's equation in R^3, since in this case, the double layer potential

$$u(\mathbf{x}) = \frac{1}{4\pi}\int_{\partial D} \psi(\mathbf{y}) \frac{\partial}{\partial \nu(\mathbf{y})} \frac{1}{|\mathbf{x} - \mathbf{y}|}\, ds(\mathbf{y}) \qquad (\mathbf{x} \in R^3\backslash\partial D)$$

and the single layer potential

$$u(\mathbf{x}) = \frac{1}{4\pi}\int_{\partial D} \psi(\mathbf{y}) \frac{1}{|\mathbf{x} - \mathbf{y}|}\, ds(\mathbf{y}) \qquad (\mathbf{x} \in R^3\backslash\partial D)$$

for $D \subset R^3$ satisfy the same discontinuity properties as the corresponding potentials in R^2. Hence the Dirichlet and Neumann boundary value problems can be reduced to the problem of solving Fredholm integral equations, where in the present case the kernels are no longer continuous but instead have a weak singularity of the above form (cf. Colton and Kress).

5.3 APPLICATIONS TO THE DIRICHLET AND NEUMANN PROBLEMS

In Section 5.1 we formulated integral equations which, if soluble, yield the solution of the Dirichlet and Neumann problems for Laplace's equation. The Fredholm alternative gives us a tool for deciding whether these equations are in fact soluble and thus a method for showing the existence of a solution to these boundary value problems for Laplace's equation. Assuming a solution exists, the method of integral equations will also provide us with a numerical method for approximating the solution of these boundary value problems (cf. Section 5.5). In this section we shall use the method of integral equations to establish the existence of a solution to the interior Dirichlet problem, the interior Neumann problem, the exterior Neumann problem, and the exterior Dirichlet problem for Laplace's equation. We shall, as usual, assume that D is a bounded, simply connected domain in R^2 such that ∂D is in class C^2, i.e., ∂D has continuous curvature.

5.3.1 The Interior Dirichlet Problem

We look for a solution $u \in C^2(D)\cap C(\overline{D})$ of

$$\Delta_2 u = 0 \quad \text{in} \quad D$$
$$u = f \quad \text{on} \quad \partial D$$

in the form of a double layer potential

$$u(\mathbf{x}) = \frac{1}{2\pi}\int_{\partial D} \psi(\mathbf{y}) \frac{\partial}{\partial \nu(\mathbf{y})} \log \frac{1}{|\mathbf{x}-\mathbf{y}|}\, ds(\mathbf{y}) \qquad (\mathbf{x} \in D), \tag{5.53}$$

where $\psi \in C(\partial D)$. Letting $\mathbf{x}$ tend to ∂D and using the jump relations for double layer potentials (see Section 5.1) gives the Fredholm integral equation

$$\psi(\mathbf{x}) - \frac{1}{\pi}\int_{\partial D} \psi(\mathbf{y}) \frac{\partial}{\partial \nu(\mathbf{y})} \log \frac{1}{|\mathbf{x}-\mathbf{y}|}\, ds(\mathbf{y}) = -2f(\mathbf{x}) \qquad (\mathbf{x} \in \partial D) \tag{5.54}$$

for the determination of $\psi(\mathbf{x})$. The existence of a unique solution to (5.54) will follow from the Fredholm alternative if we can show that the only solution of the homogeneous equation

$$\psi(\mathbf{x}) - \frac{1}{\pi}\int_{\partial D} \psi(\mathbf{y}) \frac{\partial}{\partial \nu(\mathbf{y})} \log \frac{1}{|\mathbf{x}-\mathbf{y}|}\, ds(\mathbf{y}) = 0 \qquad (\mathbf{x} \in \partial D) \tag{5.55}$$

is $\psi(\mathbf{x})$ identically zero. (Note that any square integrable solution of (5.55) is in fact continuous.) We shall now show that this is true, establishing the existence of a solution to the interior Dirichlet problem for Laplace's equation.

Let $\psi(\mathbf{x})$ satisfy (5.55) and define $u(\mathbf{x})$ by (5.53) for this $\psi(\mathbf{x})$. Then $u(\mathbf{x})$ is harmonic in D and $u(\mathbf{x}) = 0$ for $\mathbf{x} \in \partial D$. By the uniqueness of the solution to the interior Dirichlet problem, $u(\mathbf{x})$ is identically zero for $\mathbf{x} \in D$. This implies that $\partial u(\mathbf{x})/\partial \nu = 0$ for $\mathbf{x} \in \partial D$. Now consider $u(\mathbf{x})$ as given by (5.53) for $\mathbf{x} \in R^2\backslash\overline{D}$. Then $u(\mathbf{x})$ is harmonic in $R^2\backslash\overline{D}$, and by the continuity property of the normal derivative of the double layer potential $\partial u(\mathbf{x})/\partial \nu = 0$ for $\mathbf{x} \in \partial D$ (letting $\mathbf{x}$ approach ∂D from $R^2\backslash\overline{D}$). Furthermore, as $\mathbf{x}$ tends to infinity we see from (5.53) that $u(\mathbf{x}) = 0(1/r)$ and $\partial u(\mathbf{x})/\partial r = 0(1/r^2)$ where $r = |\mathbf{x}|$. Hence, from Green's formula

$$\iint_{\Omega_R\backslash D} |\nabla u|^2\, d\mathbf{x} = \int_{\partial\Omega_R} u \frac{\partial u}{\partial \nu}\, ds = 0\left(\frac{1}{R^2}\right),$$

where $\Omega_R = \{\mathbf{x}\colon |\mathbf{x}| \le R\}$, $D \subset \Omega_R$. Letting R tend to infinity, we see that $u(\mathbf{x})$ is equal to a constant for $\mathbf{x} \in R^2\backslash D$. But, since $u(\mathbf{x})$ tends to zero as $|\mathbf{x}|$ tends to infinity, we know that this constant is zero. Hence, $u(\mathbf{x})$ as defined by (5.53) vanishes in D

and $R^2\backslash\overline{D}$, and the discontinuity property (cf. (5.6))

$$u^+(\mathbf{x}) - u^-(\mathbf{x}) = -\psi(\mathbf{x}) \qquad (\mathbf{x} \in \partial D)$$

of the double layer potential now implies that $\psi(\mathbf{x}) = 0$ for $\mathbf{x} \in \partial D$.

5.3.2 The Interior Neumann Problem

We look for a solution $u \in C^2(D) \cap C^1(\overline{D})$ of

$$\Delta_2 u = 0 \qquad \text{in} \qquad D$$

$$\frac{\partial u}{\partial \nu} = g \qquad \text{on} \qquad \partial D$$

in the form of a single layer potential

$$u(\mathbf{x}) = \frac{1}{2\pi} \int_{\partial D} \psi(\mathbf{y}) \log \frac{1}{|\mathbf{x} - \mathbf{y}|}\, ds(\mathbf{y}) \qquad (\mathbf{x} \in D), \tag{5.56}$$

where $\psi \in C(\partial D)$. Note that if $u(\mathbf{x})$ is a solution of the interior Neumann problem, so is $u(\mathbf{x}) + c$ for any constant c. Letting $\mathbf{x}$ tend to ∂D and using the jump relations for the normal derivative of the single layer potential (see Section 5.1), we arrive at the Fredholm integral equation

$$\psi(\mathbf{x}) + \frac{1}{\pi} \int_{\partial D} \psi(\mathbf{y}) \frac{\partial}{\partial \nu(\mathbf{x})} \log \frac{1}{|\mathbf{x} - \mathbf{y}|}\, ds(\mathbf{y}) = 2g(\mathbf{x}) \qquad (\mathbf{x} \in \partial D) \tag{5.57}$$

for the determination of $\psi(\mathbf{x})$. The existence of a solution to (5.57) will follow from the Fredholm alternative if we can show that every solution of the adjoint homogeneous equation

$$\psi(\mathbf{x}) + \frac{1}{\pi} \int_{\partial D} \psi(\mathbf{y}) \frac{\partial}{\partial \nu(\mathbf{y})} \log \frac{1}{|\mathbf{x} - \mathbf{y}|}\, ds(\mathbf{y}) = 0 \qquad (\mathbf{x} \in \partial D) \tag{5.58}$$

is orthogonal to $g(\mathbf{x})$. We note again that any square integrable solution of (5.58) is in fact continuous. We now show that this is true and establish the existence of a solution to the interior Neumann problem for Laplace's equation. Recall from equation (4.4) that a solution to the interior Neumann problem can exist only if

$$\int_{\partial D} g\, ds = 0, \tag{5.59}$$

so we assume that this is true.

Let $\psi(\mathbf{x})$ satisfy (5.58) and consider the harmonic function $u(\mathbf{x})$ for $\mathbf{x} \in R^2\backslash\overline{D}$ defined by

$$u(\mathbf{x}) = \frac{1}{2\pi} \int_{\partial D} \psi(\mathbf{y}) \frac{\partial}{\partial \nu(\mathbf{y})} \log \frac{1}{|\mathbf{x} - \mathbf{y}|}\, ds(\mathbf{y}) \qquad (\mathbf{x} \in R^2\backslash\overline{D}). \tag{5.60}$$

Without loss of generality we can assume that $\psi(\mathbf{x})$ is real valued. Then, letting $\mathbf{x}$ tend to ∂D from $R^2\backslash\overline{D}$ and using the discontinuity properties of double layer potentials, we see from (5.58) that $u(\mathbf{x}) = 0$ for $\mathbf{x} \in \partial D$. Since $u(\mathbf{x})$ as defined by (5.60) tends to zero as $|\mathbf{x}|$ tends to infinity, we can conclude from the maximum principle that $u(\mathbf{x})$ is identically zero for $\mathbf{x} \in R^2\backslash\overline{D}$. The continuity of the normal derivative of (5.60) as $\mathbf{x}$ crosses ∂D now implies that for $\mathbf{x} \in D$, $u(\mathbf{x})$ is harmonic and $\partial u(\mathbf{x})/\partial\nu = 0$ for $\mathbf{x} \in \partial D$. Hence, from Section 4.3, $u(\mathbf{x})$ equals a constant in D. The discontinuity property of the double layer potential now implies that $\psi(\mathbf{x})$ is equal to a constant. Conversely, from the Gauss formula (5.1), we see that $\psi(\mathbf{x})$ equal to a constant is in fact a solution of the homogeneous adjoint equation (5.58). From (5.59) we can now conclude that every solution of (5.58) is orthogonal to $g(\mathbf{x})$ and therefore (5.57) has a solution.

We draw the reader's attention to how neatly the above method of integral equations deals with the potentially awkward problem of nonuniqueness of a solution to the interior Neumann problem.

5.3.3 The Exterior Neumann Problem

We look for a solution $u \in C^2(R^2\backslash\overline{D})\cap C^1(R^2\backslash D)$ of

$$\Delta_2 u = 0 \qquad \text{in} \qquad R^2\backslash\overline{D}$$

$$\frac{\partial u}{\partial \nu} = g \qquad \text{on} \qquad \partial D \tag{5.61}$$

$u(\mathbf{x})$ is regular at infinity

where $u(\mathbf{x})$ is regular at infinity if $\lim_{r\to\infty} |ru(r,\theta)| < \infty$ and $\lim_{r\to\infty} |r^2\,(\partial/\partial r)u(r,\theta)| < \infty$ uniformly for $0 \le \theta \le 2\pi$, where (r,θ) are polar coordinates. Note that from Green's formula, we know that if a solution to (5.61) exists then

$$\int_{\partial D} g\,ds = \int_{\partial D} \frac{\partial u}{\partial \nu}\,ds = \int_{\partial\Omega_R} \frac{\partial u}{\partial \nu}\,ds = 0\left(\frac{1}{R}\right),$$

where $\Omega_R = \{\mathbf{x}: |\mathbf{x}| \le R\}$. Letting R tend to infinity, we see that a necessary condition for the existence of a solution of the exterior Neumann problem is that

$$\int_{\partial D} g\,ds = 0. \tag{5.62}$$

We now look for a solution of (5.61) in the form of a single layer potential

$$u(\mathbf{x}) = \frac{1}{2\pi}\int_{\partial D} \psi(\mathbf{y}) \log\frac{1}{|\mathbf{x}-\mathbf{y}|}\,ds(\mathbf{y}) \qquad (\mathbf{x} \in R^2\backslash\overline{D}), \tag{5.63}$$

where $\psi \in C(\partial D)$. If $\psi \in C(\partial D)$ satisfies

$$\int_{\partial D} \psi\,ds = 0, \tag{5.64}$$

then $u(\mathbf{x})$, as defined by (5.63), will be regular at infinity. Letting $\mathbf{x}$ tend to ∂D in

(5.63) and using the jump relations for the normal derivative of the single layer potential, we see that $\psi(\mathbf{x})$ must satisfy the Fredholm integral equation

$$\psi(\mathbf{x}) - \frac{1}{\pi}\int_{\partial D} \psi(\mathbf{y}) \frac{\partial}{\partial \nu(\mathbf{x})} \log \frac{1}{|\mathbf{x}-\mathbf{y}|}\, ds(\mathbf{y}) = -2g(\mathbf{x}) \qquad (\mathbf{x} \in \partial D). \qquad \textbf{(5.65)}$$

By the Fredholm alternative, a unique solution will exist provided the homogeneous equation

$$\psi(\mathbf{x}) - \frac{1}{\pi}\int_{\partial D} \psi(\mathbf{y}) \frac{\partial}{\partial \nu(\mathbf{x})} \log \frac{1}{|\mathbf{x}-\mathbf{y}|}\, ds(\mathbf{y}) = 0 \qquad (\mathbf{x} \in \partial D) \qquad \textbf{(5.66)}$$

has only the trivial solution. Note again that any square integrable solution of (5.66) is in fact continuous.

Let $\psi(\mathbf{x})$ be a solution of the homogeneous equation. Then from (5.66) and the Gauss formula (5.1) we have

$$\begin{aligned} 0 &= \int_{\partial D} \psi(\mathbf{x})\, ds - \frac{1}{\pi}\int_{\partial D} \psi(\mathbf{y}) \int_{\partial D} \frac{\partial}{\partial \nu(\mathbf{x})} \log \frac{1}{|\mathbf{x}-\mathbf{y}|}\, ds(\mathbf{x}) ds(\mathbf{y}) \\ &= 2\int_{\partial D} \psi(\mathbf{x})\, ds, \end{aligned}$$

so (5.64) is satisfied. For this $\psi(\mathbf{x})$, define the harmonic function $u(\mathbf{x})$ by (5.63). Then $\partial u(\mathbf{x})/\partial \nu = 0$ for $\mathbf{x} \in \partial D$, and from the analysis in Section 5.3.1 we see that this implies that $u(\mathbf{x})$ is identically zero for $\mathbf{x} \in R^2 \backslash \overline{D}$. In particular, the condition of regularity at infinity makes the solution of the exterior Neumann problem unique in contrast to the nonuniqueness of the interior Neumann problem. By the continuity of the single layer potential, $u(\mathbf{x})$ is harmonic in D and continuously assumes zero boundry data on ∂D. Hence, $u(\mathbf{x})$ is identically zero for $\mathbf{x} \in D$. Since by (5.12)

$$\frac{\partial u^+}{\partial \nu}(\mathbf{x}) - \frac{\partial u^-}{\partial \nu}(\mathbf{x}) = \psi(\mathbf{x}) \qquad (\mathbf{x} \in \partial D),$$

we now conclude that $\psi(\mathbf{x})$ is identically zero. Hence, we conclude that there exists a unique solution to (5.65).

Let $\psi(\mathbf{x})$ be this unique solution. Then from (5.62) and (5.65) we have

$$\begin{aligned} \int_{\partial D} \psi(\mathbf{x})\, ds - \frac{1}{\pi}\int_{\partial D} \psi(\mathbf{y}) \int_{\partial D} \frac{\partial}{\partial \nu(\mathbf{x})} \log \frac{1}{|\mathbf{x}-\mathbf{y}|}\, ds(\mathbf{x}) ds(\mathbf{y}) &= -2\int_{\partial D} g(\mathbf{x})\, ds \\ &= 0, \end{aligned}$$

or, using the Gauss formula (5.1),

$$2\int_{\partial D} \psi(\mathbf{x})\, ds = 0.$$

Thus (5.64) is satisfied, so (5.63) with $\psi(\mathbf{x})$ the unique solution of (5.65) provides the unique solution of the exterior Neumann problem.

5.3.4 The Exterior Dirichlet Problem

We look for a solution $u \in C^2(R^2\backslash\overline{D})\cap C(R^2\backslash D)$ of

$$\Delta_2 u = 0 \quad \text{in} \quad R^2\backslash\overline{D}$$

$$u = f \quad \text{on} \quad \partial D \tag{5.67}$$

$u(\mathbf{x})$ is bounded at infinity.

We first want to show that the solution, if it exists, is unique. Define $v(r,\theta)$ by $v(r,\theta) = u(1/r,\theta)$, where (r,θ) are polar coordinates, and let D' be the image of $R^2\backslash\overline{D}$ under the mapping $(r,\theta) \to (1/r,\theta)$. Then it can be directly verified that $v(\mathbf{x}) = v(r,\theta)$ is harmonic in $D'\backslash\{0\}$, and $v(\mathbf{x}) = f(\mathbf{x})$ for $\mathbf{x} \in \partial D'$. We want to show that the origin is a removable singularity, and $v(\mathbf{x})$ is harmonic for $\mathbf{x} \in D'$. To do this, let $\Omega_r = \{\mathbf{x}: |\mathbf{x}| < r\}$ be a disk contained in D'. Then $v \in C^2(\partial\Omega_r)$, and from Fourier's Theorem (cf. Section 1.4),

$$v(r,\theta) = \sum_{n=-\infty}^{\infty} a_n(r)\theta^{in\theta}$$

$$a_n(r) = \frac{1}{2\pi}\int_{-\pi}^{\pi} v(r,\theta)e^{-in\theta}\,d\theta.$$

Differentiating under the integral sign and integrating by parts using

$$\Delta_2 v = \frac{\partial^2 v}{\partial r^2} + \frac{1}{r}\frac{\partial v}{\partial r} + \frac{1}{r^2}\frac{\partial^2 v}{\partial\theta^2} = 0$$

shows that

$$a_n'' + \frac{1}{r}a_n' - \frac{n^2}{r^2}a_n = 0,$$

and hence

$$a_0(r) = a_0 + b_0 \log r$$

$$a_n(r) = a_n r^{|n|} + b_n r^{-|n|} \qquad (n = \pm 1, \pm 2, \ldots),$$

where the a_n and b_n are constants. Since $v(\mathbf{x})$ is bounded at the origin, $b_n = 0$ for $n = 0, \pm 1, \pm 2, \ldots$, so

$$v(r,\theta) = \sum_{n=-\infty}^{\infty} a_n r^{|n|} e^{in\theta}.$$

Hence, $v(\mathbf{x})$ is harmonic in Ω_r, and the origin is a removable singularity. We can now conclude from the uniqueness of the solution to the interior Dirichlet problem that the solution of the exterior Dirichlet problem (5.67) is unique.

In a first attempt to show the existence of a solution to the exterior Dirichlet problem, we look for a solution in the form of a double layer potential

$$u(\mathbf{x}) = \frac{1}{2\pi}\int_{\partial D} \psi(\mathbf{y})\frac{\partial}{\partial\nu(\mathbf{y})}\log\frac{1}{|\mathbf{x}-\mathbf{y}|}\,ds(\mathbf{y}) \qquad (\mathbf{x}\in R^2\backslash\overline{D}), \tag{5.68}$$

where $\psi \in C(\partial D)$. The jump relations for double layer potentials imply that $\psi(\mathbf{x})$ satisfies the Fredholm integral equation

$$\psi(\mathbf{x}) + \frac{1}{\pi}\int_{\partial D} \psi(\mathbf{y})\frac{\partial}{\partial\nu(\mathbf{y})}\log\frac{1}{|\mathbf{x}-\mathbf{y}|}\,ds(\mathbf{y}) = 2f(\mathbf{x}) \qquad (\mathbf{x}\in \partial D). \tag{5.69}$$

But the Gauss formula (5.1) implies that $\psi(\mathbf{x})$ equal to a constant is a solution of the homogeneous equation, so by the Fredholm alternative there exists a nontrivial real valued solution $\phi(\mathbf{x})$ of the homogeneous adjoint equation. Thus a solution of (5.69) will not exist unless

$$\int_{\partial D} f\phi\, ds = 0.$$

Since in general we cannot assume that $f(\mathbf{x})$ satisfies this condition, we must modify our approach. In particular, the solution of the exterior Dirichlet problem cannot in general be represented in the form of a double layer potential.

As a second attempt to solve the exterior Dirichlet problem, we look for a solution in the form

$$u(\mathbf{x}) = \frac{1}{2\pi}\int_{\partial D} \psi(\mathbf{y})\left[\frac{\partial}{\partial\nu(\mathbf{y})}\log\frac{1}{|\mathbf{x}-\mathbf{y}|} + 1\right] ds(\mathbf{y}) \qquad (\mathbf{x}\in R^2\backslash\overline{D}) \tag{5.70}$$

where $\psi \in C(\partial D)$. Then $u(\mathbf{x})$ as defined by (5.70) is harmonic for $\mathbf{x}\in R^2\backslash\overline{D}$, bounded at infinity, and $u(\mathbf{x}) = f(\mathbf{x})$ for $\mathbf{x}\in\partial D$ provided $\psi(\mathbf{x})$ satisfies the Fredholm integral equation

$$\psi(\mathbf{x}) + \frac{1}{\pi}\int_{\partial D} \psi(\mathbf{y})\left[\frac{\partial}{\partial\nu(\mathbf{y})}\log\frac{1}{|\mathbf{x}-\mathbf{y}|} + 1\right] ds(\mathbf{y}) = 2f(\mathbf{x}) \qquad (\mathbf{x}\in \partial D). \tag{5.71}$$

By the Fredholm alternative, a unique solution to (5.71) will exist, provided the homogeneous equation

$$\psi(\mathbf{x}) + \frac{1}{\pi}\int_{\partial D} \psi(\mathbf{y})\left[\frac{\partial}{\partial\nu(\mathbf{y})}\log\frac{1}{|\mathbf{x}-\mathbf{y}|} + 1\right] ds(\mathbf{y}) = 0 \qquad (\mathbf{x}\in \partial D) \tag{5.72}$$

has only the trivial solution. We note that any square integrable solution of (5.72) is continuous.

Let $\psi(\mathbf{x})$ be a solution of (5.72) and define $u(\mathbf{x})$ by (5.70) for this $\psi(\mathbf{x})$. Then $u(\mathbf{x}) = 0$ for $\mathbf{x}\in\partial D$, and by the uniqueness of the solution to the exterior Dirichlet problem, $u(\mathbf{x}) = 0$ for $\mathbf{x} \in R^2 \backslash \overline{D}$. Letting $\mathbf{x}$ tend to infinity in (5.70) now shows that

$$\int_{\partial D} \psi \, ds = 0, \tag{5.73}$$

so (5.72) takes the form

$$\psi(\mathbf{x}) + \frac{1}{\pi} \int_{\partial D} \psi(\mathbf{y}) \frac{\partial}{\partial \nu(\mathbf{y})} \log \frac{1}{|\mathbf{x} - \mathbf{y}|} \, ds(\mathbf{y}) = 0 \qquad (\mathbf{x} \in \partial D).$$

But this is identical to (5.58), whose only solution is $\psi(\mathbf{x})$ equal to a constant. But from (5.73) we see that this constant must be zero. Hence, (5.72) has only the trivial solution, and a unique solution to (5.71) exists. The existence of a solution to the exterior Dirichlet problem is now established.

5.4 HILBERT-SCHMIDT THEORY AND EIGENVALUE PROBLEMS

We now examine the eigenvalue problem for the integral equation

$$\phi(s) - \lambda \int_0^l K(s,t)\phi(t) \, dt = 0 \tag{5.74}$$

where the kernel is **symmetric,** i.e., $K(s,t)$ is real valued and $K(s,t) = K(t,s)$. Hence we want to consider the operator equation

$$\phi - \lambda \mathbf{K}\phi = 0, \tag{5.75}$$

where $\mathbf{K}$ is self-adjoint, i.e.,

$$\mathbf{K}^* = \mathbf{K}. \tag{5.76}$$

We are concerned with square integrable solutions of (5.75) and assume that the kernel $K(s,t)$, in addition to being symmetric, is not identically zero and has at most a weak singularity at $s = t$ of the form

$$K(s,t) = \frac{\kappa(s,t)}{|s - t|^\alpha} \qquad (0 \leq \alpha < 1/2),$$

where $\kappa(s,t)$ is continuous. Under these assumptions, $K(s,t)$ is square integrable, and there exists a positive constant A such that

$$\int_0^l |K(s,t)|^2 \, dt = \int_0^l |K(s,t)|^2 \, ds \leq A < \infty.$$

In addition, these assumptions imply that every square integrable solution of (5.75) is in fact continuous.

This section considers only one-dimensional integral equations, but we want to emphasize that all of our results and proofs are identical for n-dimensional integral equations, where the kernel $K(s,t)$ satisfies

$$K(\mathbf{s},\mathbf{t}) = \frac{\kappa(\mathbf{s},\mathbf{t})}{|\mathbf{s} - \mathbf{t}|^{\alpha}} \qquad (0 \leq \alpha < n/2;\ \mathbf{s},\mathbf{t} \in D \subset R^n),$$

where $\kappa(\mathbf{s},\mathbf{t})$ is continuous. This includes the eigenvalue problem for Laplace's equation where $K(\mathbf{s},\mathbf{t})$ is Green's function for Laplace's equation in D (cf. Section 5.1).

DEFINITION 14

Suppose there exists a λ such that (5.75) has a nontrivial solution ϕ. Then λ is called an **eigenvalue** of (5.75) with corresponding **eigenfunction** ϕ.

THEOREM 40

The eigenvalues of (5.75) are real.

■ ***Proof.*** Suppose $\phi = \lambda\mathbf{K}\phi$ where $\|\phi\|$ is not zero. Then $\lambda(\mathbf{K}\phi,\phi) = (\lambda\mathbf{K}\phi,\phi) = (\phi,\phi)$. But $(\mathbf{K}\phi,\phi) = (\phi,\mathbf{K}\phi) = \overline{(\mathbf{K}\phi,\phi)}$, and hence $(\mathbf{K}\phi,\phi)$ is real valued. We can now conclude that λ is real. ■

THEOREM 41

Let λ_1 and λ_2 be distinct eigenvalues of (5.75) and ϕ_1 and ϕ_2 eigenfunctions corresponding to λ_1 and λ_2, respectively. Then ϕ_1 and ϕ_2 are orthogonal, i.e., $(\phi_1,\phi_2) = 0$.

■ ***Proof.*** We have $(\phi_1,\phi_2) = (\lambda_1\mathbf{K}\phi_1,\phi_2) = \lambda_1(\mathbf{K}\phi_1,\phi_2) = \lambda_1(\phi_1,\mathbf{K}\phi_2)$. Since $\|\phi_2\|$ is not zero, λ_2 is not zero, and hence, using the previous theorem, $(\phi_1,\mathbf{K}\phi_2) = (1/\lambda_2)(\phi_1,\lambda_2\mathbf{K}\phi_2) = (1/\lambda_2)(\phi_1,\phi_2)$. Thus $(\phi_1,\phi_2) = \lambda_1/\lambda_2\ (\phi_1,\phi_2)$, and so $(\phi_1,\phi_2) = 0$. ■

By taking the real and imaginary parts of (5.75) and noting that λ and $K(s,t)$ are real, we can assume without loss of generality that the eigenfunctions of (5.75) are real valued. We can furthermore assume that the eigenfunctions have norm one, i.e., eigenfunctions corresponding to distinct eigenvalues form an orthonormal set.

THEOREM 42

The eigenvalues of (5.75) are isolated and can accumulate only at infinity.

■ ***Proof.*** Let ϕ_i, $i = 1,\ldots,n$, be eigenfunctions of (5.75) with eigenvalues λ_i, i.e., $\phi_i = \lambda_i\mathbf{K}\phi_i$. By the above results we have that if the λ_i are distinct eigenvalues then

$$(\phi_i,\phi_j) = \delta_{ij},$$

where δ_{ij} is the Kronecker delta function. Hence

$$0 \leq \int_0^l \left[K(s,t) - \sum_{i=1}^n \frac{\phi_i(s)\phi_i(t)}{\lambda_i} \right]^2 dt = \int_0^l |K(s,t)|^2 \, dt - \sum_{i=1}^n \frac{|\phi_i(s)|^2}{\lambda_i^2}$$

and integrating with respect to s gives

$$\sum_{i=1}^n \frac{1}{\lambda_i^2} \leq B^2 = \int_0^l \int_0^l |K(s,t)|^2 \, dt ds, \tag{5.77}$$

which holds for any n distinct eigenvalues $\lambda_1,\lambda_2,\ldots,\lambda_n$. However, if $\phi_1,\phi_2,\ldots,\phi_n$ are distinct eigenfunctions corresponding to the same eigenvalue λ, and we use the Gram-Schmidt process to orthonormalize these eigenfunctions, then we see that (5.77) is also valid for $\lambda = \lambda_1 = \lambda_2 = \ldots = \lambda_n$, i.e., $n \leq \lambda^2 B^2$. This implies that no eigenvalue can have more than a finite multiplicity. Furthermore, if $|\lambda_i| \leq M$, then $n \leq M^2 B^2$, so there are only a finite number of eigenvalues in any interval $[-M,M]$. Hence, (5.75) has at most countably many eigenvalues. If the set of eigenvalues is indeed infinite, we know from (5.77) that

$$\sum_{i=1}^\infty \frac{1}{\lambda_i^2} \leq B^2,$$

and hence

$$\lim_{i\to\infty} |\lambda_i| = \infty. \qquad \blacksquare$$

By using the Gram-Schmidt process as indicated in the proof of the above theorem, we can always assume that distinct eigenfunctions corresponding to the same eigenvalue are orthonormal. Hence by orthonormalizing the eigenfunctions corresponding to distinct eigenvalues we can always assume without loss of generality that the eigenfunctions of (5.75) form a (countable) orthonormal set.

We now need to show that there exists at least one eigenvalue of the integral equation (5.75). To this end we need the next theorem.

DEFINITION 15

A set of functions $\{\psi_j(s)\}$ is said to be **equicontinuous** on an interval $[0,l]$ if for every $\epsilon > 0$ there exists a number $\delta = \delta(\epsilon)$ independent of j such that

$$|\psi_j(s_1) - \psi_j(s_2)| < \epsilon$$

for $s_1,s_2 \in [0,l]$, $|s_1 - s_2| < \delta$.

DEFINITION 16

A set of functions $\{\psi_j(s)\}$ defined on $[0,l]$ is said to be **uniformly bounded** if there exists a constant M independent of j such that

$$\max_{0\leq s\leq l} |\psi_j(s)| \leq M.$$

THEOREM 43 Arzela-Ascoli Theorem

Let $\{\psi_j(s)\}$ be a set of uniformly bounded and equicontinuous functions defined on an interval $[0,l]$. Then there exists a subsequence of $\{\psi_j(s)\}$ that is uniformly convergent on $[0,l]$.

■ ***Proof.*** Let $s_1,s_2,s_3,\ldots$ be a dense sequence of points on $[0,l]$ and let $\{\psi_{1n}(s)\}$ be a subsequence of $\{\psi_j(s)\}$ such that $\lim_{n\to\infty} \psi_{1n}(s_1)$ exists. Now let $\{\psi_{2n}(s)\}$ be a subsequence of $\{\psi_{1n}(s)\}$ such that $\lim_{n\to\infty} \psi_{2n}(s_2)$ exists; repeating this process we have an array of functions $\psi_{mn}(s)$ such that each row is a subsequence of the previous row and the mth row converges at $s_1,s_2,\ldots,s_m$ as n tends to infinity. The existence of the subsequences follows from the uniform boundedness of $\{\psi_j(s)\}$.

$$\begin{array}{cccc} \psi_{11} & \psi_{12} & \psi_{13} & \cdots \\ \psi_{21} & \psi_{22} & \psi_{23} & \cdots \\ \psi_{31} & \psi_{32} & \psi_{33} & \cdots \\ \vdots & \vdots & \vdots & \cdots \end{array}$$

Hence the subsequence $\{\psi_{nn}(s)\}$ converges for each point $s = s_1,s_2,\ldots$. Note that we can choose the $\psi_{mn}(s)$ such that $\psi_{nn}(s)$ does not equal $\psi_{mm}(s)$ for all s unless $n = m$.

We now show that the sequence $\{\psi_{nn}(s)\}$ converges uniformly for $s \in [0,l]$. Let $\epsilon > 0$ be given and δ be as in the definition of equicontinuity. Choose L large enough such that any point $s \in [0,l]$ has a distance less than δ from at least one of the points $s_1,s_2,\ldots,s_L$ and choose N such that

$$|\psi_{jj}(s_l) - \psi_{kk}(s_l)| < \epsilon$$

for $j,k > N$ and $l = 1,\ldots,L$. By the definition of equicontinuity and the triangle inequality,

$$|\psi_{jj}(s) - \psi_{kk}(s)| \leq |\psi_{jj}(s) - \psi_{jj}(s_l)| + |\psi_{jj}(s_l) - \psi_{kk}(s_l)| + |\psi_{kk}(s_l) - \psi_{kk}(s)| < 3\epsilon$$

for all $s \in [0,l]$, and this implies the uniform convergence of the sequence $\{\psi_{nn}(s)\}$. ■

We note that this theorem and its proof remain valid if $\{\psi_j(s)\}$ is replaced by a sequence $\{\psi_j(\mathbf{s})\}$ defined on any closed, bounded subset of R^n.

With the aid of the Arzela-Ascoli theorem we can now show that there exists at least one eigenvalue of the integral equation (5.75).

THEOREM 44

There exists at least one eigenvalue of the integral equation (5.75).

■ ***Proof.*** We first define a bilinear form $F(\phi,\psi)$ by

$$F(\phi,\psi) = (\phi,\mathbf{K}\psi)$$

and the associated quadratic form by

$$F(\phi) = F(\phi,\phi).$$

Note that since $K = K^*$, $F(\phi,\psi)$ satisfies

$$F(\phi,\psi) = \overline{F(\psi,\phi)}.$$

We now define $|\lambda|$ by

$$\begin{aligned}\frac{1}{|\lambda|} &= \sup \frac{|F(\phi)|}{\|\phi\|^2} \\ &= \sup_{\|\phi\|=1} |F(\phi)|,\end{aligned} \tag{5.78}$$

where the supremum is taken over the class of square integrable functions. If we define the norm of the operator $\mathbf{K}$ by

$$\|\mathbf{K}\| = \sup \frac{\|\mathbf{K}\phi\|}{\|\phi\|} = \sup_{\|\phi\|=1} \|\mathbf{K}\phi\|,$$

we see from the Schwarz inequality and (5.77) that

$$\begin{aligned}\|\mathbf{K}\phi\|^2 &= \int_0^l \left[\int_0^l K(s,t)\phi(t)\,dt\right]^2 ds \\ &\leq \int_0^l \int_0^l |K(s,t)|^2\,dt\,ds \int_0^l |\phi(t)|^2\,dt = B^2\|\phi\|^2,\end{aligned}$$

and

$$|F(\phi)| = |(\phi,\mathbf{K}\phi)| \leq \|\phi\|\,\|\mathbf{K}\phi\| \leq \|\mathbf{K}\|\,\|\phi\|^2. \tag{5.79}$$

Hence, the above suprema exist. We shall now show that either $|\lambda|$ or $-|\lambda|$ is an eigenvalue.

From (5.78) and (5.79) we know that $1/|\lambda| \leq \|\mathbf{K}\|$. To establish the reverse inequality, we set $\psi = \mathbf{K}\phi$ and

$$\mu^2 = \frac{\|\mathbf{K}\phi\|}{\|\phi\|},$$

where ϕ is such that $\mathbf{K}\phi$ is not the zero function. (We shall soon see that this is always possible.) Then

$$\|\mathbf{K}\phi\|^2 = (\phi,\mathbf{K}\psi) = F\left(\mu\phi, \frac{1}{\mu}\psi\right),$$

and since for this choice of ψ we have

$$F\left(\mu\phi \pm \frac{1}{\mu}\psi\right) = \mu^2 F(\phi) \pm 2F(\phi,\psi) + \frac{1}{\mu^2}F(\psi),$$

we deduce that (cf. Exercise 11 of Chapter 1)

$$\begin{aligned}\|\mathbf{K}\phi\|^2 &= \frac{1}{4}\left[F\left(\mu\phi + \frac{1}{\mu}\psi\right) - F\left(\mu\phi - \frac{1}{\mu}\psi\right)\right] \\ &\leq \frac{1}{4|\lambda|}\left[\left\|\mu\phi + \frac{1}{\mu}\psi\right\|^2 + \left\|\mu\phi - \frac{1}{\mu}\psi\right\|^2\right] \\ &= \frac{1}{2|\lambda|}\left[\mu^2\|\phi\|^2 + \frac{1}{\mu^2}\|\psi\|^2\right].\end{aligned}$$

Because of our special choice of ψ and μ, we have

$$\|\mathbf{K}\phi\|^2 \leq \frac{1}{|\lambda|}\|\mathbf{K}\phi\|\,\|\phi\|.$$

Therefore, we can now conclude that

$$\|\mathbf{K}\| \leq \frac{1}{|\lambda|}$$

and hence

$$\|\mathbf{K}\| = \frac{1}{|\lambda|}. \tag{5.80}$$

We now see that $1/|\lambda| > 0$ unless $K(s,t)$ is identically zero. Indeed, suppose $\|\mathbf{K}\| = 0$. Then $\mathbf{K}\phi = 0$ for every square integrable function ϕ, and setting $\phi(t) = K(s,t)$ for fixed s shows that

$$\int_0^l |K(s,t)|^2\,dt = 0$$

for all $s \in [0,l]$. This implies that $K(s,t)$ is identically zero. Hence, from now on we assume that $1/|\lambda| > 0$.

Returning now to (5.78), we assume without loss of generality that for ϕ an extremal function, $F(\phi) > 0$. Let $\{\phi_j\}$ denote a sequence such that $\|\phi_j\| = 1$ and

$$\frac{1}{\lambda} = \lim_{j\to\infty} F(\phi_j) > 0. \tag{5.81}$$

(If $F(\phi) < 0$, this inequality is reversed. This has no effect on our analysis.) Define the sequence $\{\psi_j\}$ by

$$\psi_j = \mathbf{K}\phi_j.$$

Then, by the Schwarz inequality,

$$|\psi_j(s_1) - \psi_j(s_2)|^2 \le \int_0^l |K(s_1,t) - K(s_2,t)|^2\,dt,$$

from which we can easily deduce that the set $\{\psi_j(s)\}$ is equicontinuous. The Schwarz inequality also implies that

$$|\psi_j(s)|^2 \le \int_0^l |K(s,t)|^2\,dt \le A,$$

so the set $\{\psi_j(s)\}$ is uniformly bounded. By the Arzela-Ascoli theorem, there exists a uniformly convergent subsequence of $\{\psi_j(s)\}$, which we again denote by $\{\psi_j(s)\}$. Hence,

$$\lim_{j\to\infty} \psi_j(s) = \psi(s)$$

exists uniformly and since the uniform limit of continuous functions is continuous, we can conclude that $\psi(s)$ is continuous on $[0,l]$. We now show that ψ is an eigenfunction of (5.75) with eigenvalue λ. (We can of course then normalize ψ such that $\|\psi\| = 1$.)

By (5.79) and (5.80) we have

$$|F(\phi_j)| \le \|\mathbf{K}\phi_j\| \le \frac{1}{|\lambda|},$$

and hence from (5.81),

$$\|\psi\| = \lim_{j\to\infty} \|\mathbf{K}\phi_j\| = \frac{1}{|\lambda|} > 0, \tag{5.82}$$

so ψ is not the zero function. Now define the operator $\mathbf{R}$ by

$$\mathbf{R}\phi = \frac{1}{\lambda}\phi - \mathbf{K}\phi$$

for all square integrable functions ϕ. Our task will be complete if we can show that

$$\lim_{j\to\infty} \mathbf{R}\psi_j = 0. \tag{5.83}$$

But $\mathbf{R}\psi_j = \mathbf{R}\mathbf{K}\phi_j = \mathbf{K}\mathbf{R}\phi_j$, and by the Schwarz inequality,

$$|\mathbf{R}\psi_j(s)|^2 \le \|\mathbf{R}\phi_j\|^2 \int_0^l |K(s,t)|^2\,dt \le \|\mathbf{R}\phi_j\|^2 A$$

so (5.83) now follows from the identity

$$\|\mathbf{R}\phi_j\|^2 = \left\|\frac{1}{\lambda}\phi_j - \psi_j\right\|^2 = \frac{1}{\lambda^2} - 2\frac{1}{\lambda}F(\phi_j) + \|\psi_j\|^2,$$

which from (5.81) and (5.82) tends to zero as j tends to infinity. This completes the proof of the theorem. ∎

We now want to find other eigenvalues and eigenfunctions. Suppose we have already found n eigenvalues $\lambda_1,\lambda_2,\ldots,\lambda_n$ and n corresponding orthonormal eigenfunctions $\phi_1,\phi_2,\ldots,\phi_n$. Define the symmetric kernel $K_n(s,t)$ by

$$K_n(s,t) = K(s,t) - \sum_{i=1}^{n} \frac{\phi_i(s)\phi_i(t)}{\lambda_i} . \tag{5.84}$$

If $K_n(s,t)$ is not identically zero, then the optimization problem

$$\frac{1}{|\lambda_{n+1}|} = \sup \frac{(\phi,\mathbf{K}_n\phi)}{\|\phi\|^2}$$

where

$$\mathbf{K}_n\phi = \int_0^l K_n(s,t)\phi(t)\,dt$$

provides, as above, an eigenvalue λ_{n+1} and a (normalized) eigenfunction ϕ_{n+1} of the integral equation

$$\phi = \lambda\mathbf{K}_n\phi. \tag{5.85}$$

We shall show that ϕ_{n+1} is orthogonal to $\phi_1,\phi_2,\ldots,\phi_n$ and is therefore an eigenfunction of the original integral equation (5.75). To show this, we see from (5.84) that

$$\mathbf{K}_n\phi_i = \mathbf{K}\phi_i - \frac{\phi_i}{\lambda_i} = 0$$

for $i = 1,2,\ldots,n$. From (5.85) we now see that

$$(\phi_i,\phi_{n+1}) = \lambda_{n+1}(\phi_i,\mathbf{K}_n\phi_{n+1}) = \lambda_{n+1}(\mathbf{K}_n\phi_i,\phi_{n+1}) = 0$$

for $i = 1,2,\ldots,n$. This process can either be continued indefinitely or until a point is reached where $K_n(s,t)$ is identically zero. In the latter case (5.84) implies that $K(s,t)$ is a degenerate kernel, whereas in the former case there exists an infinite sequence of eigenvalues

$$0 < |\lambda_1| \le |\lambda_2| \le \ldots \le |\lambda_n| \le \ldots$$

with associated (orthonormal) eigenfunctions

$$\phi_1,\phi_2,\ldots,\phi_n,\ldots$$

We now show that, in the sense of mean square convergence of the partial sums,

$$K(s,t) = \sum_{i=1}^{\infty} \frac{\phi_i(s)\phi_i(t)}{\lambda_i} . \tag{5.86}$$

By the orthogonality of the eigenfunctions ϕ_i and the inequality (5.77) we can conclude that the series on the right side of (5.86) converges in the mean to a square integrable function defined on $[0,l] \times [0,l]$. To establish (5.86) we observe that

$$\left\| K - \sum_{i=1}^{\infty} \frac{\phi_i\phi_i}{\lambda_i} \right\| = \left\| K_n - \sum_{i=n+1}^{\infty} \frac{\phi_i\phi_i}{\lambda_i} \right\| \leq \|K_n\| + \left\| \sum_{i=n+1}^{\infty} \frac{\phi_i\phi_i}{\lambda_i} \right\|, \tag{5.87}$$

where

$$\|\psi\| = \int_0^l \int_0^l |\psi(s,t)|^2 \, dsdt$$

for ψ a square integrable function defined on $[0,l] \times [0,l]$. But the identity (5.80) applied to $\mathbf{K}_n$ shows that

$$\|\mathbf{K}_n\| = \frac{1}{|\lambda_{n+1}|}$$

and by the convergence of the series in (5.86) and the fact that $|\lambda_{n+1}|$ tends to infinity as n tends to infinity, we can deduce from (5.87) that (5.86) is valid in the sense of mean square convergence.

In preparation for the final theorems of this chapter, we shall now introduce the idea of completeness of a set $\{\phi_i(s)\}$ of square integrable orthonormal functions defined on $[0,l]$. Let $f(s)$ be a square integrable function defined on $[0,l]$ and define the Fourier coefficient c_i by

$$c_i = (f,\phi_i) \qquad (i = 1,2,\ldots).$$

Then, for any integer n,

$$\int_0^l \left| f(s) - \sum_{i=1}^{n} c_i\phi_i(s) \right|^2 ds = \int_0^l |f(s)|^2 \, ds - \sum_{i=1}^{n} |c_i|^2,$$

i.e.,

$$\sum_{i=1}^{n} |c_i|^2 \leq \int_0^l |f(s)|^2 \, ds.$$

Letting n tend to infinity gives **Bessel's inequality**

$$\sum_{i=1}^{\infty} |c_i|^2 \leq \int_0^l |f(s)|^2 \, ds.$$

If $\sum_{i=1}^{n} c_i\phi_i(s)$ converges in the mean to $f(s)$, then we have **Parseval's equality** (generalizing the results of Section 1.4)

$$\sum_{i=1}^{\infty} |c_i|^2 = \int_0^l |f(s)|^2 \, ds.$$

Conversely, Parseval's equality implies that $\Sigma_{i=1}^{n} c_i\phi_i(s)$ converges to $f(s)$. If this is true for every square integrable function $f(s)$, the set $\{\phi_i(s)\}$ is **complete**.

If the set $\{\phi_i(s)\}$ is complete and

$$(f,\phi_i) = \int_0^l f(s)\overline{\phi_i(s)}\,ds = 0 \tag{5.88}$$

for every integer i, then $f(s) = 0$ almost everywhere (cf. Section 5.1), since in this case $c_i = 0$ for every i. Conversely, suppose (5.88) implies that $f(s) = 0$ almost everywhere. We show that in this case the set $\{\phi_i(s)\}$ is complete by supposing the contrary. In that case, there exists a square integrable function $f(s)$ such that

$$\sum_{i=1}^{\infty} |c_i|^2 < \int_0^l |f(s)|^2\,ds.$$

Consider the functions $\psi_j(s)$ defined by

$$\psi_j(s) = \sum_{i=1}^{j} c_i\phi_i(s) \qquad (j = 1,2,\ldots).$$

Then for $j > k$,

$$\|\psi_j - \psi_k\| = \sum_{i=k+1}^{j} |c_i|^2$$

and since $\Sigma_{i=1}^{\infty} |c_i|^2$ is convergent, $\{\psi_j(s)\}$ is a Cauchy sequence. Remember that Cauchy sequences in the space of square integrable functions are convergent in the sense of mean square convergence to a square integrable function (cf. Weir) and hence we can conclude that there exists a square integrable function $\psi(s)$ such that

$$\lim_{j\to\infty} \|\psi_j - \psi\| = 0.$$

But then

$$\int_0^l [f(s) - \psi(s)]\overline{\phi_i(s)}\,ds = 0 \qquad (i = 1,2,\ldots)$$

and

$$\int_0^l |f(s) - \psi(s)|^2\,ds = \int_0^l |f(s)|^2\,ds - \sum_{i=1}^{\infty} |c_i|^2 > 0,$$

and hence $f(s) - \psi(s)$ is not almost everywhere equal to zero, violating our hypothesis. This establishes our result.

Suppose that the set $\{\phi_i(s)\}$ is not orthonormal, but (5.88) implies that $f(s) = 0$ almost everywhere. Then the Gram-Schmidt procedure implies that the set obtained by orthonormalizing the set $\{\phi_i(s)\}$ is complete. Hence, $\{\phi_i(s)\}$ is complete in the sense that for every $\epsilon > 0$ there exist constants $c_1, c_2, \ldots, c_n$, $c_i = c_i(\epsilon)$, such that

$$\int_0^l \left| f(s) - \sum_{i=1}^{n} c_i \phi_i(s) \right|^2 ds < \epsilon.$$

Finally, we note that there is nothing special about the interval $[0,l]$; any other compact interval or contour in R^2 will do.

We now want to determine when a given function $f(s)$ defined on $[0,l]$ can be expanded in a series

$$f(s) = \sum_{i=1}^{\infty} (\phi_i, f)\phi_i(s)$$

of orthonormalized eigenfunctions of (5.75). A first result in this direction is the following theorem:

THEOREM 45

$(f,\phi_i) = 0$ for all of the eigenfunctions ϕ_i of (5.75) if and only if $\mathbf{K}f = 0$. Hence, for $\{\phi_i\}$ to be complete it is necessary and sufficient that $\mathbf{K}f = 0$ implies that $f = 0$.

■ ***Proof.*** Let f be such that $\mathbf{K}f = 0$. Then

$$(f,\phi_i) = \lambda_i(f,\mathbf{K}\phi_i) = \lambda_i(\mathbf{K}f,\phi_i) = 0.$$

On the other hand, if $(f,\phi_i) = 0$ for every ϕ_i then $\mathbf{K}f = \mathbf{K}_n f$ where the kernel $K_n(s,t)$ of $\mathbf{K}_n$ is defined by (5.84). Then

$$\|\mathbf{K}f\| = \|\mathbf{K}_n f\| \leq \|\mathbf{K}_n\| \|f\|,$$

and since $\|\mathbf{K}_n\| = 1/|\lambda_{n+1}|$ we know that $\|\mathbf{K}_n\|$ tends to zero as n tends to infinity. Hence, $\|\mathbf{K}f\| = 0$, implying that $\mathbf{K}f = 0$. The theorem now follows from the above discussion on complete orthonormal sets. ■

THEOREM 46 Hilbert-Schmidt Theorem

Let f be a square integrable function defined on $[0,l]$ such that there exists a square integrable function g where $f = \mathbf{K}g$. Then f has an absolutely and uniformly convergent expansion

$$f = \sum_{i=1}^{\infty} (\phi_i, f)\phi_i = \sum_{i=1}^{\infty} \frac{(\phi_i, g)}{\lambda_i} \phi_i \tag{5.89}$$

in terms of the orthonormalized eigenfunctions of $\mathbf{K}$.

■ ***Proof.*** We first show that (5.89) is absolutely and uniformly convergent. This follows from the fact that by the Schwarz inequality we have

$$\left[\sum_{i=m}^{n} |(\phi_i, f)\phi_i| \right]^2 = \left[\sum_{i=m}^{n} \left| (\phi_i, g) \frac{\phi_i}{\lambda_i} \right| \right]^2 \leq \sum_{i=m}^{n} |(\phi_i, g)|^2 \sum_{i=m}^{n} \frac{\phi_i^2}{\lambda_i^2},$$

and hence by Bessel's inequality and the inequality preceding (5.77), the partial sums of (5.89) form a uniformly convergent Cauchy sequence.

We now must show that the series in (5.89) converges to f. We first note that since $K(s,t)$ is continuous, f and the ϕ_i, $i = 1,2,\ldots$, are continuous. Hence, it suffices to show that the series converges to f in the sense of mean square convergence. But this follows from the inequality

$$\left\| f - \sum_{i=1}^{n} (\phi_i, f)\phi_i \right\| = \left\| \mathbf{K}g - \sum_{i=1}^{n} (\phi_i, g)\frac{\phi_i}{\lambda_i} \right\| = \|\mathbf{K}_n g\| \le \|\mathbf{K}_n\| \, \|g\|,$$

since $\|\mathbf{K}_n\| = 1/|\lambda_{n+1}|$ which tends to zero as n tends to infinity. The theorem is now proved. ■

As an example of the application of the above theory to partial differential equations, consider the Dirichlet eigenvalue problem

$$\begin{aligned} &\Delta_2 u + \lambda u = 0 \quad \text{in} \quad D \\ &u = 0 \quad \text{on} \quad \partial D, \end{aligned} \tag{5.90}$$

where D is a bounded, simply connected domain with C^2 boundary ∂D. We have already shown (see Section 5.1) that (5.90) is equivalent to the integral equation

$$u(\mathbf{x}) = \lambda \iint_D \frac{1}{2\pi} G(\mathbf{x},\mathbf{y})u(\mathbf{y})\, d\mathbf{y}, \tag{5.91}$$

where $G(\mathbf{x},\mathbf{y})$ is the Green's function for Laplace's equation in D. From the theory developed in this section (applied to two-dimensional integral equations) we can conclude that there exists a countable set of eigenvalues $0 < \lambda_1 \le \lambda_2 \le \ldots \le \lambda_n \le \ldots$ and orthonormalized eigenfunctions $u_1(\mathbf{x}), u_2(\mathbf{x}), \ldots$. From Exercise 18 of this chapter, we see that the kernel $(1/2\pi)G(\mathbf{x},\mathbf{y})$ is not degenerate, and hence there exist an infinite number of eigenvalues and eigenfunctions. Note also that all of the eigenvalues are positive since if $\lambda = 0$, $u(\mathbf{x})$ is identically zero by the maximum principle for Laplace's equation and if $\lambda = -k^2 < 0$, then by Green's first identity, using the fact that $u \in C^1(\overline{D})$ (cf. Vekua),

$$\iint_D (k^2|u(\mathbf{x})|^2 + |\nabla u(\mathbf{x})|^2)\, d\mathbf{x} = 0,$$

which implies that $u(\mathbf{x})$ is again identically zero.

We shall now show that any function $v \in C^2(\overline{D})$ such that $v(\mathbf{x}) = 0$ for $\mathbf{x} \in \partial D$ can be expanded in a uniformly convergent series of the $u_i(\mathbf{x})$, i.e.,

$$v(\mathbf{x}) = \sum_{i=1}^{\infty} (u_i, v)u_i(\mathbf{x}).$$

To show this it suffices by the Hilbert-Schmidt theorem to show that there exists a square integrable function $g(\mathbf{x})$ defined on D such that

$$v(\mathbf{x}) = \frac{1}{2\pi} \iint_D G(\mathbf{x},\mathbf{y}) g(\mathbf{y})\, d\mathbf{y}.$$

But from the identity (see Chapter 4)

$$\begin{aligned} v(\mathbf{x}) &= \frac{1}{2\pi} \int_{\partial D} \left(\frac{\partial v}{\partial \nu}(\mathbf{y}) G(\mathbf{x},\mathbf{y}) - v(\mathbf{y}) \frac{\partial}{\partial \nu(\mathbf{y})} G(\mathbf{x},\mathbf{y}) \right) ds(\mathbf{y}) \\ &\quad - \frac{1}{2\pi} \iint_D G(\mathbf{x},\mathbf{y}) \Delta_2 v(\mathbf{y})\, d\mathbf{y} \\ &= -\frac{1}{2\pi} \iint_D G(\mathbf{x},\mathbf{y}) \Delta_2 v(\mathbf{y})\, d\mathbf{y}, \end{aligned} \tag{5.92}$$

we have $g(\mathbf{x}) = -\Delta_2 v(\mathbf{x})$ satisfying the necessary conditions.

We close this section by deriving a monotonicity property of the first eigenvalue for the Dirichlet eigenvalue problem (5.90). This result will be used later when we study the inverse scattering problem in Chapter 6. We begin by considering a real valued function $u \in C^4(\overline{D})$ such that $u(\mathbf{x}) = \Delta_2 u(\mathbf{x}) = 0$ for $\mathbf{x}$ on ∂D. Then, if $u_i(\mathbf{x})$ is an eigenfunction of (5.90) associated with the eigenvalue λ_i, $\|u_i\| = 1$, setting $v(\mathbf{x}) = \Delta_2 u(\mathbf{x})$ yields

$$\Delta_2 u(\mathbf{x}) = \sum_{i=1}^{\infty} (u_i, \Delta_2 u) u_i(\mathbf{x}) = -\sum_{i=1}^{\infty} \frac{a_i}{\lambda_i} u_i(\mathbf{x}), \tag{5.93}$$

where $a_i = (u_i, \Delta_2^2 u)$ and in the last equality we used (5.89). Similarly,

$$u(\mathbf{x}) = \sum_{i=1}^{\infty} (u_i, u) u_i(\mathbf{x}) = -\sum_{i=1}^{\infty} \frac{(u_i, \Delta_2 u)}{\lambda_i} u_i(\mathbf{x}) = \sum_{i=1}^{\infty} \frac{a_i}{\lambda_i^2} u_i(\mathbf{x}). \tag{5.94}$$

Since $u(\mathbf{x}) = 0$ for $\mathbf{x}$ on ∂D, Green's first identity gives

$$\iint_D |\nabla u|^2\, d\mathbf{x} = -\iint_D u \Delta_2 u\, d\mathbf{x}, \tag{5.95}$$

so (5.93)–(5.95) imply that

$$\iint_D |\nabla u|^2\, d\mathbf{x} = \sum_{i=1}^{\infty} \frac{a_i^2}{\lambda_i^3}$$

$$\iint_D |u|^2\, d\mathbf{x} = \sum_{i=1}^{\infty} \frac{a_i^2}{\lambda_i^4}.$$

Thus, the **Rayleigh quotient**

$$R(u) = \frac{\displaystyle\iint_D |\nabla u|^2 \, d\mathbf{x}}{\displaystyle\iint_D |u|^2 \, d\mathbf{x}} \tag{5.96}$$

is always greater than or equal to the first eigenvalue λ_1 for $u(\mathbf{x})$ satisfying the above assumptions. By a suitable approximation process (cf. Garabedian 1964), this conclusion remains valid for functions $u(\mathbf{x})$ that are continuous in $\overline{D}$, piecewise continuously differentiable in $\overline{D}$, and vanish on ∂D. In particular, the first eigenfunction $u_1(\mathbf{x})$ minimizes the Rayleigh quotient, i.e., for all u in the above class of functions, $R(u) \geq \lambda_1$ and $R(u_1) = \lambda_1$.

Now let $D \subset D_1$. Then if $u(\mathbf{x})$ minimizes the Rayleigh quotient over D we can continue $u(\mathbf{x})$ to D_1 by defining $u(\mathbf{x})$ to be zero for $\mathbf{x} \in D_1 \backslash D$. Then the Rayleigh quotient of $u(\mathbf{x})$ over D_1 is the same as the Rayleigh quotient of $u(\mathbf{x})$ over D, i.e., the minimum value of the Rayleigh quotient over D_1 is less than or equal to the minimum value of the Rayleigh quotient over D. Since the minimum value of the Rayleigh quotient is the first eigenvalue, we have the following theorem:

THEOREM 47

Suppose $D \subset D_1$. Let λ_1 be the first Dirichlet eigenvalue of the Laplacian in D and μ_1 the first Dirichlet eigenvalue of the Laplacian in D_1. Then $\mu_1 \leq \lambda_1$.

For further developments on the properties of eigenvalues and eigenfunctions for Laplace's equation we refer the reader to Bandle, Courant and Hilbert 1953, and Garabedian 1964.

5.5 THE NUMERICAL SOLUTION OF FREDHOLM INTEGRAL EQUATIONS OF THE SECOND KIND

We close this chapter by indicating how the Fredholm integral equation of the second kind

$$\phi(s) - \int_0^l K(s,t)\phi(t)\, dt = f(s) \qquad (0 \leq s \leq l) \tag{5.97}$$

can be solved numerically. Without loss of generality, we have set $\lambda = 1$. We assume that $K(s,t)$ is continuous for s, t in the square $0 \leq s \leq l$, $0 \leq t \leq l$, and that $f(s)$ is continuous for $0 \leq s \leq l$. Our aim is to use a simple quadrature formula (the trapezoidal rule) to reduce the integral equation (5.97) to a system of algebraic equations. For more sophisticated quadrature schemes and alternative methods for approximating the solution of (5.97) we refer the reader to Atkinson, Delves and Walsh, and Linz. We note that the proof of the Fredholm alternative shows that (5.97) can be reduced to a system of algebraic equations by approximating the kernel $K(s,t)$ by a degenerate kernel. However, this procedure is nontrivial as well as time-consuming, so in this section we shall present a more straightforward approach known as the **Nyström method**.

We begin by deriving the trapezoidal rule for approximating the integral $\int_0^l f(t)\,dt$, where $f \in C[0,l]$. Let $h = l/n$, $t_k = kh$, where n is a positive integer. Then for each $k = 0,1,2,\ldots,n-1$, approximate $f(t)$ on $[t_k,t_{k+1}]$ by a linear function that agrees with $f(t)$ at the endpoints $t = t_k$ and $t = t_{k+1}$. Let $p(t)$ denote the corresponding piecewise linear function defined over the full interval $[0,l]$:

$$p(t) = \frac{t_{k+1} - t}{h} f(t_k) + \frac{t - t_k}{h} f(t_{k+1}) \qquad (t_k \le t \le t_{k+1}).$$

Then

$$\int_{t_k}^{t_{k+1}} p(t)\,dt = \frac{h}{2}(f(t_k) + f(t_{k+1})),$$

and hence

$$\begin{aligned}\int_0^l p(t)\,dt &= \frac{h}{2}\sum_{k=0}^{n-1}(f(t_k) + f(t_{k+1})) \\ &= \frac{h}{2}\left(f(0) + 2\sum_{k=1}^{n-1} f(t_k) + f(l)\right).\end{aligned} \tag{5.98}$$

Error estimates for

$$\left|\int_0^l (f(t) - p(t))\,dt\right|$$

can be found in Linz.

We now use the trapezoidal rule (5.98) to reduce (5.97) to a system of algebraic equations. Let t_k be as above and let $s_j = t_j$. Then we replace (5.97) by the algebraic system

$$\tilde{\phi}(s_j) - \frac{h}{2}\left[K(s_j,0)\tilde{\phi}(0) + 2\sum_{k=1}^{n-1} K(s_j,t_k)\tilde{\phi}(t_k) + K(s_j,l)\tilde{\phi}(l)\right] = f(s_j) \tag{5.99}$$

for $j = 0,1,2,\ldots,n-1$. This is a system of n linear equations for the n unknowns $\tilde{\phi}(s_j)$, $j = 0,1,2,\ldots,n-1$. Having determined $\tilde{\phi}(s_j)$, $j = 0,1,2,\ldots,n-1$, we can now write a continuous approximation $\tilde{\phi}(s)$ to $\phi(s)$ by using the trapezoidal rule again to approximate the integral in (5.97):

$$\tilde{\phi}(s) = f(s) + \frac{h}{2}\left[K(s,0)\tilde{\phi}(0) + 2\sum_{k=1}^{n-1} K(s,t_k)\tilde{\phi}(t_k) + K(s,l)\tilde{\phi}(l)\right]. \tag{5.100}$$

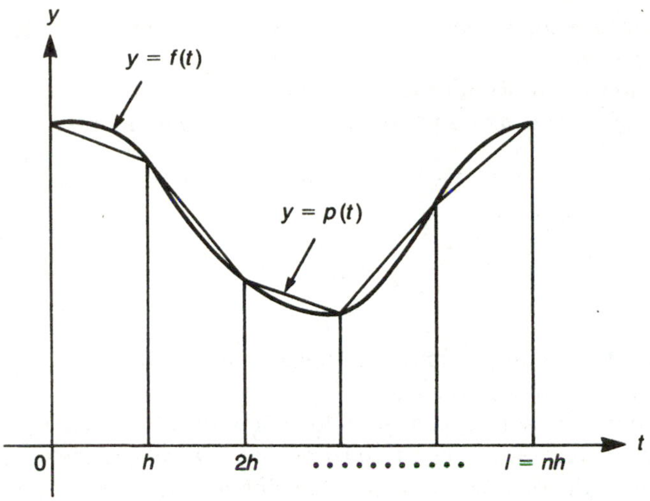

Figure 5.4

For an error analysis of this procedure, we refer the reader to Atkinson, Delves and Walsh, and Linz.

Exercises

1. Show that if you use a double layer potential to solve the interior Dirichlet problem for a disk of radius a, you arrive at the integral equation

$$\psi(s) + \frac{1}{2\pi a}\int_{\partial D} \psi(s)\, ds = -2f(s),$$

where $f(s)$ is the Dirichlet boundary data. Solve this equation for $\psi(s)$ and substitute back into the double layer potential to obtain Poisson's formula for the solution of Dirichlet's problem in a disk.

2. Show that the solution to the exterior Dirichlet problem can be represented in the form

$$u(\mathbf{x}) = \frac{1}{2\pi}\int_{\partial D} \psi(\mathbf{y})\,\frac{\partial}{\partial \nu(\mathbf{y})}\left[\log\frac{1}{|\mathbf{x}-\mathbf{y}|} + \log\left(\frac{|\mathbf{y}|}{R}\left|\mathbf{x} - \frac{R^2\mathbf{y}}{|\mathbf{y}|^2}\right|\right)\right] ds(\mathbf{y}),$$

where $\{\mathbf{x}\colon |\mathbf{x}| \le R\} \subset D$.

3. **(a)** Show that there exists at most one solution $u \in C^2(D) \cap C^1(\overline{D})$ of the Robin problem

$$\Delta_2 u = 0 \qquad \text{in} \qquad D$$

$$\frac{\partial u}{\partial \nu} + u = f \qquad \text{on} \qquad \partial D,$$

where D is a bounded, simply connected domain with C^2 boundary ∂D and unit outward normal $\boldsymbol{\nu}$.

(b) Show that a solution to the Robin problem exists by representing the solution as a single layer potential and applying the Fredholm alternative.

4. **(a)** Use elementary calculus arguments to show that there exists at most one solution $u \in C^2(D) \cap C(\overline{D})$ of

$$\Delta_2 u + q(x)u = 0 \qquad \text{in} \qquad D$$

$$u = f \qquad \text{on} \qquad \partial D,$$

where D is a bounded, simply connected domain with C^2 boundary ∂D, and $q \in C(\overline{D})$ is nonpositive in D.

(b) Use Green's function for Laplace's equation to rewrite the above boundary value problem as a Fredholm integral equation over D, where the nonhomogeneous term is the (known) harmonic function assuming the boundary data $f(\mathbf{x})$ for $\mathbf{x} \in \partial D$. Use the Fredholm alternative to deduce the existence of a solution to this integral equation and hence the existence of a solution to the boundary value problem in part (a).

5. Show that

$$\Delta_2 u + \lambda u = 0 \qquad \text{in} \qquad D$$

$$u = f \qquad \text{on} \qquad \partial D,$$

where $f \in C(\partial D)$ and λ is a positive constant, is soluble provided

$$\int_{\partial D} f \frac{\partial \phi}{\partial \nu} \, ds = 0$$

for every solution $\phi(\mathbf{x})$ of

$$\Delta_2 \phi + \lambda \phi = 0 \qquad \text{in} \qquad D$$

$$\phi = 0 \qquad \text{on} \qquad \partial D.$$

Here D is again a bounded, simply connected domain with C^2 boundary ∂D and unit outward normal $\boldsymbol{\nu}$.

6. Show that the Volterra integral equation

$$\phi(s) - \int_0^s K(s,t)\, \phi(t)\, dt = f(s) \qquad (0 \leq s \leq l)$$

with continuous kernel $K(s,t)$ can always be solved by successive approximations regardless of the magnitude of

$$\int_0^l \int_0^l |K(x,t)|^2 \, dt \, ds.$$

7. Consider the integral equation of the first kind

$$\int_0^1 K(s,t)\,\phi(t)\,dt = f(s),$$

where $K(s,t)$ is a degenerate kernel. What can one say about the existence and uniqueness of solutions to this equation?

8. Let $K(s,t)$ be continuous and symmetric. Show that the integral equation

$$\phi(s) - \lambda \int_0^l K(s,t)\,\phi(t)\,ds = f(s) \qquad (0 \le s \le l)$$

can be solved by successive approximations provided $|\lambda| < \lambda_1$, where λ_1 is the smallest eigenvalue of the homogeneous equation.

9. Let $K(s,t)$ be symmetric and continuous and define the operator $\mathbf{K}$ by

$$\mathbf{K}\phi = \int_0^l K(s,t)\,\phi(t)\,dt.$$

Show that if λ is not an eigenvalue, then the Fredholm integral equation

$$\phi - \lambda\mathbf{K}\phi = f$$

has the solution

$$\phi = f + \lambda \sum_{i=1}^{\infty} \frac{(\phi_i, f)}{\lambda_i - \lambda}\,\phi_i$$

in terms of the eigenvalues λ_i and eigenfunctions ϕ_i of the operator $\mathbf{K}$.

10. Let $\mathbf{K}$, λ_i, and ϕ_i be as in Exercise 9. Show that there exists a unique solution to

$$\mathbf{K}\phi = f$$

if and only if the homogeneous equation has only the trivial solution and $\sum_{i=1}^{\infty} \lambda_i^2 |(\phi_i, f)|^2 < \infty$.

11. Solve the Fredholm integral equations

(a) $$\phi(s) = s + \frac{1}{2}\int_{-1}^{1} (s + t)\phi(t)\,dt.$$

(Answer: $\phi(s) = (3s + 1)/2$.)

(b) $$\phi(s) = 1 + \int_0^{2\pi} \cos(s - t)\,\phi(t)\,dt.$$

(Answer: $\phi(s) = 1$.)

12. For what values of λ does the equation

$$\phi(s) - \lambda \int_0^1 \frac{s-t}{(s+t)(t+1)} \phi(t)\, dt = 0$$

have a nonzero solution? For what values of λ does the equation

$$\phi(s) - \lambda \int_0^1 \frac{s-t}{(s+t)(t+1)} \phi(t)\, dt = s$$

have solutions? What are the solutions?

13. Determine the eigenvalues and eigenfunctions of the integral equation

$$\phi(s) - \lambda \int_0^1 st\, \phi(t)\, dt = 0.$$

State explicitly what the Hilbert-Schmidt theorem says in this case.

14. Show that the integral equation

$$\phi(s) - \lambda \int_0^\pi \sin s \cos t\, \phi(t)\, dt = 0$$

has no eigenvalues.

15. **(a)** Show that $K(s,t) = \log[1 - \cos(s-t)]$ can be expanded in the form

$$K(s,t) = -\log 2 - 2\sum_{n=1}^{\infty} \frac{\cos ns \cos nt}{n} - 2\sum_{n=1}^{\infty} \frac{\sin ns \sin nt}{n}.$$

(b) Show that the eigenvalues of

$$\phi(s) - \lambda \int_0^{2\pi} K(s,t)\phi(t)\, dt = 0$$

are $\lambda_0 = -1/(2\pi \log 2)$, $\lambda_n = -n/2\pi$, $n = 1,2,\ldots$, with eigenfunctions $\phi_0(s) = c_0$, $\phi_n(s) = c_1 \cos ns + c_2 \sin ns$ where c_0, c_1, c_2 are constants.

16. Show that if

$$K(s,t) = \sum_{n=1}^{\infty} \frac{\sin (n+1)s \sin nt}{n^2} \qquad (0 \le s,t \le \pi),$$

then the integral equation

$$\phi(s) - \lambda \int_0^\pi K(s,t)\phi(t)\, dt = 0$$

has no eigenvalues.

17. Consider the **Lalesco-Picard equation**

$$\phi(s) - \lambda \int_{-\infty}^{\infty} e^{-|s-t|}\, \phi(t)\, dt = 0.$$

Show that if $\phi(s)$ is twice continuously differentiable, then $\phi(s)$ is a solution of the differential equation

$$\phi'' + (2\lambda - 1)\, \phi = 0.$$

Deduce that there exists no square integrable solution of the integral equation for any value of λ, i.e., eigenvalues do not exist. Why doesn't the theory developed in this chapter apply?

18. Show that Green's function for Laplace's equation is not a degenerate kernel.

19. Show that if $\Omega = \{\mathbf{x}\colon |\mathbf{x}| < 1\}$, then all of the eigenvalues of

$$\Delta_2 u + \lambda u = 0 \quad \text{in} \quad \Omega$$

$$u = 0 \quad \text{on} \quad \partial\Omega$$

are given by $\lambda_{mn} = k_{mn}^2$ where k_{mn} is the nth root of the Bessel function $J_m(k)$. In particular, show that $\lambda_1 = k_{01}^2$.

20. Let D be a bounded, simply connected domain with C^2 boundary ∂D, and consider the eigenvalue problem

$$\Delta_2 u + \lambda u = 0 \quad \text{in} \quad D$$

$$\frac{\partial u}{\partial \nu} = 0 \quad \text{on} \quad \partial D,$$

where $\boldsymbol{\nu}$ is the unit outward normal to ∂D. Show that the eigenvalues are real and nonnegative. Find all the eigenvalues and eigenfunctions when D is the unit circle.

21. Let D be a bounded, simply connected domain with C^2 boundary ∂D, and consider the eigenvalue problem

$$\Delta_2 u + \lambda u = 0 \quad \text{in} \quad D$$

$$\frac{\partial u}{\partial \nu} + hu = 0 \quad \text{on} \quad \partial D,$$

where $h = h(\mathbf{x})$ is a positive function defined for $\mathbf{x} \in D$. Show that the eigenvalues are real and positive and that eigenvalues corresponding to different eigenvalues are orthogonal. Find all the eigenvalues and eigenfunctions when h is a constant and D is the unit circle.

22. **(a)** Determine the eigenvalues and eigenfunctions of

$$\phi(s) - \lambda \int_0^1 K(s,t)\phi(t)\,dt = 0,$$

where

$$K(s,t) = \begin{cases} s(1-t) & (s \le t) \\ t(1-s) & (s \ge t). \end{cases}$$

(Answer: $\lambda_n = \pi^2 n^2$, $\phi_n(s) = \sin n\pi s$.**)**

(b) Show that if $f \in C^2[0,1]$, $f(0) = f(1) = 0$, then $f(s)$ can be represented in the form

$$f(s) = \int_0^1 K(s,t)h(t)\,dt,$$

where $h \in C[0,1]$ and $K(s,t)$ is as in part (a). Deduce from this that $f(s)$ can be expanded in the uniformly convergent Fourier series

$$f(s) = \sum_{n=1}^{\infty} a_n \sin n\pi s,$$

where

$$a_n = 2\int_0^1 f(s) \sin n\pi s\,ds.$$

23. Use the Rayleigh quotient applied to the trial function

$$u(x,y) = \frac{30}{\pi^5} xy(\pi - x)(\pi - y)$$

to obtain an estimate for the first eigenvalue λ_1 of

$$\Delta_2 u + \lambda u = 0 \quad \text{in} \quad D$$

$$u = 0 \quad \text{on} \quad \partial D,$$

where D is the square $D = \{(x,y) : 0 < x < \pi,\ 0 < y < \pi\}$. Use the method of separation of variables to obtain the exact value of λ_1.

(Answer: $R(u) = 2.03$, $\lambda_1 = 2$.**)**

*Chapter 6

Scattering Theory

This chapter is devoted to mathematical problems connected with the scattering of time harmonic acoustic and electromagnetic waves and hence with the theory of Bessel functions. Aside from the trigonometric, exponential, logarithmic, and other elementary functions encountered in calculus, Bessel functions are probably the most widely used special functions of mathematical physics. Unfortunately, an elementary treatment such as that presented in Chapter 2 is the only familiarity with them that most students acquire. This defect in mathematical education presents a barrier to many areas of classical mathematical physics, particularly the study of the scattering of acoustic and electromagnetic waves. We wish to remedy this defect, so early in this chapter we present sufficient information on Bessel functions for the reader to proceed to most areas of application, especially our subsequent development of mathematical methods in scattering theory. We begin by presenting the basic facts on the gamma function that are needed for the development of the theory of Bessel functions, and then follow this by a discussion of the analytic properties of solutions to Bessel's differential equation. Of particular importance is the asymptotic behavior of Bessel functions, and to this end we derive in Section 6.2.5 the method of stationary phase for integrals having an oscillatory integrand and then apply this general result to the various integral representations for Bessel functions obtained in Section 6.2.4. The reader should note that our presentation touches only the surface of existing knowledge. For a comprehensive treatment see Watson.

We then use the theory of Bessel functions with our previously derived results on Fredholm integral equations to examine the scattering of acoustic and electromagnetic waves by an infinite cylinder of arbitrary cross section. We show that under appropriate assumptions both of these problems lead to exterior boundary value problems for the Helmholtz equation and then proceed to use the method of integral equations and potential theory to solve these boundary value problems. Although the approach is similar to that used in Chapter 5 for Laplace's equation, significant differences occur in the case of the Helmholtz equation because of the presence of interior eigenvalues. We conclude this chapter by considering the inverse scattering problem of determining the shape of the scattering obstacle from a knowledge of the scattered field at large distances from the obstacle. This provides another example (cf. Sections 3.1, 3.3, and 3.6) of an improperly posed problem that nevertheless arises in a realistic physical situation; indeed this problem is central to recent developments in radar, sonar, medical imaging, and nondestructive testing. The inverse scattering problem is not amenable to the use of linear methods of analysis, and provides yet another example (cf. Sections 2.6, 3.6, and 3.8) of the need to examine nonlinear problems in order to investigate the physical world.

6.1 THE GAMMA FUNCTION

The **gamma function** $\Gamma(z)$ is defined by

$$\Gamma(z) = \int_0^\infty e^{-t} t^{z-1} dt \tag{6.1}$$

for Re $z > 0$. Writing $\Gamma(z)$ in the form

$$\Gamma(z) = \int_0^1 e^{-t} t^{z-1} dt + \int_1^\infty e^{-t} t^{z-1} dt, \tag{6.2}$$

we note that the integrand in each integral is analytic in z and continuous in z and t for Re $z > 0$, $0 < t < \infty$. The first integral is uniformly convergent for Re $z \geq \delta > 0$, while the second integral is uniformly convergent for Re $z \leq A < \infty$. Hence, we can conclude from Morera's theorem (Section 1.5.2) and an interchange of orders of integration that $\Gamma(z)$ is analytic in the half plane Re $z > 0$. For Re $z > 0$ we can write

$$\begin{aligned} \int_0^1 e^{-t} t^{z-1} dt &= \int_0^1 t^{z-1} \left[\sum_{k=0}^\infty \frac{(-1)^k}{k!} t^k \right] dt \\ &= \sum_{k=0}^\infty \frac{(-1)^k}{k!} \int_0^1 t^{k+z-1} dt \\ &= \sum_{k=0}^\infty \frac{(-1)^k}{k!} \frac{1}{z+k}; \end{aligned} \tag{6.3}$$

it is easily verified that the orders of integration and summation can be interchanged. Since for $|z + k| \geq \delta > 0$, $k = 0,1,2,\ldots$, the above series is majorized by

$$\sum_{k=0}^{\infty} \frac{1}{k!\delta},$$

we can conclude that (6.3) defines a meromorphic function of z (i.e., analytic except for singularities that are poles) with simple poles at $z = 0,-1,-2,\ldots$ and thus provides the analytic continuation of the integral on the left side of (6.3) into the left half plane. Since the second integral in (6.2) is an entire function of z, we can conclude that $\Gamma(z)$ is a meromorphic function of z with simple poles at $z = 0,-1,-2,\ldots$ with residue $(-1)^n/n!$ at the pole $z = -n$.

THEOREM 48

(a) $\Gamma(z + 1) = z\Gamma(z) \qquad (z \neq 0,-1,-2,\ldots).$

(b) $\Gamma(z)\Gamma(1 - z) = \dfrac{\pi}{\sin\pi z} \qquad (z \neq 0,\pm1,\pm2,\ldots).$

■ ***Proof***

(a) Assume Re $z > 0$. Then

$$\Gamma(z + 1) = \int_0^\infty e^{-t}t^z dt = -e^{-t}t^z\Big|_0^\infty + z\int_0^\infty e^{-t}t^{z-1}dt = z\Gamma(z),$$

and for $z \neq 0,-1,-2,\ldots$ the result follows by the principle of analytic continuation.

(b) Assume $0 < \text{Re } z < 1$. Then

$$\Gamma(z)\Gamma(1 - z) = \int_0^\infty\int_0^\infty e^{-(s+t)}s^{-z}t^{z-1}ds\,dt.$$

Setting $u = s + t$, $v = t/s$, we have

$$\Gamma(z)\Gamma(1 - z) = \int_0^\infty\int_0^\infty e^{-u}v^{z-1}\frac{du\,dv}{1 + v} = \int_0^\infty \frac{v^{z-1}}{1 + v}dv.$$

To evaluate this last integral consider

$$\int_C \frac{z^{\alpha-1}}{1 + z}dz \qquad (0 < \alpha < 1),$$

where C is shown in Figure 6.1 and $z^{\alpha-1}$ is defined on the complex plane cut from 0 to ∞, i.e., $z^{\alpha-1} = e^{(\alpha-1)\log z}$ where $z = re^{i\theta}$, $0 \leq \theta < 2\pi$.

The only pole of the integrand is at $z = -1$ with residue $e^{(\alpha-1)\pi i}$. It is easily verified that the integrals around the circles of radius ρ and radius R go to zero as ρ tends to zero and R tends to infinity respectively. Hence by the residue theorem,

$$\int_0^\infty \frac{z^{\alpha-1}}{1 + z}dz + \int_\infty^0 \frac{(ze^{2\pi i})^{\alpha-1}}{1 + z}dz = 2\pi i e^{(\alpha-1)\pi i},$$

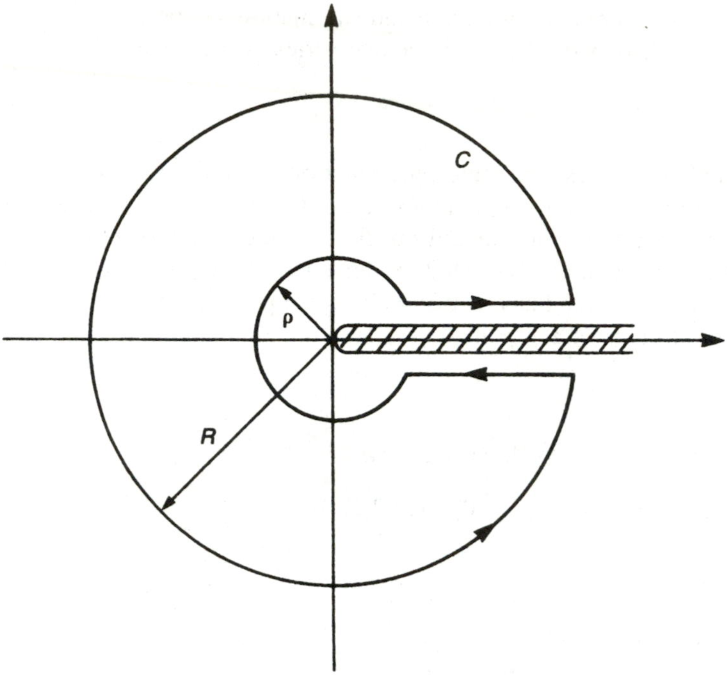

Figure 6.1

or

$$\int_0^\infty \frac{z^{\alpha-1}}{1+z}\,dz = \frac{\pi}{\sin \alpha\pi}\,.$$

Part (b) of the theorem now follows for $0 < \text{Re } z < 1$, and by analytic continuation for $z \neq 0, \pm 1, \pm 2, \ldots$. ■

We now give some applications of Theorem 48.

COROLLARY 1

(a) $\Gamma(n+1) = n! \qquad (n = 0,1,2,\ldots)$.
(b) $\Gamma(\frac{1}{2}) = \sqrt{\pi}$

$$\Gamma(n+\tfrac{1}{2}) = \frac{1\cdot 3\cdot 5 \ldots (2n-1)}{2^n}\sqrt{\pi} \qquad (n = 1,2,\ldots)\,.$$

■ ***Proof***

(a) This follows from part (a) of Theorem 48, induction, and the fact that $\Gamma(1) = 1$.

(b) $\Gamma(\frac{1}{2}) = \int_0^\infty e^{-t}t^{-1/2}dt = 2\int_0^\infty e^{-u^2}du = \sqrt{\pi}$ and the expression for $\Gamma(n+\frac{1}{2})$ now follows from part (a) of the above theorem. ■

COROLLARY 2

$1/\Gamma(z)$ is an entire function of z.

■ ***Proof.*** $\Gamma(n) \neq 0$ for $n = 0, \pm 1, \pm 2, \ldots$, since $\Gamma(n) = (n-1)!$ for $n = 1, 2, \ldots$, and $\Gamma(n) = \infty$ for $n = 0, -1, -2, \ldots$. If any other value of z were a zero of $\Gamma(z)$, then from part (b) of Theorem 48 it would have to be a pole of $\Gamma(1-z)$, which is impossible. Hence $\Gamma(z)$ has no zeros in the complex plane, and the derivative of $1/\Gamma(z)$ exists for each z and is equal to $\Gamma'(z)/\Gamma^2(z)$. So $1/\Gamma(z)$ is analytic for all z in the complex plane and is an entire function of z. ■

COROLLARY 3

$$\frac{1}{\Gamma(z)} = \frac{1}{2\pi i}\int_C e^t t^{-z}\,dt,$$

where C is shown in Figure 6.2, and $t^{-z} = e^{-z \log t}$, where the complex plane is cut along the negative real axis.

■ ***Proof.*** Let $-\infty < z < 0$. Then as a tends to zero the integral around the circle of radius a tends to zero. Hence, in the limit,

$$\frac{1}{2\pi i}\int_C e^t t^{z-1}\,dt = -\frac{1}{2\pi i}\int_\infty^0 e^{-\rho}(\rho e^{-\pi i})^{-z}\,d\rho - \frac{1}{2\pi i}\int_0^\infty e^{-\rho}(\rho e^{\pi i})^{-z}\,d\rho$$

$$= \frac{(e^{\pi i z} - e^{-\pi i z})}{2\pi i}\int_0^\infty e^{-\rho}\rho^{-z}\,d\rho$$

$$= \frac{\sin \pi z}{\pi}\Gamma(1-z) = \frac{1}{\Gamma(z)}$$

from part (b) of the above theorem. The validity of the identity for all complex z now follows by analytic continuation. ■

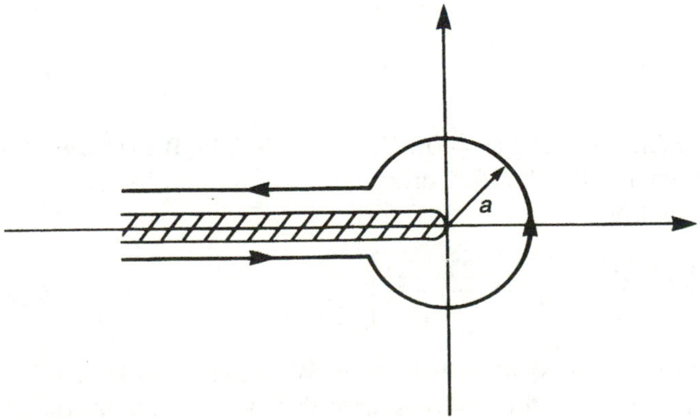

Figure 6.2

6.2 BESSEL FUNCTIONS

6.2.1 Definitions

Bessel's differential equation is

$$y'' + \frac{1}{z}y' + \left(1 - \frac{\nu^2}{z^2}\right)y = 0, \tag{6.4}$$

where we assume the parameter ν to be real. By direct calculation and the ratio test we see that

$$J_\nu(z) = \sum_{k=0}^{\infty} \frac{(-1)^k}{\Gamma(k+1)\Gamma(k+\nu+1)}\left(\frac{z}{2}\right)^{\nu+2k} \tag{6.5}$$

is a solution of (6.4) that is analytic in the complex plane (cut along $(-\infty,0]$ if $\nu \neq 0,\pm1,\pm2,\ldots$). $J_\nu(z)$ is called the **Bessel function of the first kind** of order ν. For negative integral $\nu = -n$, $n = 1,2,\ldots$, the first n terms of the series (6.5) vanish, and we have

$$J_{-n}(z) = \sum_{k=n}^{\infty} \frac{(-1)^k}{k!(k-n)!}\left(\frac{z}{2}\right)^{-n+2k} = \sum_{s=0}^{\infty} \frac{(-1)^{n+s}}{(n+s)!s!}\left(\frac{z}{2}\right)^{n+2s} = (-1)^n J_n(z).$$

On the other hand, if ν is not an integer $J_\nu(z)$ and $J_{-\nu}(z)$ are linearly independent solutions of Bessel's equation, since as z tends to zero,

$$\begin{aligned} J_\nu(z) &= \frac{1}{\Gamma(1+\nu)}\left(\frac{z}{2}\right)^{\nu} + O(z^{\nu+2}) \\ J_{-\nu}(z) &= \frac{1}{\Gamma(1-\nu)}\left(\frac{z}{2}\right)^{-\nu} + O(z^{-\nu+2}). \end{aligned} \tag{6.6}$$

Hence, for ν not an integer, a solution of Bessel's equation that is independent of $J_\nu(z)$ is given by

$$Y_\nu(z) = \frac{J_\nu(z)\cos\nu\pi - J_{-\nu}(z)}{\sin\nu\pi} \tag{6.7}$$

valid for z in the complex plane cut along $(-\infty,0]$. $Y_\nu(z)$ is called the **Bessel function of the second kind,** or Neumann function, of order ν.

For ν an integer, we define

$$Y_n(z) = \lim_{\nu\to n} Y_\nu(z) = \frac{1}{\pi}\left[\frac{\partial J_\nu(z)}{\partial\nu} - (-1)^n \frac{\partial J_{-\nu}(z)}{\partial\nu}\right]_{\nu=n}. \tag{6.8}$$

$Y_n(z)$ is clearly a solution of Bessel's equation for $\nu = n$. We want to show that $J_n(z)$ and $Y_n(z)$ are linearly independent. To this end, we note that because the residue of

$\Gamma(z)$ at the simple pole $z = -n$, $n = 0,1,2,\ldots$, is $(-1)^n/n!$, we have

$$\lim_{z\to n} \frac{\Gamma'(z)}{\Gamma^2(z)} = (-1)^{n+1}n! \qquad (n = 0,1,2,\ldots).$$

Hence from (6.5) and (6.8) we have for $n = 1,2,3,\ldots$ that as z tends to zero

$$Y_n(z) = -\frac{(n-1)!}{\pi}\left(\frac{z}{2}\right)^{-n} + \begin{cases} O(z\log z) & (n=1) \\ O(z^{-n+2}) & (n>1), \end{cases} \tag{6.9}$$

whereas for $n = 0$

$$Y_0(z) = \frac{2}{\pi}\log\frac{z}{2} + O(1) \tag{6.10}$$

as z tends to zero. Noting that from (6.8)

$$Y_{-n}(z) = (-1)^n Y_n(z)$$

for all integers n, we can now conclude that $J_n(z)$ and $Y_n(z)$ are linearly independent for all integers $n = 0,\pm1,\pm2,\ldots$. From the definition of $Y_n(z)$ it follows that $Y_n(z)$ is analytic in the complex plane cut along $(-\infty,0]$.

Finally, we define the **Hankel functions** $H_\nu^{(1)}(z)$ and $H_\nu^{(2)}(z)$ by

$$\begin{aligned} H_\nu^{(1)}(z) &= J_\nu(z) + iY_\nu(z) \\ H_\nu^{(2)}(z) &= J_\nu(z) - iY_\nu(z) \end{aligned} \tag{6.11}$$

for ν an arbitrary real number and z any point of the complex plane cut along $(-\infty,0]$. Clearly, the Hankel functions are linearly independent and each is linearly independent of $J_\nu(z)$.

6.2.2 Wronskians

If $y_1(z)$ and $y_2(z)$ are two solutions of Bessel's equation, the Wronskian

$$W(y_1(z),y_2(z)) = \begin{vmatrix} y_1(z) & y_2(z) \\ y_1'(z) & y_2'(z) \end{vmatrix}$$

can be computed as follows. Since

$$(zy_1')' + \left(z - \frac{\nu^2}{z}\right)y_1 = 0$$

$$(zy_2')' + \left(z - \frac{\nu^2}{z}\right)y_2 = 0,$$

we subtract the first equation multiplied by y_2 from the second equation multiplied by y_1 and find that

$$\frac{d}{dz}\left[zW(y_1(z),y_2(z))\right] = 0,$$

i.e.,

$$W(y_1(z), y_2(z)) = \frac{C}{z},$$

where C is a constant. This constant can be evaluated from the relation

$$C = \lim_{z \to 0} zW(y_1(z), y_2(z))$$

where we make use of (6.6), (6.7), and (6.11), noting that (6.6) can be differentiated with respect to z. For example, choosing $y_1(z) = J_\nu(z)$ and $y_2(z) = J_{-\nu}(z)$ where ν is not an integer and using (6.6), we have

$$C = \lim_{z \to 0} \frac{-2\nu}{\Gamma(1+\nu)\Gamma(1-\nu)}[1 + O(z^2)] = -\frac{2 \sin \nu\pi}{\pi},$$

i.e.,

$$W(J_\nu(z), J_{-\nu}(z)) = -\frac{2 \sin \nu\pi}{\pi z}. \qquad \textbf{(6.12)}$$

The validity of (6.12) for integral ν follows by continuity and we find that $W(J_n(z), J_{-n}(z)) = 0$ as expected, since $J_n(z)$ and $J_{-n}(z)$ are linearly dependent. In the same way we find that

$$\begin{aligned} W(J_\nu(z), Y_\nu(z)) &= \frac{2}{\pi z} \\ W(J_\nu(z), H_\nu^{(1)}(z)) &= \frac{2i}{\pi z} \\ W(H_\nu^{(1)}(z), H_\nu^{(2)}(z)) &= -\frac{4i}{\pi z} \end{aligned} \qquad \textbf{(6.13)}$$

and so on.

6.2.3 A Generating Function for $J_n(z)$

Consider the function $\exp[(1/2)z(t - (1/t))]$. For fixed z this is an analytic function of t in any annulus $0 < \rho_1 \le |t| \le \rho_2$. It therefore has a Laurent expansion

$$\exp\left[\frac{1}{2}z\left(t - \frac{1}{t}\right)\right] = \sum_{n=-\infty}^{\infty} a_n(z)t^n$$

where

$$a_n(z) = \frac{1}{2\pi i}\int_C t^{-n-1} \exp\left[\frac{1}{2}z\left(t - \frac{1}{t}\right)\right]dt \qquad \textbf{(6.14)}$$

and C is a positively oriented simple closed curve enclosing the origin. Setting $t = 2u/z$ and deforming C gives

$$a_n(z) = \frac{1}{2\pi i}\left(\frac{z}{2}\right)^n \int_{|u|=1} u^{-n-1} \exp\left[u - \frac{z^2}{4u}\right] du$$
$$= \frac{1}{2\pi i} \sum_{k=0}^{\infty} \frac{(-1)^k}{k!}\left(\frac{z}{2}\right)^{n+2k} \int_{|u|=1} u^{-n-k-1} e^u du,$$

and computing the residue,

$$a_n(z) = \sum_{k=0}^{\infty} \frac{(-1)^k}{k!\Gamma(k+n+1)}\left(\frac{z}{2}\right)^{n+2k}.$$

Hence,

$$a_n(z) = J_n(z),$$

i.e.,

$$\exp\left[\frac{1}{2} z\left(t - \frac{1}{t}\right)\right] = \sum_{n=-\infty}^{\infty} J_n(z) t^n \qquad (t \neq 0). \tag{6.15}$$

We now give two applications of (6.14) and (6.15). Setting $t = e^{i\theta}$ in (6.14) gives

$$J_n(z) = \frac{1}{2\pi} \int_{-\pi}^{\pi} \exp\,(iz \sin\theta - in\theta) d\theta$$

or

$$J_n(z) = \frac{1}{\pi} \int_0^{\pi} \cos\,(z \sin\theta - n\theta) d\theta. \tag{6.16}$$

The integral representation (6.16) is often much more useful than the series expansion (6.5). Setting $z = kr$ and $t = ie^{i\theta}$ in (6.15) gives

$$\exp\,[ikr\cos\theta] = \sum_{n=-\infty}^{\infty} i^n J_n(kr) e^{in\theta}. \tag{6.17}$$

This is the Bessel function expansion of the "plane wave" $\exp\,[ikr\cos\theta]$ and is very useful in scattering theory (cf. Section 6.3).

6.2.4 Integral Representations

We begin by generalizing (6.16) to $J_\nu(z)$ for arbitrary real ν. To this end, we recall the representation

$$\frac{1}{\Gamma(k+\nu+1)} = \frac{1}{2\pi i} \int_C e^s s^{-(k+\nu+1)} ds,$$

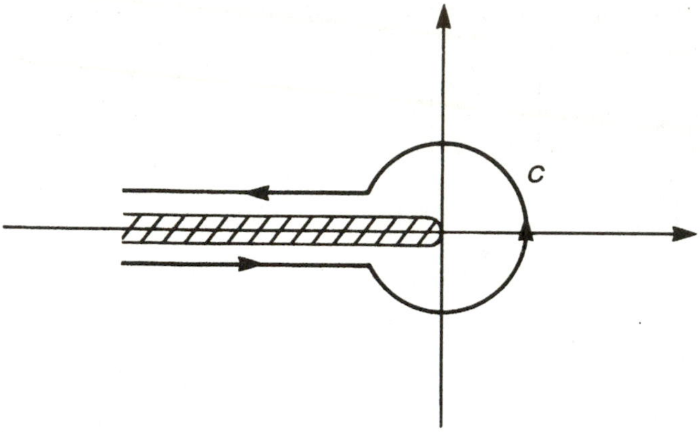

Figure 6.3

where C is shown in Figure 6.3. Substituting this into the series expansion of $J_\nu(z)$ and interchanging orders of integration and summation (which can be easily shown to be valid) we have

$$\begin{aligned} J_\nu(z) &= \sum_{k=0}^{\infty} \frac{(-1)^k}{\Gamma(k+1)} \left(\frac{z}{2}\right)^{\nu+2k} \frac{1}{2\pi i} \int_C e^s s^{-(k+\nu+1)} ds \\ &= \left(\frac{z}{2}\right)^{\nu} \frac{1}{2\pi i} \int_C e^s s^{-\nu-1} \sum_{k=0}^{\infty} \frac{(-1)^k}{\Gamma(k+1)}{}^k \left(\frac{z^2}{4s}\right)^k ds \\ &= \left(\frac{z}{2}\right)^{\nu} \frac{1}{2\pi i} \int_C s^{-\nu-1} \exp\left(s - \frac{z^2}{4s}\right) ds. \end{aligned} \tag{6.18}$$

Assuming z to be positive and setting $s = zt/2$, we can write (6.18) in the form (after deforming the contour back to C)

$$J_\nu(z) = \frac{1}{2\pi i} \int_C t^{-\nu-1} \exp\left[\frac{1}{2} z\left(t - \frac{1}{t}\right)\right] dt. \tag{6.19}$$

By analytic continuation, (6.19) is valid for $\operatorname{Re} z > 0$. Setting $t = \rho e^{i\theta}$ and choosing the circular part of C to have radius one, we have

$$\begin{aligned} J_\nu(z) &= \frac{1}{\pi} \int_0^{\pi} \cos(z \sin\theta - \nu\theta) d\theta \\ &\quad - \frac{\sin \nu\pi}{\pi} \int_1^{\infty} \rho^{-\nu-1} \exp\left[-\frac{1}{2} z\left(\rho - \frac{1}{\rho}\right)\right] d\rho \qquad (\operatorname{Re} z > 0), \end{aligned}$$

or, setting $\rho = e^{\alpha}$,

$$\begin{aligned} J_\nu(z) &= \frac{1}{\pi} \int_0^{\pi} \cos(z \sin\theta - \nu\theta) d\theta \\ &\quad - \frac{\sin \nu\pi}{\pi} \int_0^{\infty} \exp[-z \sinh\alpha - \nu\alpha] d\alpha \end{aligned} \tag{6.20}$$

valid for $\operatorname{Re} z > 0$. Note that for $\nu = n$ we recover our previous integral representation for $J_n(z)$.

Now assume $\operatorname{Re} z > 0$ and ν is nonintegral. Then from (6.20) and the definition of $Y_\nu(z)$ we have

$$\begin{aligned} Y_\nu(z) = \frac{\cot \nu\pi}{\pi} &\int_0^\pi \cos(z \sin\theta - \nu\theta)\,d\theta \\ &- \frac{\cos \nu\pi}{\pi} \int_0^\infty \exp[-z \sinh\alpha - \nu\alpha]\,d\alpha \\ &- \frac{\csc \nu\pi}{\pi} \int_0^\pi \cos(z\sin\theta + \nu\theta)\,d\theta \\ &- \frac{1}{\pi}\int_0^\infty \exp[-z\sinh\alpha + \nu\alpha]\,d\alpha, \end{aligned}$$

or, replacing θ by $\pi - \theta$ in the third integral on the right side and performing some simple calculations,

$$\begin{aligned} Y_\nu(z) = \frac{1}{\pi} &\int_0^\pi \sin(z\sin\theta - \nu\theta)\,d\theta \\ &- \frac{1}{\pi}\int_0^\infty e^{-z\sinh\alpha}(e^{\nu\alpha} + e^{-\nu\alpha}\cos\nu\pi)\,d\alpha \qquad (\operatorname{Re} z > 0). \end{aligned} \tag{6.21}$$

Since both sides are entire functions of ν, it follows by analytic continuation that (6.21) is valid for all real ν and not just nonintegral values of ν.

From (6.20), (6.21), and the definition of the Hankel functions we can now immediately write down integral representations of $H_\nu^{(1)}(z)$ and $H_\nu^{(2)}(z)$ for $\operatorname{Re} z > 0$. However, since we shall have no need of these representations in what follows, we shall not present these formulas here.

6.2.5 Asymptotic Expansions for Positive z

Let $z = x > 0$. Then since $\sinh\alpha \geq \alpha$ for $0 \leq \alpha < \infty$, we easily see from (6.20) and (6.21) that

$$\begin{aligned} J_\nu(x) &= \frac{1}{\pi}\int_0^\pi \cos(x\sin\theta - \nu\theta)\,d\theta + O\left(\frac{1}{x}\right) \\ Y_\nu(x) &= \frac{1}{\pi}\int_0^\pi \sin(x\sin\theta - \nu\theta)\,d\theta + O\left(\frac{1}{x}\right) \end{aligned} \tag{6.22}$$

as x tends to infinity. Hence, in order to determine the asymptotic behavior of $J_\nu(x)$ and $Y_\nu(x)$ (and hence $H_\nu^{(1)}(x)$ and $H_\nu^{(2)}(x)$) for large positive x it suffices to determine how

$$\int_0^\pi \exp(ix\sin\theta - i\nu\theta)\,d\theta = \int_0^\pi e^{ix\sin\theta}e^{-i\nu\theta}\,d\theta$$

behaves as x tends to infinity, assuming this decays slower than $O(1/x)$. To this end we have the following theorem:

THEOREM 49 Method of Stationary Phase

Consider the integral

$$f(x) = \int_a^b e^{ix\phi(t)} g(t)dt,$$

where $\phi(t)$ and $g(t)$ are analytic in some complex neighborhood of $[a,b]$, $\phi(t)$ is real valued on $[a,b]$, and there exists only one point t_0 in $[a,b]$ such that $\phi'(t_0) = 0$. Assume further that $t_0 \in (a,b)$ and $\phi''(t_0)$ does not equal zero. Then as x tends to infinity,

$$f(x) = \sqrt{\frac{2\pi}{x|\phi''(t_0)|}}\, g(t_0) \exp\left(ix\phi(t_0) \pm i\frac{\pi}{4}\right) + O\left(\frac{1}{x}\right),$$

where the plus sign applies if $\phi''(t_0) > 0$ and the minus sign if $\phi''(t_0) < 0$.

■ ***Proof.*** The proof is based on showing that it suffices to consider an integral of a particularly simple form.

We begin by observing that it suffices to consider

$$\int_{t_0-\delta}^{t_0+\delta} e^{ix\phi(t)} g(t)dt \tag{6.23}$$

for any $\delta > 0$. This follows because the remaining intervals of integration are such that $\phi'(t)$ does not vanish, and hence if $[\alpha,\beta]$ is such an interval,

$$\int_\alpha^\beta e^{ix\phi(t)} g(t)dt = \int_\alpha^\beta \frac{g(t)}{ix\phi'(t)} \frac{d}{dt}(e^{ix\phi(t)})dt$$
$$= \frac{g(t)}{ix\phi'(t)} e^{ix\phi(t)}\Bigg|_\alpha^\beta - \frac{1}{ix}\int_\alpha^\beta e^{ix\phi(t)} \frac{d}{dt}\left(\frac{g(t)}{\phi'(t)}\right)dt = O\left(\frac{1}{x}\right)$$

as x tends to infinity.

We next note that instead of (6.23) it suffices to consider the integral

$$e^{ix\phi(t_0)} \int_{-\delta_1}^{\delta_2} e^{\pm ixt^2} h(t)dt, \tag{6.24}$$

where

$$h(0) = g(t_0)\sqrt{\frac{2}{|\phi''(t_0)|}}$$

and δ_1, δ_2 are arbitrarily small positive numbers. To see this, we make the change of variables in (6.23) defined by

$$\tau = \sqrt{\pm\frac{\phi(t) - \phi(t_0)}{(t - t_0)^2}}\,(t - t_0),$$

where we choose the plus sign if $\phi''(t_0) > 0$ and the minus sign if $\phi''(t_0) < 0$. Then

$$\phi(t) = \phi(t_0) \pm \tau^2$$

$$\left.\frac{d\tau}{dt}\right|_{t=t_0} = \sqrt{\frac{|\phi''(t_0)|}{2}}$$

$$\delta_2 = \tau(t)|_{t=t_0+\delta}$$

$$-\delta_1 = \tau(t)|_{t=t_0-\delta},$$

and hence

$$\int_{t_0-\delta}^{t_0+\delta} e^{ix\phi(t)}g(t)dt = e^{ix\phi(t_0)}\int_{-\delta_1}^{\delta_2} e^{\pm ix\tau^2}g(t(\tau))\,\frac{dt}{d\tau}\,d\tau,$$

where

$$g(t(\tau))\left.\frac{dt}{d\tau}\right|_{\tau=0} = g(t_0)\sqrt{\frac{2}{|\phi''(t_0)|}},$$

and this is of the form (6.24).

Finally, we show that it suffices to consider

$$g(t_0)\sqrt{\frac{2}{|\phi''(t_0)|}}\,e^{ix\phi(t_0)}\int_{-\delta_1}^{\delta_2} e^{\pm ixt^2}dt. \tag{6.25}$$

This can be seen by rewriting (6.24) in the form

$$h(0)e^{ix\phi(t_0)}\int_{-\delta_1}^{\delta_2} e^{\pm ixt^2}dt + e^{ix\phi(t_0)}\int_{-\delta_1}^{\delta_2} e^{\pm ixt^2}t\,p(t)dt,$$

where

$$p(t) = \frac{h(t) - h(0)}{t}.$$

Note that by our strong assumptions on $\phi(t)$ and $g(t)$, we know that $h(t)$ and $p(t)$ are analytic in a neighborhood of $[-\delta_1,\delta_2]$. Without loss of generality, consider the case when the plus sign appears in the integrand. Then, integrating by parts, we have

$$\int_{-\delta_1}^{\delta_2} e^{ixt^2} t p(t) dt = \frac{1}{2ix} p(t) e^{ixt^2} \Big|_{-\delta_1}^{\delta_2} - \frac{1}{2ix} \int_{-\delta_1}^{\delta_2} p'(t) e^{ixt^2} dt = O\left(\frac{1}{x}\right).$$

Hence, ignoring terms of order $1/x$, we need only consider the asymptotic behavior of (6.25), i.e., the integral

$$\int_{-\delta_1}^{\delta_2} e^{\pm ixt^2} dt. \tag{6.26}$$

To determine the asymptotic behavior of (6.26), we first evaluate the **Fresnel integral**

$$\int_0^{\infty} e^{\pm ixt^2} dt.$$

To this end, consider

$$\int_C e^{iz^2} dz,$$

where C is the contour shown in Figure 6.4.

Then, by Cauchy's theorem,

$$\int_C e^{iz^2} dz = 0.$$

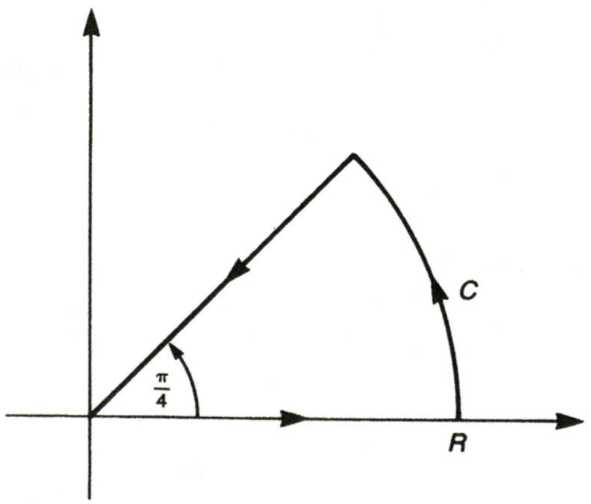

Figure 6.4

By the techniques used in the proof of Jordan's lemma (cf. Section 1.5.3), the integral along the circular arc vanishes as R tends to infinity and hence

$$\int_0^\infty e^{it^2}dt = e^{i\pi/4}\int_0^\infty e^{-t^2}dt = \frac{1}{2}\sqrt{\pi}\,e^{i\pi/4}.$$

Taking complex conjugates now shows that

$$\int_0^\infty e^{\pm it^2}dt = \frac{1}{2}\sqrt{\pi}\,e^{\pm i\pi/4}. \tag{6.27}$$

We now write

$$\int_{-\delta_1}^{\delta_2} e^{\pm ixt^2}dt = \int_{-\delta_1}^{0} e^{\pm ixt^2}dt + \int_0^{\delta_2} e^{\pm ixt^2}dt \tag{6.28}$$

and first consider the second integral on the right hand side. Then

$$\begin{aligned}\int_0^{\delta_2} e^{\pm ixt^2}dt &= \int_0^\infty e^{\pm ixt^2}dt - \int_{\delta_2}^\infty te^{\pm ixt^2}\frac{1}{t}\,dt\\ &= \frac{1}{\sqrt{x}}\int_0^\infty e^{\pm i\tau^2}d\tau - \left(\pm\frac{e^{ixt^2}}{2ixt}\bigg|_{\delta_2}^{\infty} \pm \frac{1}{2ix}\int_{\delta_2}^\infty \frac{e^{\pm ixt^2}}{t^2}\,dt\right)\\ &= \frac{1}{2}\sqrt{\frac{\pi}{x}}\,e^{\pm i\pi/4} + O\!\left(\frac{1}{x}\right).\end{aligned}$$

A similar calculation applies to the first integral on the right hand side of (6.28) and hence

$$\int_{-\delta_1}^{\delta_2} e^{\pm ixt^2}dt = \sqrt{\frac{\pi}{x}}\,e^{\pm i\pi/4} + O\!\left(\frac{1}{x}\right). \tag{6.29}$$

Equations (6.25) and (6.29) now imply the theorem. ■

Now consider the integral

$$\int_0^\pi e^{ix\sin\theta}e^{-i\nu\theta}d\theta.$$

Then $\phi(\theta) = \sin\theta$ and $\phi'(\theta) = 0$ only for $\theta = \pi/2$. By the method of stationary phase, as x tends to infinity,

$$\int_0^\pi e^{ix\sin\theta}e^{-i\nu\theta}d\theta = \sqrt{\frac{2\pi}{x}}\exp\left[i\left(x - \frac{1}{2}\nu\pi - \frac{1}{4}\pi\right)\right] + O\!\left(\frac{1}{x}\right).$$

Hence from (6.22) and the definition of the Hankel functions, as $x > 0$ tends to infinity,

$$\begin{aligned}
J_\nu(x) &= \sqrt{\frac{2}{\pi x}}\cos\left(x - \frac{1}{2}\nu\pi - \frac{1}{4}\pi\right) + O\left(\frac{1}{x}\right)\\
Y_\nu(x) &= \sqrt{\frac{2}{\pi x}}\sin\left(x - \frac{1}{2}\nu\pi - \frac{1}{4}\pi\right) + O\left(\frac{1}{x}\right)\\
H_\nu^{(1)}(x) &= \sqrt{\frac{2}{\pi x}}\exp\left[i\left(x - \frac{1}{2}\nu\pi - \frac{1}{4}\pi\right)\right] + O\left(\frac{1}{x}\right)\\
H_\nu^{(2)}(x) &= \sqrt{\frac{2}{\pi x}}\exp\left[-i\left(x - \frac{1}{2}\nu\pi - \frac{1}{4}\pi\right)\right] + O\left(\frac{1}{x}\right).
\end{aligned} \tag{6.30}$$

A more careful analysis (cf. Lebedev) would show that the error terms in (6.30) are in fact $O(1/x^{3/2})$; however, we shall not need this sharper result.

In the method of stationary phase, the regularity conditions on $\phi(t)$ and $g(t)$ can be weakened to $\phi \in C^2[a,b]$ and $g \in C[a,b]$. If the stationary point t_0 (a point such that $\phi'(t_0) = 0$) occurs at one of the endpoints of $[a,b]$, then the asymptotic behavior is one-half the value stated in the theorem. Finally, if several stationary points occur in the interval $[a,b]$, we break the integral into a finite sum of integrals each containing one stationary point and apply the method of stationary phase to each one of these integrals and then sum the results. For details, see Erdélyi.

From (6.30) we see that both $J_\nu(x)$ and $Y_\nu(x)$ have an infinite number of zeros on the positive x axis. We have previously made use of this fact in Chapter 2. From the series expansion of $J_\nu(z)$ we see that

$$\frac{d}{dz}\left(z^\nu J_\nu(z)\right) = z^\nu J_{\nu-1}(z),$$

and hence for $\nu > 0$ the first (positive) zero of $J_\nu(z)$ is greater than the first (positive) zero of $J_{\nu-1}(z)$. In particular, the smallest (positive) zero of $J_n(z)$ for $n = 0,1,\ldots,$ is the first (positive) zero of $J_0(z)$. We shall make use of this fact later on in this chapter.

6.2.6 Addition Formulas

Let $\mathbf{x},\mathbf{y} \in R^2$. Then for $\mathbf{y}$ fixed we see that for $\mathbf{x}$ not equal to $\mathbf{y}$, $u(\mathbf{x}) = H_0^{(1)}(|\mathbf{x} - \mathbf{y}|)$ is a solution of the Helmholtz equation

$$\Delta_2 u + u = 0$$

satisfying the Sommerfeld radiation condition

$$\lim_{r\to\infty} \sqrt{r}\left(\frac{\partial u}{\partial r} - iu\right) = 0,$$

where $r = |\mathbf{x}|$; this follows from (6.30). In applications, we are interested in expanding $H_0^{(1)}(|\mathbf{x} - \mathbf{y}|)$ in a series of Bessel or Hankel functions. Such expansions are called addition formulas.

Let (r,θ) be the polar coordinates of $\mathbf{x}$ and (ρ,ϕ) the polar coordinates of $\mathbf{y}$. Then

$$|\mathbf{x} - \mathbf{y}| = \sqrt{r^2 - 2r\rho\cos(\theta - \phi) + \rho^2}.$$

Let $r > \rho$ and set $\psi = \theta - \phi$. Then for fixed r and ρ we have from Fourier's theorem

$$H_0^{(1)}(\sqrt{r^2 - 2r\rho\cos\psi + \rho^2}) = \frac{1}{2}a_0 + \sum_{n=1}^{\infty} a_n(r,\rho)\cos n\psi, \tag{6.31}$$

where

$$a_n(r,\rho) = \frac{2}{\pi}\int_0^{\pi} H_0^{(1)}(|\mathbf{x} - \mathbf{y}|)\cos n\psi\, d\psi. \tag{6.32}$$

Now keep only ρ fixed. Then differentiating under the integral sign and integrating by parts using the fact that $H_0^{(1)}(\sqrt{r^2 - 2r\rho\cos\psi + \rho^2})$ satisfies

$$\Delta_2 u + u = \frac{\partial^2 u}{\partial r^2} + \frac{1}{r}\frac{\partial u}{\partial r} + \frac{1}{r^2}\frac{\partial^2 u}{\partial \psi^2} + u = 0,$$

we see that

$$a_n'' + \frac{1}{r}a_n' + \left(1 - \frac{n^2}{r^2}\right)a_n = 0, \tag{6.33}$$

i.e., $a_n(r,\rho)$, as a function of r with ρ fixed, is a solution of Bessel's equation. Since $H_0^{(1)}(\sqrt{r^2 - 2r\rho\cos\psi + \rho^2})$ satisfies the Sommerfeld radiation condition uniformly with respect to ψ, we can conclude from (6.32) that, as a function of r, $a_n(r,\rho)$ also satisfies the Sommerfeld radiation condition. From (6.30) and (6.33), we see that

$$a_n(r,\rho) = c_1 H_n^{(1)}(r), \tag{6.34}$$

where $c_1 = c_1(\rho)$. The same analysis (but keeping r fixed and noting that, as a function of $\mathbf{y}$, $H_0^{(1)}(|\mathbf{x} - \mathbf{y}|)$ is bounded at the origin) shows that

$$a_n(r,\rho) = c_2 J_n(\rho), \tag{6.35}$$

where $c_2 = c_2(r)$. Equations (6.34) and (6.35) imply that

$$a_n(r,\rho) = cH_n^{(1)}(r)J_n(\rho), \tag{6.36}$$

where c is a constant independent of r and ρ.

We now need to evaluate the constant c. To do this, we first note that

$$|\mathbf{x} - \mathbf{y}| = \sqrt{r^2 - 2r\rho\cos\psi + \rho^2} = r\sqrt{1 - 2\frac{\rho}{r}\cos\psi + \frac{\rho^2}{r^2}}$$
$$= r - \rho\cos\psi + O\left(\frac{1}{r}\right),$$

and hence from (6.30) and (6.32) we have

$$a_n(r,\rho) = \frac{2}{\pi}\sqrt{\frac{2}{\pi r}}\, e^{i(r-\pi/4)} \int_0^{\pi} e^{-i\rho\cos\psi} \cos n\psi \, d\psi + O\left(\frac{1}{r}\right). \tag{6.37}$$

But from (6.17) and the fact that $J_{-n}(\rho) = (-1)^n J_n(\rho)$ we have

$$e^{-i\rho\cos\psi} = \sum_{n=-\infty}^{\infty} (-i)^n J_n(\rho) e^{in\psi} = J_0(\rho) + 2\sum_{n=1}^{\infty} (-i)^n J_n(\rho) \cos n\psi, \tag{6.38}$$

and hence from (6.37) and (6.38) we see that

$$a_n(r,\rho) = 2\sqrt{\frac{2}{\pi r}}\, e^{i(r-\pi/4)}(-i)^n J_n(\rho) + O\left(\frac{1}{r}\right). \tag{6.39}$$

But from (6.30) and (6.36),

$$a_n(r,\rho) = c\sqrt{\frac{2}{\pi r}}\, e^{i(r-\pi/4)}(-i)^n J_n(\rho) + O\left(\frac{1}{r}\right), \tag{6.40}$$

so $c = 2$. Hence from (6.31) and (6.36) we see that

$$H_0^{(1)}\left(\sqrt{r^2 - 2r\rho\cos\psi + \rho^2}\right)$$
$$= J_0(\rho) H_0^{(1)}(r) + 2\sum_{n=1}^{\infty} J_n(\rho) H_n^{(1)}(r) \cos n\psi \qquad (r > \rho). \tag{6.41}$$

If $r < \rho$, then, since $|\mathbf{x} - \mathbf{y}|$ is symmetric in $\mathbf{x}$ and $\mathbf{y}$, (6.41) is valid with r and ρ interchanged. Finally, we can obtain addition formulas for $J_0(|\mathbf{x} - \mathbf{y}|)$, $Y_0(|\mathbf{x} - \mathbf{y}|)$, and $H_0^{(2)}(|\mathbf{x} - \mathbf{y}|)$ by taking the real and imaginary parts or the complex conjugate of (6.41).

6.3 THE SCATTERING OF ACOUSTIC WAVES

Consider the scattering of an acoustic wave by a cylinder of arbitrary cross section. If we assume that the wave motion is in a homogeneous isotropic medium with speed of sound c, then the wave motion can be determined from a velocity potential $U(\mathbf{x},t)$ where, in the linearized theory, $U(\mathbf{x},t)$ satisfies the wave equation (see Chapter 1)

$$\Delta_3 U = \frac{1}{c^2}\frac{\partial^2 U}{\partial t^2}.$$

If we further assume that $\mathbf{x} = (x,y,z)$, where the z axis coincides with the axis of the cylinder and the incident wave propagates in a direction that is perpendicular to the cylinder, then $U(\mathbf{x},t)$ is independent of z and $U(\mathbf{x},t)$ satisfies

$$\Delta_2 U = \frac{1}{c^2}\frac{\partial^2 U}{\partial t^2}.$$

Hence, for time harmonic acoustic waves of the form

$$U(\mathbf{x},t) = u(\mathbf{x})e^{-i\omega t} \qquad (\mathbf{x} \in R^2)$$

with frequency $\omega > 0$, $u(\mathbf{x})$ satisfies the Helmholtz equation

$$\Delta_2 u + k^2 u = 0$$

in the exterior of D where D denotes the cross section of the cylinder and the wave number $k > 0$ is defined by $k^2 = \omega^2/c^2$. For a "sound-soft" cylinder the pressure on the boundary ∂D vanishes, i.e.,

$$u(\mathbf{x}) = 0 \qquad (\mathbf{x} \in \partial D),$$

whereas for a "sound-hard" cylinder (no waves penetrating the cylinder),

$$\frac{\partial u}{\partial \nu}(\mathbf{x}) = 0 \qquad (\mathbf{x} \in \partial D),$$

where $\boldsymbol{\nu}$ is the unit outward normal to ∂D.

Assume now that the incident field is the **plane wave**

$$U^i(\mathbf{x},t) = \exp\,[i(k\boldsymbol{\alpha} \cdot \mathbf{x} - \omega t)],$$

where $\boldsymbol{\alpha}$ is a unit vector with no z component, for example $\boldsymbol{\alpha} = (1,0,0)$. Then the total field $u(\mathbf{x})$ (where as discussed above, $\mathbf{x} \in R^2$ and we have factored out the term $e^{-i\omega t}$) is of the form

$$u(\mathbf{x}) = u^i(\mathbf{x}) + u^s(\mathbf{x}),$$

where, in polar coordinates (r,θ), the incident field is

$$u^i(r,\theta) = \exp\,(ikr\cos\theta)$$

and the scattered field $u^s(r,\theta)$ satisfies the **Sommerfeld radiation condition**

$$\lim_{r\to\infty} \sqrt{r}\left(\frac{\partial u^s}{\partial r} - iku^s\right) = 0$$

uniformly in θ. This condition guarantees that, for large r, $U^s(\mathbf{x},t) = u^s(\mathbf{x})e^{-i\omega t}$ is an outgoing cylindrical wave:

$$U^s(\mathbf{x},t) \sim \frac{F(\theta)}{\sqrt{r}} \exp\,[i(kr - \omega t)]$$

for some function $F(\theta)$, where $F(\theta)$ is known as the **far field pattern** of the scattered wave.

We now summarize our discussion of the scattering of acoustic waves by a cylinder. To solve the scattering problem of a plane, time harmonic wave incident upon an infinite cylinder, we must find a complex valued function $u^s(\mathbf{x})$, which we now relabel $u(\mathbf{x})$, such that

$$\Delta_2 u + k^2 u = 0 \qquad \text{in} \qquad R^2\backslash\overline{D} \qquad \textbf{(6.}$$

$$u(r,\theta) = -\exp\,[ikr\cos\theta] \qquad \text{on} \qquad \partial D$$

$$\frac{\partial u}{\partial \nu}(r,\theta) = -\frac{\partial}{\partial \nu}\exp\,[ikr\cos\theta] \qquad \text{on} \qquad \partial D \tag{6.44}$$

$$\lim_{r\to\infty}\sqrt{r}\left(\frac{\partial u}{\partial r} - iku\right) = 0, \tag{6.45}$$

where $D \subset R^2$ is the cross section of the cylinder. The existence and uniqueness of a solution to (6.42)–(6.45) for arbitrarily shaped cylinders will be discussed in Section 6.5 using the method of integral equations. Here we shall consider only the case of a sound-soft circular cylinder of radius a, i.e., $D = \{\mathbf{x}\colon |\mathbf{x}| < a\}$, and solve this problem by the method of separation of variables.

In polar coordinates, (6.42) becomes

$$\frac{\partial^2 u}{\partial r^2} + \frac{1}{r}\frac{\partial u}{\partial r} + \frac{1}{r^2}\frac{\partial^2 u}{\partial \theta^2} + k^2 u = 0.$$

Looking for solutions in the form $u_n(r,\theta) = R(r)\Phi(\theta)$, we easily find that

$$R'' + \frac{1}{r}R' + \left(k^2 - \frac{n^2}{r^2}\right)R = 0$$

$$\Phi'' + n^2\Phi = 0,$$

where the separation parameter is chosen to be an integer n so that $u_n(r,\theta)$ is periodic in θ of period 2π. Then

$$R(r) = a_n H_n^{(1)}(kr) + b_n H_n^{(2)}(kr)$$

$$\Phi(\theta) = c_n \cos n\theta + d_n \sin n\theta,$$

where a_n, b_n, c_n, d_n are constants. Since we want $u_n(r,\theta)$ to satisfy the Sommerfeld radiation condition (6.45), we know that $b_n = 0$ from the asymptotic behavior of the Hankel functions. Furthermore, from the boundary condition (6.43) we see that the solution of (6.42), (6.43), (6.45) must be an even function of θ, so we can set $d_n = 0$. We thus look for a solution of (6.42), (6.43), (6.45) in the form (noting that $H_n^{(1)}(kr) = (-1)^n H_n^{(1)}(kr)$)

$$u(r,\theta) = \sum_{n=0}^{\infty} \alpha_n H_n^{(1)}(kr)\cos n\theta, \tag{6.46}$$

where the α_n are constants to be determined.

It remains to satisfy the boundary condition (6.43). But from (6.38) we see

$$[ikr\cos\theta] = J_0(kr) + 2\sum_{n=1}^{\infty} i^n J_n(kr)\cos n\theta \tag{6.47}$$

and hence (6.43), (6.46), and (6.47) imply that

$$\alpha_0 = -\frac{J_0(ka)}{H_0^{(1)}(ka)} \tag{6.48}$$

$$\alpha_n = -2i^n \frac{J_n(ka)}{H_n^{(1)}(ka)} \qquad (n \geq 1),$$

assuming that $H_n^{(1)}(ka)$ does not vanish. But if $H_n^{(1)}(ka) = 0$, so does $H_n^{(2)}(ka)$, since $H_n^{(2)}(ka)$ is just the complex conjugate of $H_n^{(1)}(ka)$, and this contradicts the fact that the Wronskian of $H_n^{(1)}(kr)$ and $H_n^{(2)}(kr)$ is not zero (cf. Section 6.2.2). Hence $H_n^{(1)}(ka)$ does not vanish and from (6.46) and (6.48) we have formally

$$u(r,\theta) = -\left[\frac{J_0(ka)}{H_0^{(1)}(ka)} + 2\sum_{n=1}^{\infty} i^n \frac{J_n(ka)}{H_n^{(1)}(ka)} H_n^{(1)}(kr)\cos n\theta\right] \qquad (r \geq a). \tag{6.49}$$

Note that (6.49) is not very useful for high frequencies (large values of k) since in this case the series is very slowly convergent.

It must now be verified directly that (6.49) is in fact the solution of (6.42), (6.43), (6.45). However, there are some difficulties in verifying that it satisfies the Sommerfeld radiation condition since the asymptotic expansions obtained in Section 6.2.5 do not hold uniformly with respect to n. We now present one approach for verifying that (6.49) is the desired solution.

We first note that from the series representation of $J_n(kr)$, for $n \geq 0$ and r in compact subsets of $(0,\infty)$ we have

$$|J_n(kr)| = O\left(\frac{(kr)^n}{2^n n!}\right), \tag{6.50}$$

where the order estimate is independent of r and n. From (6.41) we have for $r > \rho$,

$$H_n^{(1)}(kr)J_n(k\rho) = \frac{1}{\pi}\int_0^{\pi} H_0^{(1)}(k\sqrt{r^2 - 2r\rho\cos\psi + \rho^2})\cos n\psi\, d\psi. \tag{6.51}$$

By analytic continuation, this representation is valid for all complex ρ such that $|\rho| < r$, and this fact and the absolute integrability of $H_0^{(1)}(kr\sqrt{-2i\cos\psi})$ imply that there exists a positive constant M independent of n and r such that for $n \geq 0$ and r in compact subsets of $(0,\infty)$ we have

$$|H_n^{(1)}(kr)J_n(ikr)| \leq M. \tag{6.52}$$

But since from its power series expansion

$$|J_n(ikr)| \geq \frac{(kr)^n}{2^n n!},$$

we have for $n \geq 0$ and r in compact subsets of $(0,\infty)$

$$|H_n^{(1)}(kr)| = O\left(\frac{2^n n!}{(kr)^n}\right). \tag{6.}$$

, we have previously noted,

$$\frac{d}{dz}[z^\nu J_\nu(z)] = z^\nu J_{\nu-1}(z),$$

and from the definition of the Neumann function $Y_\nu(z)$ we see that

$$\frac{d}{dz}[z^\nu Y_\nu(z)] = z^\nu Y_{\nu-1}(z),$$

and hence from the definition of the Hankel function $H_\nu^{(1)}(z)$ we have

$$\frac{d}{dz}[z^\nu H_\nu^{(1)}(z)] = z^\nu H_{\nu-1}^{(1)}(z).$$

Therefore relations analogous to (6.50) and (6.53) are valid for the derivatives of $J_n(kr)$ and $H_n^{(1)}(kr)$. We can now conclude that the series (6.49) and its derivatives with respect to r and θ are absolutely and uniformly convergent on compact subsets of $R^2\backslash(0,0)$.

Using the addition formula (6.41) we can now represent (6.49) in the form

$$u(\mathbf{x}) = \frac{i}{4}\int_{|\mathbf{y}|=a}\left(u(\mathbf{y})\frac{\partial}{\partial\nu}H_0^{(1)}(k|\mathbf{x}-\mathbf{y}|) - \frac{\partial u}{\partial\nu}(\mathbf{y})H_0^{(1)}(k|\mathbf{x}-\mathbf{y}|)\right)ds(\mathbf{y}), \qquad \textbf{(6.54)}$$

where (r,θ) are the polar coordinates of $\mathbf{x}$, $|\mathbf{x}| > a$, $\boldsymbol{\nu}$ is the unit outward normal to $|\mathbf{y}| = a$, and we have used the formula (6.13) for the Wronskian of $J_n(z)$ and $H_n^{(1)}(z)$. Since $H_0^{(1)}(k|\mathbf{x}-\mathbf{y}|)$ satisfies the Sommerfeld radiation condition we can now conclude from (6.54) that $u(\mathbf{x})$ does also. The fact that $u(\mathbf{x})$ is a solution of the Helmholtz equation (6.42) also follows from (6.54), and the fact that $u(\mathbf{x})$ satisfies the boundary condition (6.43) follows from the uniform convergence of the series (6.49). We have now shown that (6.49) is the solution of (6.42), (6.43), (6.45).

We note in conclusion that the problem of the scattering of acoustic waves by a sound-soft sphere can be obtained in a similar manner as the above case of scattering by a cylinder. In this case, the expansion corresponding to (6.49) involves Bessel functions of half-integer order and Legendre polynomials (cf. Exercise 18). The problem of a sound-hard scattering obstacle can also be treated in a similar way.

6.4 MAXWELL'S EQUATIONS

In the previous section we used acoustic scattering as a motivation for studying ex-
rior boundary value problems for the Helmholtz equation. We now want to show the same problem arises in the scattering of electromagnetic waves by a perfectly cting infinite cylinder.

nsider electromagnetic wave propagation in a homogeneous, isotropic, non- medium in R^3 with electric permittivity ε and magnetic permeability μ.

The electromagnetic wave with frequency $\omega > 0$ is described by the electric and magnetic fields

$$\mathbf{E}(\mathbf{x},t) = \varepsilon^{-1/2}\mathbf{E}(\mathbf{x})e^{-i\omega t}$$
$$\mathbf{H}(\mathbf{x},t) = \mu^{-1/2}\mathbf{H}(\mathbf{x})e^{-i\omega t}.$$

From the time dependent form of **Maxwell's equations**

$$\operatorname{curl}\mathbf{E} + \mu\frac{\partial \mathbf{H}}{\partial t} = 0$$

$$\operatorname{curl}\mathbf{H} - \varepsilon\frac{\partial \mathbf{E}}{\partial t} = 0$$

we see that the space dependent fields $\mathbf{E}(\mathbf{x})$ and $\mathbf{H}(\mathbf{x})$ satisfy the time harmonic form of Maxwell's equations

$$\begin{aligned}\operatorname{curl}\mathbf{E} - ik\mathbf{H} &= 0\\ \operatorname{curl}\mathbf{H} + ik\mathbf{E} &= 0,\end{aligned} \tag{6.55}$$

where the wave number k is given by $k = \omega\sqrt{\varepsilon\mu}$. Hence, the mathematical description of the scattering of time harmonic electromagnetic waves by an obstacle D leads to boundary value problems for (6.55). In particular, if D is perfectly conducting then we require the tangential component of the electric field to vanish on ∂D. In addition, if $\mathbf{E}^i(\mathbf{x})$ denotes the incident field and $\mathbf{E}^s(\mathbf{x})$ the scattered field, then $\mathbf{E}^s(\mathbf{x})$ must be an "outgoing" wave that satisfies the **Silver-Müller radiation condition**

$$\lim_{r\to\infty} r(\mathbf{H}\times\mathbf{e}_r - \mathbf{E}) = 0, \tag{6.56}$$

where $\mathbf{e}_r$ denotes the unit vector in the radial direction, and on ∂D

$$\boldsymbol{\nu}\times\mathbf{E} = 0 \tag{6.57}$$

where $\mathbf{E}(\mathbf{x}) = \mathbf{E}^i(\mathbf{x}) + \mathbf{E}^s(\mathbf{x})$, and $\boldsymbol{\nu}$ is the unit outward normal to ∂D.

Now consider the scattering due to an infinite cylinder with axis in the z direction, $\mathbf{x} = (x,y,z) \in R^3$, and assume that the incident wave propagates in a direction perpendicular to the cylinder. Then $\mathbf{E}(\mathbf{x}) = (E_1(\mathbf{x}),E_2(\mathbf{x}),E_3(\mathbf{x}))$ and $\mathbf{H}(\mathbf{x}) = (H_1(\mathbf{x}),H_2(\mathbf{x}),H_3(\mathbf{x}))$ are independent of z, so (6.55) reduces to the two groups of equations

$$\begin{aligned}\frac{\partial E_3}{\partial y} &= ikH_1\\ \frac{\partial E_3}{\partial x} &= -ikH_2\\ \frac{\partial H_2}{\partial x} - \frac{\partial H_1}{\partial y} &= -ikE_3\end{aligned} \tag{6.58}$$

and

$$\frac{\partial E_2}{\partial x} - \frac{\partial E_1}{\partial y} = ikH_3$$

$$\frac{\partial H_3}{\partial y} = -ikE_1 \tag{6.59}$$

$$\frac{\partial H_3}{\partial x} = ikE_2,$$

where the first group involves only $H_1(\mathbf{x})$, $H_2(\mathbf{x})$, and $E_3(\mathbf{x})$, and the second group involves only $E_1(\mathbf{x})$, $E_2(\mathbf{x})$, and $H_3(\mathbf{x})$. Let $\mathbf{e}_x$, $\mathbf{e}_y$, $\mathbf{e}_z$ denote the unit vectors in the x, y, z directions. Then if $H_1(\mathbf{x})$, $H_2(\mathbf{x})$, and $E_3(\mathbf{x})$ satisfy (6.58), $\mathbf{E}(\mathbf{x}) = E_3(\mathbf{x})\mathbf{e}_z$ and $\mathbf{H}(\mathbf{x}) = H_1(\mathbf{x})\mathbf{e}_x + H_2(\mathbf{x})\mathbf{e}_y$ specify an electromagnetic wave polarized perpendicular to the z axis (i.e., $\mathbf{E} \times \mathbf{H} \cdot \mathbf{e}_z = 0$). Similarly, (6.59) specifies an electromagnetic wave polarized parallel to the z axis. The general solution of (6.58) and (6.59) is obtained by adding the solutions of (6.58) and (6.59) together.

By eliminating $H_1(\mathbf{x})$ and $H_2(\mathbf{x})$ from (6.58) we see that $E_3(\mathbf{x})$ satisfies

$$\Delta_2 E_3 + k^2 E_3 = 0 \tag{6.60}$$

and if $E_3(\mathbf{x})$ can be determined we can recover $H_1(\mathbf{x})$ and $H_2(\mathbf{x})$ from (6.58). Similarly, from (6.59) we see that $H_3(\mathbf{x})$ satisfies

$$\Delta_2 H_3 + k^2 H_3 = 0, \tag{6.61}$$

and if $H_3(\mathbf{x})$ can be determined, we can recover $E_1(\mathbf{x})$ and $E_2(\mathbf{x})$ from (6.59). Hence, the problem of finding $\mathbf{E}(\mathbf{x})$ and $\mathbf{H}(\mathbf{x})$ reduces to the problem of solving the Helmholtz equation for $E_3(\mathbf{x})$ and $H_3(\mathbf{x})$.

For $\mathbf{E}(\mathbf{x})$ and $\mathbf{H}(\mathbf{x})$ to satisfy (6.56), $E_3(\mathbf{x})$ and $H_3(\mathbf{x})$ must satisfy the Sommerfeld radiation condition. It remains to determine the boundary conditions satisfied by $E_3(\mathbf{x})$ and $H_3(\mathbf{x})$. If $(\mathbf{E}(\mathbf{x}),\mathbf{H}(\mathbf{x}))$ is polarized perpendicular to the z axis such that $E_1(\mathbf{x}) = E_2(\mathbf{x}) = H_3(\mathbf{x}) = 0$, then from (6.57) we see that

$$E_3(\mathbf{x}) = 0 \quad \text{on} \quad \partial D, \tag{6.62}$$

and $E_3^s(\mathbf{x}) = -E_3^i(\mathbf{x})$ on ∂D. On the other hand, if $(\mathbf{E}(\mathbf{x}),\mathbf{H}(\mathbf{x}))$ is polarized parallel to the z axis such that $H_1(\mathbf{x}) = H_2(\mathbf{x}) = E_3(\mathbf{x}) = 0$, then from (6.57) we see that $\mathbf{E}(\mathbf{x}) = |\mathbf{E}(\mathbf{x})|\boldsymbol{\nu}$ on ∂D. But from (6.59) we have

$$\nabla H_3(\mathbf{x}) = ik(E_2(\mathbf{x})\mathbf{e}_x - E_1(\mathbf{x})\mathbf{e}_y),$$

and hence $\nabla H_3(\mathbf{x}) \cdot E(\mathbf{x}) = 0$, i.e.,

$$\frac{\partial H_3}{\partial \nu}(\mathbf{x}) = 0 \quad \text{on} \quad \partial D. \tag{6.63}$$

summary, the problem of the scattering of time harmonic electromagnetic an infinite cylinder can be reduced to solving (6.60), (6.62) or (6.61), the incident field is polarized perpendicular or parallel to the z axis $E_3(\mathbf{x})$ and $H_3(\mathbf{x})$ satisfy the Sommerfeld radiation condition. If the incident field has arbitrary polarization, then one must decompose it into its components that are polarized perpendicular and parallel to the z axis, solve (6.60)–(6.63) such that $E_3(\mathbf{x})$ and $H_3(\mathbf{x})$ satisfy the Sommerfeld radiation condition, and then add the two solutions.

6.5 SCATTERING BY A CYLINDER OF ARBITRARY CROSS SECTION

We shall consider the problem of the scattering of a time harmonic plane wave by an infinite cylinder of arbitrary cross section D. We make the assumption that D is bounded, simply connected, has C^2 boundary ∂D with unit outward normal $\boldsymbol{\nu}$, and, without loss of generality, that D contains the origin. Then, as discussed in Sections 6.3 and 6.4, we must solve the problem

$$\Delta_2 u + k^2 u = 0 \qquad \text{in} \qquad R^2 \backslash \overline{D} \tag{6.64}$$

$$u(\mathbf{x}) = -u^i(\mathbf{x}) \qquad \text{on} \qquad \partial D \tag{6.65}$$

or

$$\frac{\partial u}{\partial \nu}(\mathbf{x}) = -\frac{\partial u^i}{\partial \nu}(\mathbf{x}) \qquad \text{on} \qquad \partial D \tag{6.66}$$

$$\lim_{r\to\infty} \sqrt{r}\left(\frac{\partial u}{\partial r} - iku\right) = 0, \tag{6.67}$$

where $u^i(r,\theta) = \exp\,[ikr\cos\theta]$, (6.65) corresponds to an electromagnetic wave polarized perpendicular to the cylinder or, in acoustics, to a sound-soft cylinder, and (6.66) corresponds to an electromagnetic wave polarized parallel to the cylinder or, in acoustics, to a sound-hard cylinder. Equation (6.67) is assumed to hold uniformly in θ, where (r,θ) are polar coordinates. Our aim is to establish the uniqueness and existence of a (complex valued) solution to (6.64)–(6.67) in a manner that is suitable for numerical computation. For (6.64), (6.65), (6.67) we require that $u \in C^2(R^2\backslash\overline{D}) \cap C(R^2\backslash D)$ and for (6.64), (6.66), (6.67) we want $u \in C^2(R^2\backslash\overline{D}) \cap C^1(R^2\backslash D)$. However, it can be shown that if $u \in C^2(R^2\backslash\overline{D}) \cap C(R^2\backslash D)$ then in fact $u \in C^2(R^2\backslash\overline{D}) \cap C^1(R^2\backslash D)$ (cf. Colton and Kress). Hence, we shall always assume that any solution of (6.64) and (6.67) is such that $u \in C^2(R^2\backslash\overline{D}) \cap C^1(R^2\backslash D)$.

DEFINITION 17

The (radiating) **fundamental solution** to the Helmholtz equation (6.64) is

$$\Phi(\mathbf{x},\mathbf{y}) = \frac{i}{4} H_0^{(1)}(k|\mathbf{x}-\mathbf{y}|).$$

Note that $\Phi(\mathbf{x},\mathbf{y})$ satisfies the Sommerfeld radiation condition with respect to both $\mathbf{x}$ and $\mathbf{y}$, and as $|\mathbf{x}-\mathbf{y}|$ tends to zero we have from Section 6.2.1 that

$$\Phi(\mathbf{x},\mathbf{y}) = \frac{1}{2\pi}\log\frac{1}{|\mathbf{x}-\mathbf{y}|} + O(1).$$

Theorem 50 generalizes the result previously obtained for D equal to a disk see Section 6.3).

THEOREM 50

Let $u(\mathbf{x})$ be a solution of the Helmholtz equation in the exterior of D satisfying the Sommerfeld radiation condition. Then for $\mathbf{x} \in R^2 \backslash \overline{D}$,

$$u(\mathbf{x}) = \int_{\partial D} \left(u(\mathbf{y}) \frac{\partial}{\partial \nu(\mathbf{y})} \Phi(\mathbf{x},\mathbf{y}) - \frac{\partial u}{\partial \nu}(\mathbf{y}) \Phi(\mathbf{x},\mathbf{y}) \right) ds(\mathbf{y}).$$

■ ***Proof.*** Let $\mathbf{x} \in R^2 \backslash \overline{D}$ and circumscribe it with a disk

$$\Omega_{x,\epsilon} = \{\mathbf{y}: |\mathbf{x} - \mathbf{y}| < \epsilon\},$$

where $\Omega_{x,\epsilon} \subset R^2 \backslash \overline{D}$. Let Ω be a disk of radius R centered at the origin and containing D and $\Omega_{x,\epsilon}$ in its interior. Then from Green's second identity we have

$$\int_{\partial D + \partial \Omega_{x,\epsilon} + \partial \Omega} \left(u(\mathbf{y}) \frac{\partial}{\partial \nu(\mathbf{y})} \Phi(\mathbf{x},\mathbf{y}) - \frac{\partial u}{\partial \nu}(\mathbf{y}) \Phi(\mathbf{x},\mathbf{y}) \right) ds(\mathbf{y}) = 0.$$

From the definition of the Hankel function, we have that

$$\frac{d}{dz} H_0^{(1)}(z) = -H_1^{(1)}(z),$$

and hence on $\partial \Omega_{x,\epsilon}$

$$\frac{\partial}{\partial \nu(\mathbf{y})} \Phi(\mathbf{x},\mathbf{y}) = \frac{1}{2\pi} \frac{1}{|\mathbf{x} - \mathbf{y}|} + O(|\mathbf{x} - \mathbf{y}| \log |\mathbf{x} - \mathbf{y}|). \tag{6.69}$$

Using (6.68) and (6.69) and letting ϵ tend to zero, we see (as in the case of the representation theorem for Laplace's equation, see Section 4.1) that

$$\begin{aligned} u(\mathbf{x}) = &\int_{\partial D} \left(u(\mathbf{y}) \frac{\partial}{\partial \nu(\mathbf{y})} \Phi(\mathbf{x},\mathbf{y}) - \frac{\partial u}{\partial \nu}(\mathbf{y}) \Phi(\mathbf{x},\mathbf{y}) \right) ds(\mathbf{y}) \\ &- \int_{|\mathbf{y}|=R} \left(u(\mathbf{y}) \frac{\partial}{\partial \nu(\mathbf{y})} \Phi(\mathbf{x},\mathbf{y}) - \frac{\partial u}{\partial \nu}(\mathbf{y}) \Phi(\mathbf{x},\mathbf{y}) \right) ds(\mathbf{y}); \end{aligned} \tag{6.70}$$

we remind the reader that $\boldsymbol{\nu}$ is always the unit outward normal to the boundary of e (interior) domain. Hence, to establish the theorem it suffices to show that the ral on $|\mathbf{y}| = R$ tends to zero as R tends to infinity.

Ve first show that as R tends to infinity

$$|^2 ds = O(1), \tag{6.71}$$

i.e., the integral on the left side of (6.71) is a bounded function of R. To accomplish this, we first observe that from the Sommerfeld radiation condition we have that

$$\begin{aligned} 0 &= \lim_{R\to\infty} \int_{|\mathbf{y}|=R} \left| \frac{\partial u}{\partial r} - iku \right|^2 ds \\ &= \lim_{R\to\infty} \int_{|\mathbf{y}|=R} \left(\left| \frac{\partial u}{\partial r} \right|^2 + k^2|u|^2 + 2k \,\mathrm{Im}\left(u \frac{\partial \bar{u}}{\partial r} \right) \right) ds. \end{aligned} \tag{6.72}$$

Green's first identity applied to $D_R = \Omega \backslash \overline{D}$ gives

$$\int_{|\mathbf{y}|=R} u \frac{\partial \bar{u}}{\partial r} ds = \int_{\partial D} u \frac{\partial \bar{u}}{\partial \nu} ds - k^2 \iint_{D_R} |u|^2 d\mathbf{y} + \iint_{D_R} |\mathrm{grad}\, u|^2 d\mathbf{y},$$

and hence from (6.72) we have that

$$\lim_{R\to\infty} \int_{|\mathbf{y}|=R} \left(\left| \frac{\partial u}{\partial r} \right|^2 + k^2|u|^2 \right) ds = -2k \,\mathrm{Im} \int_{\partial D} u \frac{\partial \bar{u}}{\partial \nu} ds.$$

This now implies that as R tends to infinity

$$\int_{|\mathbf{y}|=R} |u|^2 ds = O(1).$$

To complete the proof of the theorem, we now note the identity

$$\begin{aligned} \int_{|\mathbf{y}|=R} &\left(u(\mathbf{y}) \frac{\partial}{\partial \nu(\mathbf{y})} \Phi(\mathbf{x},\mathbf{y}) - \frac{\partial u}{\partial \nu}(\mathbf{y}) \Phi(\mathbf{x},\mathbf{y}) \right) ds(\mathbf{y}) = \\ &\int_{|\mathbf{y}|=R} u(\mathbf{y}) \left(\frac{\partial}{\partial r} \Phi(\mathbf{x},\mathbf{y}) - ik\Phi(\mathbf{x},\mathbf{y}) \right) ds(\mathbf{y}) \\ &- \int_{|\mathbf{y}|=R} \Phi(\mathbf{x},\mathbf{y}) \left(\frac{\partial u}{\partial r}(\mathbf{y}) - iku(\mathbf{y}) \right) ds(\mathbf{y}), \end{aligned} \tag{6.73}$$

where $\partial/\partial r$ denotes the derivative with respect to $\mathbf{y}$ in the radial direction. Applying the Schwarz inequality to each of the integrals in (6.73) and using (6.71), the facts that $\Phi(\mathbf{x},\mathbf{y}) = O(1/\sqrt{R})$ and $\Phi(\mathbf{x},\mathbf{y})$ and $u(\mathbf{y})$ satisfy the Sommerfeld radiation condition now show that

$$\lim_{R\to\infty} \int_{|\mathbf{y}|=R} \left(u(\mathbf{y}) \frac{\partial}{\partial \nu(\mathbf{y})} \Phi(\mathbf{x},\mathbf{y}) - \frac{\partial u}{\partial \nu}(\mathbf{y}) \Phi(\mathbf{x},\mathbf{y}) \right) ds(\mathbf{y}) = 0.$$

The proof of the theorem is now complete. ■

Let D be any domain and $u(\mathbf{x})$ a solution of the Helmholtz equation in D. If $\mathbf{x}$ is an interior point of D and $\Omega = \{\mathbf{y}: |\mathbf{x} - \mathbf{y}| < \epsilon\}$ a disk contained in D, then from the proof of the above theorem we see that

$$u(\mathbf{x}) = -\int_{\partial\Omega} \left(u(\mathbf{y}) \frac{\partial}{\partial \nu(\mathbf{y})} \Phi(\mathbf{x},\mathbf{y}) - \frac{\partial u}{\partial \nu}(\mathbf{y}) \Phi(\mathbf{x},\mathbf{y}) \right) ds(\mathbf{y}).$$

ɪence, since $\Phi(\mathbf{x},\mathbf{y})$ is an analytic function of x_1 and x_2 where $\mathbf{x} = (x_1,x_2)$ and $\mathbf{x}$ does not equal $\mathbf{y}$, we can conclude (as in the case of Laplace's equation, see Section 4.2) that $u(\mathbf{x})$ is analytic in Ω. Since $\mathbf{x}$ was an arbitrary interior point, we have:

THEOREM 51

Solutions $u(\mathbf{x})$, $\mathbf{x} = (x_1,x_2)$, of the Helmholtz equation are analytic functions of their independent variables x_1 and x_2.

We are now in a position to prove the following uniqueness theorem for the scattering problem (6.64)–(6.67):

THEOREM 52

Let $u(\mathbf{x})$ be a solution of the Helmholtz equation in the exterior of D satisfying the Sommerfeld radiation condition and the boundary condition $u(\mathbf{x}) = 0$ or $\partial u(\mathbf{x})/\partial \nu = 0$ for $\mathbf{x}$ on ∂D. Then $u(\mathbf{x})$ is identically zero in the exterior of D.

■ ***Proof.*** Let Ω be a disk centered at the origin and containing D in its interior. Then from Green's second identity, the reality of the wave number k, and the fact that $u(\mathbf{x}) = \overline{u(\mathbf{x})} = 0$ or $\partial u(\mathbf{x})/\partial \nu = \partial \overline{u(\mathbf{x})}/\partial \nu = 0$ for $\mathbf{x}$ on ∂D, we have

$$\int_{\partial\Omega} \left(\bar{u}\frac{\partial u}{\partial r} - u\frac{\partial \bar{u}}{\partial r} \right) ds = 0. \tag{6.74}$$

But since $u(\mathbf{x})$ is infinitely differentiable (in fact analytic) for $\mathbf{x} \in R^2\backslash\overline{D}$, we have that for $\mathbf{x} \in R^2\backslash\Omega$,

$$\begin{aligned} u(r,\theta) &= \sum_{n=-\infty}^{\infty} a_n(r)e^{in\theta} \\ a_n(r) &= \frac{1}{2\pi}\int_{-\pi}^{\pi} u(r,\theta)e^{-in\theta}d\theta, \end{aligned} \tag{6.75}$$

where the series and its derivatives with respect to r are absolutely and uniformly convergent. As we have shown (cf. Section 6.2.5), $a_n(r)$ is a solution of Bessel's equation, and since $u(\mathbf{x})$ satisfies the Sommerfeld radiation condition,

$$a_n(r) = a_n H_n^{(1)}(kr), \tag{6.76}$$

where the a_n are constants. Substituting (6.75) and (6.76) into (6.74) and integrating termwise, we see from the fact that $\overline{H_n^{(1)}(kr)} = H_n^{(2)}(kr)$ and the formula for the Wronskian of $H_n^{(1)}(kr)$ and $H_n^{(2)}(kr)$ that

$$\sum_{n=-\infty}^{\infty} |a_n|^2 = 0.$$

$u(\mathbf{x})$ is identically zero for $\mathbf{x} \in R^2\backslash\Omega$, and by the analyticity of solutions to ɦoltz equation, we can conclude from the identity theorem (Section 1.5.1) identically zero for $\mathbf{x} \in R^2\backslash\overline{D}$. ■

COROLLARY

The solution of the scattering problem (6.64)–(6.67), if it exists, is unique.

■ ***Proof.*** If two solutions $u_1(\mathbf{x})$ and $u_2(\mathbf{x})$ exist, consider their difference $u(\mathbf{x}) = u_1(\mathbf{x}) - u_2(\mathbf{x})$ and conclude from the above theorem that $u(\mathbf{x})$ is identically zero, i.e., $u_1(\mathbf{x}) = u_2(\mathbf{x})$ for $\mathbf{x} \in R^2\backslash D$. ■

We now want to use the method of integral equations to establish the existence of a solution to our scattering problem. This approach also provides a numerical method for solving this problem (cf. Section 5.5).

From the asymptotic behavior of the Hankel function for large z (see equation (6.30)) and small z (see equation (6.68)), we see that the double layer potential

$$u(\mathbf{x}) = \int_{\partial D} \psi(\mathbf{y}) \frac{\partial}{\partial \nu(\mathbf{y})} \Phi(\mathbf{x},\mathbf{y}) ds(\mathbf{y}) \qquad (\mathbf{x} \in R^2\backslash\partial D) \tag{6.77}$$

with continuous density $\psi(\mathbf{y})$ satisfies the Sommerfeld radiation condition, is a solution of the Helmholtz equation in $R^2\backslash\partial D$, and satisfies the same discontinuity properties as the double layer potential for Laplace's equation discussed in Section 5.1. Similarly, the single layer potential

$$u(\mathbf{x}) = \int_{\partial D} \psi(\mathbf{y})\Phi(\mathbf{x},\mathbf{y}) ds(\mathbf{y}) \qquad (\mathbf{x} \in R^2\backslash\partial D) \tag{6.78}$$

with continuous density $\psi(\mathbf{y})$ satisfies the Sommerfeld radiation condition, is a solution of the Helmholtz equation in $R^2\backslash\partial D$, and satisfies the same discontinuity properties as the single layer potential for Laplace's equation. In particular, (6.77) solves the exterior Dirichlet problem (6.64), (6.65), (6.67) provided $\psi(\mathbf{y})$ is a solution of the integral equation

$$\psi(\mathbf{x}) + 2\int_{\partial D} \psi(\mathbf{y}) \frac{\partial}{\partial \nu(\mathbf{y})} \Phi(\mathbf{x},\mathbf{y}) ds(\mathbf{y}) = -2u^i(\mathbf{x}) \qquad (\mathbf{x} \in \partial D), \tag{6.79}$$

and (6.78) solves the exterior Neumann problem (6.64), (6.66), (6.67) provided $\psi(\mathbf{y})$ is a solution of the integral equation

$$\psi(\mathbf{x}) - 2\int_{\partial D} \psi(\mathbf{y}) \frac{\partial}{\partial \nu(\mathbf{x})} \Phi(\mathbf{x},\mathbf{y}) ds(\mathbf{y}) = 2\frac{\partial}{\partial \nu} u^i(\mathbf{x}) \qquad (\mathbf{x} \in \partial D). \tag{6.80}$$

In operator notation, we write (6.79) as

$$\psi + \mathbf{K}\psi = -2u^i \tag{6.81}$$

and (6.80) as

$$\psi - \mathbf{K}'\psi = 2\frac{\partial u^i}{\partial \nu}. \tag{6.82}$$

As in the case of Laplace's equation, the discontinuity properties of the s and double layer potentials (6.77) and (6.78) are also valid for square integrable densities if equality is interpreted as "almost everywhere" (cf. the comments after equation (5.7)).

Scattering problems for bounded obstacles $D \subset R^3$ can be reduced to Fredholm integral equations of the same form as (6.79) and (6.80) by considering potentials of the form (6.77) and (6.78) over the two-dimensional surface ∂D, where $\Phi(\mathbf{x},\mathbf{y})$ is now defined by

$$\Phi(\mathbf{x},\mathbf{y}) = \frac{1}{4\pi} \frac{\exp[ik|\mathbf{x} - \mathbf{y}|]}{|\mathbf{x} - \mathbf{y}|}$$

(cf. Section 4.8 and the remarks at the end of Section 5.2). The proof of the existence of a unique solution to these integral equations is treated exactly as in the following analysis of the scattering by an infinite cylinder (cf. Colton and Kress).

In order to apply the Fredholm alternative, we need to determine the nullspaces of $\mathbf{I} + \mathbf{K}$ and $\mathbf{I} - \mathbf{K}'$, where the nullspace is the set of continuous functions satisfying the homogeneous forms of (6.81) and (6.82) respectively. We denote the nullspace of $\mathbf{I} + \mathbf{K}$ by $N(\mathbf{I} + \mathbf{K})$ and the nullspace of $\mathbf{I} - \mathbf{K}'$ by $N(\mathbf{I} - \mathbf{K}')$ and define the sets U and V by

$$U = \left\{ u(\mathbf{x}),\ \mathbf{x} \in \partial D\colon \Delta_2 u + k^2 u = 0 \text{ in } D,\ \frac{\partial}{\partial \nu} u(\mathbf{x}) = 0 \text{ for } \mathbf{x} \in \partial D \right\}$$

$$V = \left\{ \frac{\partial}{\partial \nu} v(\mathbf{x}),\ \mathbf{x} \in \partial D\colon \Delta_2 v + k^2 v = 0 \text{ in } D,\ v(\mathbf{x}) = 0 \text{ for } \mathbf{x} \in \partial D \right\}.$$

We again assume that $u(\mathbf{x})$ and $v(\mathbf{x})$ are both in $C^1(\overline{D})$. Note that if k^2 is not an interior Neumann eigenvalue then $U = \{0\}$, and if k^2 is not an interior Dirichlet eigenvalue then $V = \{0\}$.

THEOREM 53

$N(\mathbf{I} + \mathbf{K}) = U.$

■ ***Proof.*** Let $\psi \in N(\mathbf{I} + \mathbf{K})$ and define the double layer potential by (6.77). Then $u^-(\mathbf{x}) = 0$, recalling that

$$u^-(\mathbf{x}) = \lim_{\mathbf{x} \to \partial D} u(\mathbf{x}) \qquad (\mathbf{x} \in R^2 \backslash \overline{D}),$$

and from the uniqueness of the solution to the exterior Dirichlet problem it follows that $u(\mathbf{x})$ is identically zero for $\mathbf{x} \in R^2 \backslash D$. By the continuity of the normal derivative of the double layer potential we have

$$\frac{\partial u^+}{\partial \nu}(\mathbf{x}) = \frac{\partial u^-}{\partial \nu}(\mathbf{x}) = 0,$$

...t + denotes the limit from inside D and − the limit from outside D, i.e., $u(\mathbf{x})$ is a solution of the homogeneous interior Neumann problem. Finally, since

$$\psi(\mathbf{x}) = u^-(\mathbf{x}) - u^+(\mathbf{x}) = -u^+(\mathbf{x}),$$

we have that $\psi \in U$.

Conversely, let $\psi \in U$. Then from Green's second identity we have

$$\int_{\partial D} \psi(\mathbf{y}) \frac{\partial}{\partial \nu(\mathbf{y})} \Phi(\mathbf{x},\mathbf{y}) ds(\mathbf{y}) = 0 \qquad (\mathbf{x} \in R^2 \backslash \overline{D}),$$

and letting $\mathbf{x}$ tend to ∂D we see that $\psi + \mathbf{K}\psi = 0$, i.e., $\psi \in N(\mathbf{I} + \mathbf{K})$. ■

In the same manner, we have the following result:

THEOREM 54

$N(\mathbf{I} - \mathbf{K}') = V$.

We can now conclude from the Fredholm alternative that the boundary value problem (6.64), (6.65), (6.67) is uniquely soluble provided k^2 is not an interior Neumann eigenvalue and the boundary value problem (6.64), (6.66), (6.67) is uniquely soluble provided k^2 is not an interior Dirichlet eigenvalue. It is highly desirable to remove these restrictions, particularly because for a general domain D we do not know precisely what these eigenvalues are, and in any case from physical considerations the interior eigenvalues should have nothing to do with the exterior scattering problem. To this end, we shall modify the kernel of our integral equations in a manner similar to our previous modification of the integral equation arising in the exterior Dirichlet problem for Laplace's equation (cf. Section 5.3.4). Our approach is based on the ideas of Jones, Ursell 1973, 1978, and Kleinman and Roach 1982. For a further discussion on the use of integral equation methods in scattering theory, see Colton and Kress, Kleinman and Roach 1974, and Ramm.

The idea of Jones and Ursell 1973, 1978 is to add a series of radiating waves of the form

$$\chi(\mathbf{x},\mathbf{y}) = \frac{i}{4} \sum_{n=-\infty}^{\infty} a_n H_n^{(1)}(kr) H_n^{(1)}(k\rho) e^{in(\theta-\phi)} \tag{6.83}$$

to $\Phi(\mathbf{x},\mathbf{y})$ where $\mathbf{x}$ has polar coordinates (r,θ), $\mathbf{y}$ has polar coordinates (ρ,ϕ), and the coefficients a_n are chosen such that, as a function of both $\mathbf{x}$ and $\mathbf{y}$, the series (6.83) is a solution of the Helmholtz equation (satisfying the Sommerfeld radiation condition) for $|\mathbf{x}|,|\mathbf{y}| > R$, where $B = \{\mathbf{x}\colon |\mathbf{x}| \leq R\}$ is contained in D. An appropriate choice of the a_n can easily be determined from the estimates of Section 6.3 and the fact that $H_{-n}^{(1)}(kr) = (-1)^n H_n^{(1)}(kr)$. Replacing the fundamental solution $\Phi(\mathbf{x},\mathbf{y})$ by

$$\Gamma(\mathbf{x},\mathbf{y}) = \Phi(\mathbf{x},\mathbf{y}) + \chi(\mathbf{x},\mathbf{y}),$$

we see that the modified double layer potential

$$u(\mathbf{x}) = \int_{\partial D} \psi(\mathbf{y}) \frac{\partial}{\partial \nu(\mathbf{y})} \Gamma(\mathbf{x},\mathbf{y}) ds(\mathbf{y}) \qquad (\mathbf{x} \in R^2 \backslash (\partial D \cup B)) \tag{6.84}$$

with continuous density $\psi(\mathbf{y})$ is a solution of the exterior Dirichlet problem (6.64), (6.65), (6.67) provided $\psi(\mathbf{y})$ is a solution of the integral equation

$$\psi(\mathbf{x}) + 2\int_{\partial D} \psi(\mathbf{y}) \frac{\partial}{\partial \nu(\mathbf{y})} \Gamma(\mathbf{x},\mathbf{y})ds(\mathbf{y}) = -2u^i(\mathbf{x}) \qquad (\mathbf{x} \in \partial D), \tag{6.85}$$

or, in operator notation,

$$\psi + \mathbf{K}_1\psi = -2u^i.$$

Note that $\chi(\mathbf{x},\mathbf{y})$ is continuously differentiable for $|\mathbf{x}|,|\mathbf{y}| > R$ and hence for $\mathbf{x},\mathbf{y} \in \partial D$.

THEOREM 55

The integral equation (6.85) is uniquely soluble for all $k > 0$ provided either

$$|2a_n + 1| < 1 \qquad (n = 0,\pm1,\pm2,\ldots)$$

or

$$|2a_n + 1| > 1 \qquad (n = 0,\pm1,\pm2,\ldots).$$

■ ***Proof.*** By the Fredholm alternative, we must show that $(\mathbf{I} + \mathbf{K}_1)\psi = 0$ implies that $\psi = 0$. Let $(\mathbf{I} + \mathbf{K}_1)\psi = 0$. Then $u(\mathbf{x})$, as defined by (6.84), solves the homogeneous exterior Dirichlet problem and by the uniqueness of the solution to this problem $u(\mathbf{x})$ is identically zero for $\mathbf{x} \in R^2\backslash D$. From the jump relations of the double layer potential, we have

$$\frac{\partial u^+}{\partial \nu}(\mathbf{x}) = \frac{\partial u^-}{\partial \nu}(\mathbf{x}) = 0.$$

Since $H_{-n}^{(1)}(kr) = (-1)^n H_n^{(1)}(kr)$, we can rewrite the addition formula as

$$\begin{aligned}\Phi(\mathbf{x},\mathbf{y}) &= \frac{i}{4} H_0^{(1)}(k|\mathbf{x} - \mathbf{y}|) \\ &= \frac{i}{4} \sum_{n=-\infty}^{\infty} J_n(kr)H_n^{(1)}(k\rho)e^{in(\theta-\phi)} \qquad (\rho > r),\end{aligned}$$

where $\mathbf{x} = (r,\theta)$, $\mathbf{y} = (\rho,\phi)$. Hence, from (6.83) and (6.84) we see that there exist constants α_n such that for $R_1 \le |\mathbf{x}| \le R_2$, where $R < R_1 < R_2$ and $\{\mathbf{x}\colon |\mathbf{x}| \le R_2\}$ is contained in D, we can represent $u(\mathbf{x})$ in the form

$$u(\mathbf{x}) = \sum_{n=-\infty}^{\infty} \alpha_n\{J_n(kr) + a_n H_n^{(1)}(kr)\}e^{in\theta}.$$

Using this expansion (which can be differentiated termwise with the resulting series being absolutely and uniformly convergent) and the Wronskian relations for Bessel functions (cf. Section 6.2.2), we now see from Green's second identity that

$$0 = \int_{\partial D} \left(u^+ \frac{\partial \bar{u}^+}{\partial \nu} - \bar{u}^+ \frac{\partial u^+}{\partial \nu} \right) ds = \int_{|\mathbf{x}|=R_1} \left(u \frac{\partial \bar{u}}{\partial \nu} - \bar{u} \frac{\partial u}{\partial \nu} \right) ds$$

$$= 2i \sum_{n=-\infty}^{\infty} |\alpha_n|^2 (1 - |1 + 2a_n|^2).$$

If one of the conditions of the theorem is satisfied, then $\alpha_n = 0$ for $n = 0, \pm 1, \pm 2, \ldots$ and $u(\mathbf{x})$ is identically zero for $R_1 \leq |\mathbf{x}| \leq R_2$. By the analyticity of $u(\mathbf{x})$ and the identity theorem we can now conclude that $u(\mathbf{x})$ is identically zero for $\mathbf{x} \in D \backslash B$. The jump conditions for double layer potentials now imply that $\psi(\mathbf{y}) = 0$. ■

COROLLARY

For each $k > 0$ there exists a (unique) solution to the scattering problem (6.64), (6.65), (6.67).

In the same manner, the modified single layer potential

$$u(\mathbf{x}) = \int_{\partial D} \psi(\mathbf{y}) \Gamma(\mathbf{x}, \mathbf{y}) ds(\mathbf{y}) \qquad (\mathbf{x} \in R^2 \backslash (\partial D \cup B)) \tag{6.86}$$

with continuous density $\psi(\mathbf{y})$ solves the exterior Neumann problem (6.64), (6.66), (6.67) provided $\psi(\mathbf{y})$ is a solution of the integral equation

$$\psi(\mathbf{x}) - 2 \int_{\partial D} \psi(\mathbf{y}) \frac{\partial}{\partial \nu(\mathbf{x})} \Gamma(\mathbf{x}, \mathbf{y}) ds(\mathbf{y}) = 2 \frac{\partial u^i}{\partial \nu}(\mathbf{x}) \qquad (\mathbf{x} \in \partial D). \tag{6.87}$$

Repeating essentially the same argument as used above now gives:

THEOREM 56

The integral equation (6.87) is uniquely soluble for all $k > 0$ provided that either

$$|2a_n + 1| < 1 \qquad (n = 0, \pm 1, \pm 2, \ldots)$$

or

$$|2a_n + 1| > 1 \qquad (n = 0, \pm 1, \pm 2, \ldots).$$

COROLLARY

For each $k > 0$ there exists a (unique) solution to the scattering problem (6.64), (6.66), (6.67).

For the case of the exterior Dirichlet problem, there is a somewhat simpler approach (cf. Leis, Brakhage and Werner, and Panich) for obtaining uniquely soluble integral equations for all positive values of the wave number k. In this approach, we look for a solution of (6.64), (6.65), (6.67) in the form

$$u(\mathbf{x}) = \int_{\partial D} \psi(\mathbf{y})\left\{\frac{\partial}{\partial \nu(\mathbf{y})}\Phi(\mathbf{x},\mathbf{y}) - i\Phi(\mathbf{x},\mathbf{y})\right\}ds(\mathbf{y}) \qquad (\mathbf{x} \in R^2\backslash\overline{D}). \tag{6.88}$$

From the discontinuity properties of double and single layer potentials we see that (6.88) will be a solution provided $\psi \in C(\partial D)$ is a solution of the integral equation

$$\psi(\mathbf{x}) + 2\int_{\partial D} \psi(\mathbf{y})\left\{\frac{\partial}{\partial \nu(\mathbf{y})}\Phi(\mathbf{x},\mathbf{y}) - i\Phi(\mathbf{x},\mathbf{y})\right\}ds(\mathbf{y}) = -2u^i(\mathbf{x}) \qquad (\mathbf{x} \in \partial D), \tag{6.89}$$

or, in operator notation,

$$\psi + \mathbf{K}\psi - i\mathbf{S}\psi = -2u^i,$$

where **K** is as previously defined and **S** is the single layer potential (6.78) evaluated on ∂D. The unique solubility of (6.89) will be established if we can show that the homogeneous equation has only the trivial solution $\psi(\mathbf{y})$ identically equal to zero. Hence, suppose $\psi \in C(\partial D)$ is a solution of $\psi + \mathbf{K}\psi - i\mathbf{S}\psi = 0$. Then for this ψ, $u(\mathbf{x})$ defined by (6.88) is a solution of the homogeneous exterior Dirichlet problem satisfying the Sommerfeld radiation condition and hence is identically zero for $\mathbf{x} \in R^2\backslash D$. From the discontinuity properties of single and double layer potentials, we now have that for (6.88) defined in D, $u(\mathbf{x})$ continuously assumes the boundary data

$$u^+(\mathbf{x}) = -\psi(\mathbf{x})$$

$$\frac{\partial u^+}{\partial \nu}(\mathbf{x}) = -i\psi(\mathbf{x}).$$

Hence, from Green's first identity we have that

$$i\int_{\partial D} |\psi|^2 ds = \int_{\partial D} \bar{u}\frac{\partial u}{\partial \nu}ds = \iint_D (|\nabla u|^2 - k^2|u|^2)ds. \tag{6.90}$$

Since k is real, we can now conclude that ψ is identically zero and hence by the Fredholm alternative the integral equation (6.89) is uniquely soluble.

A similar approach for solving the exterior Neumann problem runs into difficulties because the normal derivative of a double layer potential with continuous density in general does not exist on the boundary, and even if it does exist the corresponding integral operator is strongly singular. In this case, results on the existence of a solution require regularization of the integral equation, i.e., transforming the integral equation into a form for which the Fredholm alternative is applicable. For details on this procedure see Colton and Kress.

In closing, we note that all of the above results on the existence of a unique solution to (6.64)–(6.67) for $u^i(r,\theta) = \exp[ikr \cos \theta]$ are also trivially valid when

$u^i(\mathbf{x})$ is any solution of the Helmholtz equation defined in all of R^2, in particular for the plane wave moving in the direction $\boldsymbol{\alpha}$, $|\boldsymbol{\alpha}| = 1$, defined by $u^i(\mathbf{x}) = \exp[ik\mathbf{x}\cdot\boldsymbol{\alpha}]$.

6.6 THE INVERSE SCATTERING PROBLEM

6.6.1 Preliminaries

In this final section we want to consider the inverse problem to that considered in the previous section, i.e., given the scattered field at a large distance from a scattering obstacle to determine the shape of the scatterer. The importance of this inverse scattering problem in various areas of technology such as radar, sonar, medical imaging, and nondestructive testing has engendered a vast literature; for references we refer the reader to Colton 1984b and Sleeman. We again make the assumption that the (now unknown) scattering obstacle is bounded, simply connected, contains the origin, and has C^2 boundary ∂D with unit outward normal $\boldsymbol{\nu}$. For the sake of simplicity, we also assume that the boundary data is of Dirichlet type with the incident field a time harmonic plane wave moving in the direction $\boldsymbol{\alpha}$; i.e., the direct scattering problem is to find $u(\mathbf{x})$ such that

$$\Delta_2 u + k^2 u = 0 \qquad \text{in} \qquad R^2\backslash\overline{D} \tag{6.91}$$

$$u(\mathbf{x}) = -\exp[ik\mathbf{x}\cdot\boldsymbol{\alpha}] \qquad \text{on} \qquad \partial D \tag{6.92}$$

$$\lim_{r\to\infty} \sqrt{r}\left(\frac{\partial u}{\partial r} - iku\right) = 0, \tag{6.93}$$

where $\mathbf{x} \in R^2$, $\boldsymbol{\alpha} \in R^2$, $|\boldsymbol{\alpha}| = 1$, and (r,θ) are the polar coordinates of $\mathbf{x}$. The existence and uniqueness of a solution to (6.91)–(6.93) has been established in the previous section.

We now describe our inverse scattering problem more precisely. From Theorem 50 we have

$$u(\mathbf{x}) = \int_{\partial D}\left(u(\mathbf{y})\frac{\partial}{\partial\nu(\mathbf{y})}\Phi(\mathbf{x},\mathbf{y}) - \frac{\partial u}{\partial\nu}(\mathbf{y})\Phi(\mathbf{x},\mathbf{y})\right)ds(\mathbf{y}) \qquad (\mathbf{x}\in R^2\backslash\overline{D}), \tag{6.94}$$

where $\Phi(\mathbf{x},\mathbf{y})$ is the radiating fundamental solution to the Helmholtz equation. If we set $u^i(\mathbf{x}) = \exp[ik\mathbf{x}\cdot\boldsymbol{\alpha}]$, then from Green's second identity we have

$$0 = \int_{\partial D}\left(u^i(\mathbf{y})\frac{\partial}{\partial\nu(\mathbf{y})}\Phi(\mathbf{x},\mathbf{y}) - \frac{\partial u^i}{\partial\nu}(\mathbf{y})\Phi(\mathbf{x},\mathbf{y})\right)ds(\mathbf{y}) \qquad (\mathbf{x}\in R^2\backslash\overline{D}). \tag{6.95}$$

Hence, if we define the total field $\hat{u}(\mathbf{x})$ by $\hat{u}(\mathbf{x}) = u(\mathbf{x}) + u^i(\mathbf{x})$, then from (6.92), (6.94), and (6.95) we have

$$u(\mathbf{x}) = -\int_{\partial D}\frac{\partial\hat{u}}{\partial\nu}(\mathbf{y})\Phi(\mathbf{x},\mathbf{y})ds(\mathbf{y}) \qquad (\mathbf{x}\in R^2\backslash\overline{D}). \tag{6.96}$$

The asymptotic behavior of the Hankel function, the estimate

$$|\mathbf{x} - \mathbf{y}| = \sqrt{r^2 - 2r\rho \cos (\theta - \phi) + \rho^2} = r\sqrt{1 - 2\frac{\rho}{r}\cos (\theta - \phi) + \frac{\rho^2}{r^2}}$$

$$= r - \rho \cos (\theta - \phi) + O\left(\frac{1}{r}\right),$$

where (ρ,ϕ) are the polar coordinates of $\mathbf{y}$, and (6.96) imply that $u(\mathbf{x})$ has the asymptotic behavior

$$u(\mathbf{x}) = \frac{e^{ikr}}{\sqrt{r}} F(\theta;k,\alpha) + O\left(\frac{1}{r}\right), \tag{6.97}$$

where

$$F(\theta;k,\alpha) = -\frac{e^{i\pi/4}}{\sqrt{8\pi k}} \int_{\partial D} \frac{\partial \hat{u}}{\partial \nu}(\mathbf{y}) \exp[-ik\rho \cos (\theta - \phi)]ds(\mathbf{y}). \tag{6.98}$$

DEFINITION 18

The function $F(\theta;k,\alpha)$ defined by (6.98) is called the **far field pattern** of the scattered field corresponding to the incident field $u^i(\mathbf{x}) = \exp[ik\mathbf{x} \cdot \alpha]$.

We can now define our inverse scattering problem as follows:

DEFINITION 19 Inverse Scattering Problem

Given the far field pattern $F(\theta;k,\alpha)$ for $-\pi \leq \theta \leq \pi$, a fixed positive value of k, and N distinct directions $\alpha = \alpha_1,\alpha_2,\ldots,\alpha_N$, find D.

It is evident from (6.98) that the inverse scattering problem is nonlinear. However, there are even more serious difficulties. In particular, we note that $F(\theta;k,\alpha)$ is an entire analytic function of θ. But in practical applications $F(\theta;k,\alpha)$ is measured and is thus subject to noise—what we measure is in general not analytic, and even if it were there is no guarantee that it would be a far field pattern (i.e., of the form (6.98)). Hence, for a given measurement, in general no solution exists to the inverse scattering problem, i.e., the inverse scattering problem is improperly posed. (Recall that we have previously met improperly posed problems in Sections 3.1 and 3.3.) We note that for the inverse scattering problem the difficulty is not only that of nonexistence but also a lack of continuous dependence on the data. To see this, we first define the **total scattering cross section** $\sigma(k,\alpha)$ by

$$\sigma(k,\alpha) = \int_{-\pi}^{\pi} |F(\theta;k,\alpha)|^2 d\theta. \tag{6.99}$$

Now suppose ∂D is obtained from ∂D_0 by shifting ∂D_0 an infinitesimal amount $\delta\nu$

along the outer normal to ∂D_0, and let $\hat{u}_+(\mathbf{x})$ and $\hat{u}_-(\mathbf{x})$ be the total fields due to the scattering by D_0 of plane waves moving in the directions $\boldsymbol{\alpha}$ and $-\boldsymbol{\alpha}$ respectively. Let $\delta\sigma = \sigma - \sigma_0$, where $\sigma(k,\boldsymbol{\alpha})$ and $\sigma_0(k,\boldsymbol{\alpha})$ are the total scattering cross sections corresponding to D and D_0 respectively. Then a variational principle (cf. Garabedian and Colton and Kress) states that

$$\delta\sigma = \operatorname{Im} \frac{1}{k} \int_{\partial D_0} \frac{\partial \hat{u}_+}{\partial \nu} \frac{\partial \hat{u}_-}{\partial \nu} \, \delta\nu \, ds.$$

As pointed out in the introduction to Chapter 5, the solution of integral equations of the first kind does not in general depend continuously on the given data. In particular, if in the example above we assume that $\hat{u}_+(\mathbf{x})$ and $\hat{u}_-(\mathbf{x})$ are sufficiently smooth then by the Riemann-Lebesgue lemma (cf. Section 1.4),

$$\lim_{A\to\infty} \int_{\partial D_0} \frac{\partial \hat{u}_+}{\partial \nu} \frac{\partial \hat{u}_-}{\partial \nu} \sin As \, ds = 0.$$

Hence, small perturbations in the modulus of $F(\theta;k,\boldsymbol{\alpha})$ and hence in $\sigma(k,\boldsymbol{\alpha})$ can result in large perturbations of $\delta\nu$, i.e., ∂D does not depend continuously on the modulus of the far field pattern.

Although in general a solution to the inverse scattering problem does not exist for a given (measured) far field pattern, and if it does exist it does not depend continuously on the measured data, we do know that an exact knowledge of the far field pattern uniquely determines the scattering obstacle. The following theorem is due to Schiffer (cf. Colton and Kress).

THEOREM 57

The scattering obstacle D is uniquely determined by a knowledge of the far field pattern $F(\theta;k,\boldsymbol{\alpha})$ for θ on a subinterval of $[-\pi,\pi]$, fixed $k > 0$, and $\boldsymbol{\alpha}$ on an arc of the unit circle.

■ ***Proof.*** Since $F(\theta;k,\boldsymbol{\alpha})$ is an analytic function of θ, it follows from the identity theorem for analytic functions that $F(\theta;k,\boldsymbol{\alpha})$ is known for all θ in $[-\pi,\pi]$. Suppose now that there existed two obstacles D_1 and D_2 having the same far field pattern $F(\theta;k,\boldsymbol{\alpha})$. Consider first the case when D_1 and D_2 are disjoint and let $u_1(\mathbf{x})$ and $u_2(\mathbf{x})$ be the corresponding scattered fields. Expanding $F(\theta;k,\boldsymbol{\alpha})$ in a Fourier series, $u_1(\mathbf{x})$ and $u_2(\mathbf{x})$ in a Fourier-Bessel series (6.75), (6.76), and letting $r = |\mathbf{x}|$ tend to infinity, we find that the Fourier-Bessel coefficients of $u_1(\mathbf{x})$ and $u_2(\mathbf{x})$ can be determined from the Fourier coefficients of $F(\theta;k,\boldsymbol{\alpha})$, i.e., $u_1(\mathbf{x}) = u_2(\mathbf{x})$ outside a disk containing D_1 and D_2 in its interior. By analytic continuation, we can now conclude that $u(\mathbf{x}) = u_1(\mathbf{x}) = u_2(\mathbf{x})$ is a solution of the Helmholtz equation in all of R^2 satisfying the Sommerfeld radiation condition. But then for every integer n and $r > 0$,

$$a_n(r) = \frac{1}{2\pi} \int_{-\pi}^{\pi} u(r,\theta) e^{-in\theta} d\theta$$

is a solution of Bessel's equation that is analytic in a neighborhood of the origin and satisfies the Sommerfeld radiation condition (cf. Section 6.2.5). This is possible only if $a_n(r)$ is identically zero, and hence for each $r > 0$ all of the Fourier coefficients of $u(\mathbf{x})$ vanish, i.e., $u(\mathbf{x})$ is identically zero. But

$$u(\mathbf{x}) = -\exp[ik\mathbf{x} \cdot \boldsymbol{\alpha}] \qquad \text{on} \qquad \partial D_1,$$

and this is a contradiction. Hence, there cannot exist two disjoint obstacles having the same far field pattern.

Now suppose that $D_3 = D_1 \cap D_2$ is nonempty, and without loss of generality assume that D_1 is not contained in D_2. Then the above argument shows that $u(\mathbf{x})$, and hence the total field $\hat{u}(\mathbf{x})$, is a solution of the Helmholtz equation in $D_1 \backslash D_3$ and $\hat{u}(\mathbf{x})$ vanishes on the boundary of $D_1 \backslash D_3$. Hence, for $\boldsymbol{\alpha}$ on an arc of the unit circle, $\hat{u}(\mathbf{x}) = \hat{u}(\mathbf{x},\boldsymbol{\alpha})$ is an eigenfunction of the Laplacian in $D_1 \backslash D_3$, and since

$$\hat{u}(\mathbf{x},\boldsymbol{\alpha}) = \exp[ik\mathbf{x} \cdot \boldsymbol{\alpha}] + \frac{e^{ikr}}{\sqrt{r}} F(\theta;k,\boldsymbol{\alpha}) + O\left(\frac{1}{r}\right),$$

these eigenfunctions are linearly independent. But from Section 5.4 we know that the multiplicity of an eigenvalue of the Laplacian is finite. Hence we have a contradiction, i.e., there cannot exist two intersecting obstacles having the same far field pattern. This completes the proof. ■

In the proof of Theorem 57, we have glossed over the fact that the boundary of $D_1 \backslash D_3$ is in general not in class C^2, but only piecewise C^2. However, the finite multiplicity of eigenvalues of the Laplacian is also valid for domains with boundaries in this slightly wider class and we ask the reader to accept this fact on faith.

6.6.2 Herglotz Wave Functions

Our method for solving (in a sense to be described shortly) the inverse scattering problem is based on the properties of a certain class of solutions to the Helmholtz equation defined in all of R^2 called Herglotz wave functions.

DEFINITION 20

A solution of the Helmholtz equation of the form

$$v(r,\theta) = \sum_{n=-\infty}^{\infty} a_n J_n(kr) e^{in\theta}$$

such that

$$\sum_{n=-\infty}^{\infty} |a_n|^2 < \infty$$

is called a **Herglotz wave function.**

Note that by our previously obtained estimates on the Bessel function $J_n(kr)$

we can conclude that Herglotz wave functions are solutions of the Helmholtz equation in all of R^2. Our first theorem provides an integral representation of Herglotz wave functions.

THEOREM 58

A function $v(r,\theta)$ is a Herglotz wave function if and only if there exists a square integrable function $g(\phi)$ defined on $[-\pi,\pi]$ such that

$$v(r,\theta) = \int_{-\pi}^{\pi} g(\phi) \exp\,[ikr \cos(\theta - \phi)]d\phi.$$

■ ***Proof.*** Suppose $v(r,\theta)$ is a Herglotz wave function. Then

$$v(r,\theta) = \sum_{n=-\infty}^{\infty} a_n J_n(kr)e^{in\theta}, \tag{6.100}$$

where

$$\sum_{n=-\infty}^{\infty} |a_n|^2 < \infty. \tag{6.101}$$

From our discussion of orthonormal sets in Section 5.4, we see that

$$g(\phi) = \frac{1}{2\pi} \sum_{n=-\infty}^{\infty} a_n(-i)^n e^{in\phi} \tag{6.102}$$

converges in the mean-square sense and defines a square integrable function on $[-\pi,\pi]$. From the Bessel function expansion (6.17) of the plane wave

$$\exp\,[ikr \cos\theta] = \sum_{n=-\infty}^{\infty} i^n J_n(kr)e^{in\theta} \tag{6.103}$$

we now see that $v(r,\theta)$ can be represented in the form stated in the theorem.

Conversely, if $g(\phi)$ is square integrable, then by the results of Section 1.4 we see that in the sense of mean-square convergence $g(\phi)$ has a Fourier series (6.102) such that (6.101) is satisfied, and from (6.102) and (6.103) we see that $v(r,\theta)$ as defined in the statement of the theorem is a Herglotz wave function. ■

DEFINITION 21

The space of complex valued square integrable functions defined on $[-\pi,\pi]$ with inner product

$$(f,g) = \int_{-\pi}^{\pi} f(x)\overline{g(x)}dx$$

will be denoted by $L^2[-\pi,\pi]$.

DEFINITION 22

The space of Herglotz wave functions will be denoted by H. If $v(r,\theta)$ is a Herglotz wave function with the integral representation given in the above theorem, the function $g(\phi)$ is called the **Herglotz kernel** of $v(r,\theta)$.

Let D be a bounded, simply connected domain containing the origin with C^2 boundary ∂D and unit outward normal $\boldsymbol{\nu}$. Denote the first Dirichlet eigenvalue for the Laplacian in D by λ_1. Let $H|_{\partial D}$ denote the restriction of H to ∂D and denote its closure in $L^2[\partial D]$ by $\overline{H|_{\partial D}}$, where $L^2[\partial D]$ is the space of complex valued square integrable functions defined on ∂D with inner product

$$(f,g) = \int_{\partial D} f(\mathbf{x})\overline{g(\mathbf{x})}ds(\mathbf{x}).$$

Then we have the following theorem:

THEOREM 59

Let $0 < k^2 < \lambda_1$. Then $L^2[\partial D] = \overline{H|_{\partial D}}$.

■ ***Proof.*** We must show that every function in $L^2[\partial D]$ can be approximated arbitrarily closely in the mean-square sense by an element from $H|_{\partial D}$. From the Gram-Schmidt procedure and our discussion on complete sets in Section 5.4, it suffices to show that if $g \in L^2[\partial D]$ is such that for every integer $n = 0,\pm 1,\pm 2,\ldots$

$$\int_{\partial D} g(\mathbf{y})J_n(k\rho)e^{in\phi}ds(\mathbf{y}) = 0 \qquad (\mathbf{y} = \rho e^{i\phi}), \tag{6.104}$$

then $g(\mathbf{y})$ is equal to zero almost everywhere. Let Ω be a disk centered at the origin and containing D in its interior. Then for $\mathbf{x} \in R^2\backslash\Omega$ and $\mathbf{y} \in \partial D$ we have the addition formula (cf. Section 6.2.5)

$$H_0^{(1)}(k|\mathbf{x} - \mathbf{y}|) = \sum_{n=-\infty}^{\infty} J_n(k\rho)H_n^{(1)}(kr)e^{in(\phi-\theta)}, \tag{6.105}$$

where (r,θ) are the polar coordinates of $\mathbf{x}$. Hence, from (6.104) and (6.105) we see that

$$u(\mathbf{x}) = \frac{i}{4}\int_{\partial D} g(\mathbf{y})H_0^{(1)}(k|\mathbf{x} - \mathbf{y}|)ds(\mathbf{y}) \tag{6.106}$$

is identically zero for $\mathbf{x} \in R^2\backslash\Omega$. Since $u(\mathbf{x})$, as defined by (6.106), is a solution of the Helmholtz equation in the exterior of D, we can conclude by the analyticity of solutions to the Helmholtz equation that $u(\mathbf{x})$ is identically zero for $\mathbf{x} \in R^2\backslash\overline{D}$. Furthermore, $u(\mathbf{x})$ is continuous in all of R^2 and is a solution of the Helmholtz equation in D. In particular, $u(\mathbf{x}) = 0$ for $\mathbf{x} \in \partial D$, and since $0 < k^2 < \lambda_1$ we can conclude that $u(\mathbf{x})$ is identically zero for $\mathbf{x} \in D$. From the discontinuity properties of single layer potentials with square integrable densities (cf. Section 5.1) we see that

$$0 = \frac{\partial u^+}{\partial \nu}(\mathbf{x}) - \frac{\partial u^-}{\partial \nu}(\mathbf{x}) = g(\mathbf{x}),$$

and the theorem is now proved. ■

In order to state our final theorem on Herglotz wave functions we need two more definitions.

DEFINITION 23

A set S in $L^2[-\pi,\pi]$ is said to be **orthogonal** to a function $g \in L^2[-\pi,\pi]$, and we write $S \perp g$, if for every $f \in S$ $(f,g) = 0$.

Our next definition is concerned with the solution of the boundary value problem

$$\begin{aligned} \Delta_2 v + k^2 v &= 0 \quad \text{in} \quad D \\ v(\rho,\phi) &= H_0^{(2)}(k\rho) \quad \text{on} \quad \partial D. \end{aligned} \tag{6.107}$$

Note that by using the methods of Section 6.5 we can conclude that there exists a unique solution of (6.107) provided $0 < k^2 < \lambda_1$. (Rewrite (6.107) as an integral equation and apply the Fredholm alternative—see Exercise 25.) In particular, if $D = \{\mathbf{y}: |\mathbf{y}| < a\}$ the solution is the Herglotz wave function

$$v(\rho,\phi) = \frac{H_0^{(2)}(ka)}{J_0(ka)} J_0(k\rho).$$

DEFINITION 24

Let $0 < k^2 < \lambda_1$. A domain D is called a **Herglotz domain** if the unique solution of (6.107) is a Herglotz wave function.

THEOREM 60

Suppose $0 < k^2 < \lambda_1$ and D is a Herglotz domain where the solution $v(\rho,\phi)$ of (6.107) has Herglotz kernel $g(\theta)$. Let $\{\boldsymbol{\alpha}_n\}$ be a countable set of distinct unit vectors and define the set $S \subset L^2[-\pi,\pi]$ by

$$S = \{F(\theta;k,\boldsymbol{\alpha}_n) - F(\theta;k,\boldsymbol{\alpha}_1): n = 1,2,\ldots\},$$

where $F(\theta;k,\boldsymbol{\alpha}_n)$ is the far field pattern corresponding to D and the incident field $u^i(\mathbf{x}) = \exp[ik\mathbf{x}\cdot\boldsymbol{\alpha}_n]$. Then $S \perp g$.

■ ***Proof.*** Let $g(\theta)$ be as defined in the theorem and recall the far field representation (6.98)

$$F(\theta;k,\boldsymbol{\alpha}_n) = -\frac{e^{i\pi/4}}{\sqrt{8\pi k}} \int_{\partial D} \frac{\partial \hat{u}}{\partial \nu}(\mathbf{y}) \exp[-ik\rho\cos(\theta - \phi)ds(\mathbf{y}),$$

where (ρ,ϕ) are the polar coordinates of $\mathbf{y}$. Then we have

$$\begin{aligned}\int_{-\pi}^{\pi} F(\theta;k,\boldsymbol{\alpha}_n)\overline{g(\theta)}d\theta &= -\frac{e^{i\pi/4}}{\sqrt{8\pi k}}\int_{\partial D}\frac{\partial\hat{u}}{\partial\nu}(\mathbf{y})\overline{v(\rho,\phi)}ds(\mathbf{y})\\ &= -\frac{e^{i\pi/4}}{\sqrt{8\pi k}}\int_{\partial D}\frac{\partial\hat{u}}{\partial\nu}(\mathbf{y})H_0^{(1)}(k\rho)ds(\mathbf{y}),\end{aligned} \tag{6.108}$$

since $\overline{H_0^{(2)}(k\rho)} = H_0^{(1)}(k\rho)$. But since $\hat{u}(\mathbf{y}) = 0$ for $\mathbf{y} \in \partial D$,

$$\begin{aligned}\int_{\partial D}\frac{\partial\hat{u}}{\partial\nu}(\mathbf{y})H_0^{(1)}(k\rho)ds(\mathbf{y}) &= \int_{\partial D}\left(\frac{\partial\hat{u}}{\partial\nu}(\mathbf{y})H_0^{(1)}(k\rho) - \hat{u}(\mathbf{y})\frac{\partial}{\partial\nu}H_0^{(1)}(k\rho)\right)ds(\mathbf{y})\\ &= \int_{|\mathbf{y}|=b}\left(\frac{\partial\hat{u}}{\partial\nu}(\mathbf{y})H_0^{(1)}(k\rho) - \hat{u}(\mathbf{y})\frac{\partial}{\partial\nu}H_0^{(1)}(k\rho)\right)ds(\mathbf{y}),\end{aligned} \tag{6.109}$$

where D is contained in the disk $|\mathbf{x}| < b$. But on $|\mathbf{x}| = b$ we have

$$\hat{u}(\mathbf{x}) = \exp[ik\mathbf{x}\cdot\boldsymbol{\alpha}_n] + u(\mathbf{x}) = \exp[ik\mathbf{x}\cdot\boldsymbol{\alpha}_n] + \sum_{n=-\infty}^{\infty} a_n H_n^{(1)}(kr)e^{in\theta}$$

for some set of constants $\{a_n\}$. Hence, by letting b tend to zero and computing the residue, we have from (6.109) that

$$\begin{aligned}&\int_{\partial D}\frac{\partial\hat{u}}{\partial\nu}(\mathbf{y})H_0^{(1)}(k\rho)ds(\mathbf{y}) =\\ &\quad\int_{|\mathbf{y}|=b}\left(\frac{\partial}{\partial\nu}\exp[ik\mathbf{x}\cdot\boldsymbol{\alpha}_n]H_0^{(1)}(k\rho) - \exp[ik\mathbf{x}\cdot\boldsymbol{\alpha}_n]\frac{\partial}{\partial\nu}H_0^{(1)}(k\rho)\right)ds(\mathbf{y}) = -4i.\end{aligned} \tag{6.110}$$

From (6.108) and (6.110) we now have

$$\int_{-\pi}^{\pi} F(\theta;k,\boldsymbol{\alpha}_n)\overline{g(\theta)}d\theta = 4i\frac{e^{i\pi/4}}{\sqrt{8\pi k}} = -e^{-i\pi/4}\sqrt{\frac{2}{\pi k}}, \tag{6.111}$$

and therefore

$$\int_{-\pi}^{\pi}(F(\theta;k,\boldsymbol{\alpha}_n) - F(\theta;k,\boldsymbol{\alpha}_1))\overline{g(\theta)}d\theta = 0$$

for $n = 1,2,\ldots$. The theorem is now proved. ■

In the applications of the above theory of Herglotz wave functions to the inverse scattering problem we do not know D, so we do not know λ_1. However, if we make the a priori assumption that D is contained in the disk $\Omega = \{\mathbf{x}: |\mathbf{x}| < b\}$ we know from the results of Section 5.4 that $\lambda_1 \geq \mu_1$ where μ_1 is the first eigenvalue of the Laplacian in Ω. But if $w(r,\theta)$ is the first eigenfunction of the Laplacian in Ω then

$$w(r,\theta) = \sum_{n=-\infty}^{\infty} a_n J_n(\sqrt{\mu_1} r) e^{in\theta} \tag{6.112}$$

for $r < b$, where for $r < b$

$$a_n J_n(\sqrt{\mu_1} r) = \frac{1}{2\pi} \int_{-\pi}^{\pi} w(r,\theta) e^{-in\theta}. \tag{6.113}$$

Since $w(b,\theta) = 0$, letting r tend to b in (6.113) shows that

$$a_n J_n(\sqrt{\mu_1} b) = 0,$$

i.e., either $a_n = 0$ or $\sqrt{\mu_1} b$ is a positive zero of a Bessel function. We have previously established (see Section 6.2.5) that the smallest positive zero of all the Bessel functions is k_{01}, the first positive zero of $J_0(z)$. Hence, $\sqrt{\mu_1} b = k_{01}$, or

$$\mu_1 = \frac{k_{01}^2}{b^2} . \tag{6.114}$$

In particular, if we require k to satisfy

$$0 < k^2 < \frac{k_{01}^2}{b^2} \tag{6.115}$$

we can conclude that $0 < k^2 < \lambda_1$.

In this section, we have only presented a few of the basic results in the theory of Herglotz wave functions. For a more extensive development we refer the reader to Hartman and Wilcox.

6.6.3 Optimal Solutions to the Inverse Scattering Problem

We now present a method for reconstructing the shape of a scattering obstacle from an (approximate) knowledge of the far field pattern. For the sake of simplicity, we shall assume that D is starlike with respect to the origin, i.e., ∂D can be parameterized in the form

$$\rho = \rho(\phi) \qquad (-\pi \leq \phi \leq \pi), \tag{6.116}$$

where (ρ,ϕ) are the polar coordinates of a point $\mathbf{y} \in \partial D$, and $\rho(\phi)$ is a twice continuously differentiable function of ϕ of period 2π.

Our method is to find a scattering obstacle D whose far field pattern is as close as possible to the measured far field pattern $F(\theta;k,\alpha)$. The continuous dependence of D on $F(\theta;k,\alpha)$ will be obtained by imposing a priori restrictions on the class of

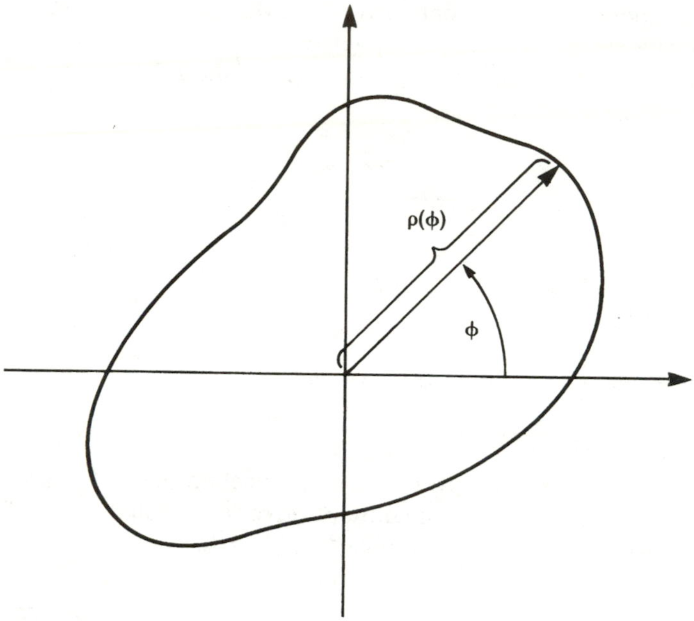

Figure 6.5

obstacles in which we assume the true scattering obstacle lies. The methods described in this section are from Colton and Monk 1985, 1986, and 1987, and Kirsch and Kress, and we refer the reader to these papers for more information as well as numerical examples.

We begin by defining and establishing some elementary properties of functionals.

DEFINITION 25

A function taking a set of functions S into the real numbers R^1 is called a **functional.**

□ **EXAMPLES 21–23.**

Let $S = C[-\pi,\pi]$, the set of complex valued continuous functions defined on $[-\pi,\pi]$. Then the following are functionals defined for $f \in S$:

(21) $F(f) = \max_{x\in[-\pi,\pi]} |f(x)|$.

(22) $F(f) = \int_{-\pi}^{\pi} \phi(x,|f(x)|)dx$, where $\phi(x,y)$ is real valued and continuous for $-\pi \le x \le \pi$ and all real y.

(23) $F(f) = |f(x_0)|$, $x_0 \in [-\pi,\pi]$.

Suppose the set of functions S is a vector space (i.e., S is closed under the operations of addition and scalar multiplication) and there exists a norm on S, i.e., to each function $f \in S$ there is a nonnegative number $\|f\|$, called the norm of f, such that

$\|f\| = 0$ if and only if $f = 0$.

$\|cf\| = |c|\,\|f\|$ for every constant c.

$\|f + g\| \le \|f\| + \|g\|$ for $f,g \in S$.

For example, if S is the set of (complex valued) square integrable functions defined on $[-\pi,\pi]$ then

$$\|f\| = \left[\int_{-\pi}^{\pi} |f(x)|^2 dx\right]^{1/2} \qquad \textbf{(6.117)}$$

defines a norm on S; if $S = C[-\pi,\pi]$ then

$$\|f\| = \max_{x\in[-\pi,\pi]} |f(x)| \qquad \textbf{(6.118)}$$

defines a norm on S. A functional F on a vector space of functions having a norm is said to be **continuous** if for every $\epsilon > 0$ there exists a $\delta > 0$ such that if $\|f - g\| < \delta$ then $|F(f) - F(g)| < \epsilon$. In particular it can be easily verified that each of the three functionals defined in the above examples are continuous in $C[-\pi,\pi]$ with the norm defined by (6.118). From now on we shall always assume that $C[-\pi,\pi]$ has the norm (6.118) associated with it. □

We now introduce the concept of compactness and prove a basic result on continuous functionals defined on a compact set.

DEFINITION 26

A set of functions K in a normed vector space S is said to be **compact** in S if every sequence of functions in K contains a subsequence which converges to some $f \in K$.

□ EXAMPLE 24

By the Arzela-Ascoli theorem (see Section 5.4), the set of bounded, equicontinuous functions defined on $[-\pi,\pi]$ is compact in $C[-\pi,\pi]$. □

□ EXAMPLE 25

Consider the set of continuously differentiable functions defined on $[-\pi,\pi]$ and denote its completion with respect to the norm

$$\|g\| = \left[\int_{-\pi}^{\pi} (|g(\theta)|^2 + |\dot{g}(\theta)|^2) d\theta\right]^{1/2}$$

by $W_2^1[-\pi,\pi]$, i.e., $W_2^1[-\pi,\pi]$ is the set of all continuously differentiable functions plus the limits of all Cauchy sequences of such functions with respect to the above norm. The space $W_2^1[-\pi,\pi]$ is called a **Sobolev space** and plays a fundamental role in the modern theory of partial differential equations. Consider the set of functions $g \in W_2^1[-\pi,\pi]$ such that $\|g\| \leq M$ for some positive constant M. Then if $\{a_n\}$ denote the Fourier coefficients of $g(\theta)$, it is easily verified that

$$\sum_{n=-\infty}^{\infty} n^2|a_n|^2 < 2\pi M.$$

In addition, by Schwarz's inequality (cf. Section 1.4)

$$|g(\theta)| = \left|\sum_{n=-\infty}^{\infty} a_n e^{in\theta}\right| \leq \sum_{n=-\infty}^{\infty} n^2|a_n|^2 \sum_{n=-\infty}^{\infty} \frac{1}{n^2}$$

and

$$|g(\theta_1) - g(\theta_2)| \leq \sum_{n=-\infty}^{\infty} n^2|a_n|^2 \sum_{n=-\infty}^{\infty} \frac{1}{n^2} |e^{in\theta_1} - e^{in\theta_2}|.$$

Thus, the Fourier series for $g(\theta)$ is uniformly convergent and $g(\theta)$ is a continuous function on $[-\pi,\pi]$. Furthermore, the set of $g \in W_2^1[-\pi,\pi]$ such that $\|g\| \leq M$ is bounded and equicontinuous and hence by the Arzela-Ascoli theorem is compact in $C[-\pi,\pi]$. □

THEOREM 61

A continuous functional defined on a compact set K of a normed vector space attains its maximum and minimum on K.

■ ***Proof.*** Let F be a continuous functional defined on K. We first show that F is bounded on K, i.e.,

$$\sup_{f\in K} |F(f)| \leq M$$

for some positive constant M. Suppose, on the contrary, that F is not bounded. Then there exists a sequence $\{f_n\}$, $f_n \in K$, such that

$$\lim_{n\to\infty} |F(f_n)| = \infty.$$

Since K is compact, there exists a convergent subsequence $\{f_{n_k}\}$, i.e.,

$$\lim_{k\to\infty} \|f_{n_k} - f_0\| = 0$$

for some $f_0 \in K$. Then for f in an arbitrarily small neighborhood of f_0, $F(f)$ will assume arbitrarily large values and this contradicts the continuity of F. Hence F is bounded on K.

Now let

$$A = \sup_{f\in K} F(f).$$

Then there exists a sequence $\{f_n\}$ such that

$$A > F(f_n) > A - \frac{1}{n}.$$

Let $\{f_{n_k}\}$ be a convergent subsequence of $\{f_n\}$, i.e.,

$$\lim_{k\to\infty} \|f_{n_k} - f_0\| = 0.$$

Then, by continuity, $F(f_0) = A$. Hence F achieves its maximum on K. The proof that F achieves its minimum on K is entirely analogous. ■

We are now in a position to reformulate the inverse scattering problem as a problem in constrained optimization, i.e., a problem of finding the minimum of a continuous functional on a compact set, where the compact set is determined by a priori constraints on the (starlike) scattering obstacle D. In particular, define the sets $U_1(M)$ and U_2 by

$$U_1(M) = \{g \in W_2^1[-\pi,\pi]: \|g\| \le M\}$$

$$U_2 = \{\rho \in C[-\pi,\pi]: \rho(\phi) \text{ is periodic of period } 2\pi,\ 0 < a \le \rho(\phi) \le b,$$
$$|\rho(\phi_1) - \rho(\phi_2)| \le C|\phi_1 - \phi_2| \bmod 2\pi\},$$

where M and C are positive constants and $W_2^1[-\pi,\pi]$ denotes the Sobolev space defined in Example 25. The a priori constraints on ∂D: $\rho = \rho(\phi)$ are incorporated in the set U_2 through the constants a, b, and C. These constraints imply that we know a priori that D contains a disk of radius a, is contained in a disk of radius b, and the boundary curve is not wildly oscillating. By the above examples, we see that $U_1(M)$ and U_2 are compact in $C[-\pi,\pi]$, and it easily follows that

$$U(M) = U_1(M) \times U_2$$

is compact in $C[-\pi,\pi] \times C[-\pi,\pi]$.

Motivated by the analysis of the previous section we require

$$0 < k^2 < \frac{k_{01}^2}{b^2}, \tag{6.119}$$

where k_{01} is the first positive zero of the Bessel function $J_0(z)$. Then for $F(\theta;k,\boldsymbol{\alpha}_n)$ a measured far field pattern corresponding to the incident field $u^i(\mathbf{x}) = \exp[ik\mathbf{x}\cdot\boldsymbol{\alpha}_n]$, $n = 1,2,\ldots,N$, we define the nonlinear optimization problem

$$\mu(F,M,N) = \min_{(g,\rho)\in U(M)} \left\{ \sum_{n=1}^{N} \left| \int_{-\pi}^{\pi} F(\theta;k,\boldsymbol{\alpha}_n)\overline{g(\theta)}\,d\theta + e^{-i\pi/4}\sqrt{\frac{2}{\pi k}} \right|^2 \right. \tag{6.120}$$
$$\left. + \int_{-\pi}^{\pi} |v(\rho(\phi),\phi) - H_0^{(2)}(k\rho(\phi))|^2 d\phi \right\},$$

where $v(\rho,\phi)$ is defined by

$$v(\rho,\phi) = \int_{-\pi}^{\pi} g(\theta) \exp[ik\rho \cos(\phi - \theta)]d\theta. \tag{6.121}$$

Note that $\mu(F,M,N) \geq 0$ and from the results of the previous section we know that if D is a Herglotz domain with (exact) far field pattern $F(\theta;k,\boldsymbol{\alpha}_n)$ and Herglotz kernel $g \in W_2^1[-\pi,\pi]$, then $\mu(F,M,N) = 0$ provided M is sufficiently large. Note also that from the compactness of $U(M)$ and the continuity of the integrals in (6.120) and (6.121) as a functional of $g(\theta)$ and $\rho(\phi)$ we can conclude that $\mu(F,M,N)$ exists.

DEFINITION 27

A function $\rho \in C[-\pi,\pi]$ will be called **admissible** if there exists a function $g \in W_2^1[-\pi,\pi]$ such that $(g,\rho) \in U(M)$ and this pair minimizes the expression in brackets in (6.120) over the set $U(M)$.

We now proceed to examine the relationships between admissible solutions and solutions of the inverse scattering problem. We first show that admissible solutions depend continuously on the measured far field data in the sense of the following theorem.

THEOREM 62

Let $\Phi(F)$ be the set of admissible solutions corresponding to the far field patterns $F(\theta;k,\boldsymbol{\alpha}_n)$, $n = 1,2,\ldots,N$. Then if for $n = 1,2,\ldots,N$, $F_j(\theta;k,\boldsymbol{\alpha}_n)$ tends to $F(\theta;k,\boldsymbol{\alpha}_n)$ in $L^2[-\pi,\pi]$, $\rho_j \in \Phi(F_j)$, there exists a convergent subsequence of $\{\rho_j\}$ and every limit point lies in $\Phi(F)$.

■ ***Proof.*** The sequence $\{\rho_j\}$ lies in a compact set, so there exists a convergent subsequence, which we again denote by $\{\rho_j\}$, and ρ_j tends to $\rho^* \in U_2$. Let $g_j(\theta)$ be such that for $F(\theta;k,\boldsymbol{\alpha}_n) = F_j(\theta;k,\boldsymbol{\alpha}_n)$, (g_j,ρ_j) minimizes the expression in brackets in (6.120). Then, after again possibly restricting ourselves to a subsequence, we see that there exists $g^* \in U_1(M)$ such that (g_j,ρ_j) tends to $(g^*,\rho^*) \in U(M)$. We want to show that

$$\mu(F,M,N) = \sum_{n=1}^{N} \left| \int_{-\pi}^{\pi} F(\theta;k,\boldsymbol{\alpha}_n)\overline{g^*(\theta)}d\theta + e^{-i\pi/4}\sqrt{\frac{2}{\pi k}}\right|^2 + \int_{-\pi}^{\pi} |v^*(\rho^*(\phi),\phi) - H_0^{(2)}(k\rho^*(\phi))|^2 d\phi,$$

where $v^*(\rho,\phi)$ is the Herglotz wave function having Herglotz kernel $g^*(\theta)$.

To this end, let $\rho^{**} \in \Phi(F)$ with associated minimizing pair (g^{**},ρ^{**}). Then we have

$$
\begin{aligned}
\mu(F,M,N) &\leq \sum_{n=1}^{N} \left| \int_{-\pi}^{\pi} F(\theta;k,\alpha_n)\overline{g^*(\theta)}d\theta + e^{-i\pi/4}\sqrt{\frac{2}{\pi k}} \right|^2 \\
&\quad + \int_{-\pi}^{\pi} |v^*(\rho^*(\phi),\phi) - H_0^{(2)}(k\rho^*(\phi))|^2 d\phi \\
&= \lim_{j\to\infty} \left\{ \sum_{n=1}^{N} \left| \int_{-\pi}^{\pi} F_j(\theta;k,\alpha_n)\overline{g_j(\theta)}d\theta + e^{-i\pi/4}\sqrt{\frac{2}{\pi k}} \right|^2 \right. \\
&\quad \left. + \int_{-\pi}^{\pi} |v_j(\rho_j(\phi),\phi) - H_0^{(2)}(k\rho_j(\phi))|^2 d\phi \right\} \\
&= \lim_{j\to\infty} \mu(F_j,M,N) \\
&\leq \lim_{j\to\infty} \left\{ \sum_{n=1}^{N} \left| \int_{-\pi}^{\pi} F_j(\theta;k,\alpha_n)\overline{g^{**}(\theta)}d\theta + e^{-i\pi/4}\sqrt{\frac{2}{\pi k}} \right|^2 \right. \\
&\quad \left. + \int_{-\pi}^{\pi} |v^{**}(\rho^{**}(\phi),\phi) - H_0^{(2)}(k\rho^{**}(\phi))|^2 d\phi \right\} \\
&= \mu(F,M,N),
\end{aligned}
\tag{6.122}
$$

where $v_j(\rho,\phi)$ and $v^{**}(\rho,\phi)$ are the Herglotz wave functions having Herglotz kernels $g_j(\theta)$ and $g^{**}(\theta)$ respectively. Hence, all the inequalities in (6.122) are in fact equalities and this implies the theorem. ∎

We now want to show that if $F(\theta;k,\alpha_n)$, $n = 1,2,\ldots,N$, are the far field patterns of a domain D such that ∂D is described by $\rho = \rho(\phi)$ where $\rho \in C^2[-\pi,\pi] \cap U_2$ then

$$\lim_{M\to\infty} \mu(F,M,N) = 0.$$

THEOREM 63

Suppose that $0 < k^2 < k_{01}^2/b^2$ and ∂D: $\rho = \rho(\phi)$ is a solution of the inverse scattering problem with far field patterns $F(\theta;k,\alpha_n)$, $n = 1,2,\ldots,N$, such that $\rho \in C^2[-\pi,\pi] \cap U_2$. Then for every integer N,

$$\lim_{M\to\infty} \mu(F,M,N) = 0.$$

∎ ***Proof.*** Since from the previous section we have that $L^2[\partial D] = \overline{H|_{\partial D}}$, we can approximate the solution of (6.107) in $L^2[\partial D]$ by a Herglotz wave function $v^*(\rho,\phi)$

with Herglotz kernel $g^*(\theta)$. Furthermore, from (6.108), (6.110), and (6.111),

$$\int_{-\pi}^{\pi} F(\theta;k,\alpha_n)\overline{g^*(\theta)}d\theta$$

is approximately equal to $-e^{-i\pi/4}\sqrt{2/\pi k}$. In both cases, the accuracy of the approximation can be as high as desired and as this accuracy is improved, $\|g^*\|$ in general increases. This observation now implies the theorem. ■

How do the above results lead us to an approximate solution of the inverse scattering problem? We begin by choosing M large and using a Newton-like procedure to determine the minimum of the functional in brackets in (6.120). For a discussion of computational methods to minimize nonlinear functionals, e.g., Newton's method, see Luenberger. Theorem 62 shows that the pair (g^*,ρ^*) for which this minimum is achieved depends continuously on the far field data $F(\theta;k,\alpha_n)$, $n = 1,2,\ldots,N$, i.e., the minimization procedure for determining our candidate $\rho^*(\phi)$ of the solution to the inverse scattering problem is stable with respect to measurements of the far field data. The accuracy of the solution $\rho^*(\phi)$ is problematic, since no error estimates are available. However, if M is large then $\mu(F,M,N)$ will be small by our second theorem, and hence if $\mu(F,M,N) < \epsilon$ for ϵ appropriately chosen we have some faith that $\rho^*(\phi)$ is a reasonably good approximation to the solution of the inverse scattering problem. In various numerical experiments it has been found that choosing ϵ to be around .01 gives good computational accuracy (cf. Colton and Monk 1986), i.e., the value of M and the initial guess in the Newton iterative procedure should be varied if necessary such that $\mu(F,M,N)$ is approximately .01 or smaller. (A poor initial guess can cause the Newton procedure to converge to a local minimum which is not an absolute minimum.) Finally, the choice of N is arbitrary, but in order to obtain a sharp minimum N should not be too small. The problem of local minima of (6.120) that are not solutions of the inverse scattering problem remains an open problem, although some results in this direction are presented in Colton and Monk 1986. The situation when the far field data $F(\theta;k,\alpha_n)$, $n = 1,2,\ldots,N$, is only known for θ in a subinterval of $[-\pi,\pi]$ can also be handled; for details we refer the reader to Ochs.

This procedure for solving the inverse scattering problem is based on finding a function $g(\theta)$ that is orthogonal to the set

$$S = \{F(\theta;k,\alpha_n) - F(\theta;k,\alpha_1): \quad n = 1,2,\ldots\},$$

where $F(\theta;k,\alpha_n)$ is the far field pattern corresponding to the scattering obstacle D and the incident field $u^i(\mathbf{x}) = \exp[ik\mathbf{x}\cdot\alpha_n]$. An alternate procedure is to work directly with the set S and to find a domain D whose far field pattern is a best fit to the measured far field pattern (cf. Angell and Kleinman, Kirsch and Kress). We shall finish this chapter by briefly describing this complementary approach, assuming again that D is starlike with respect to the origin.

From the result of Exercise 24 we see that the set

$$H_n^{(1)}(kr)e^{in\theta} \qquad (n = 0,\pm 1,\pm 2,\ldots) \tag{6.123}$$

is complete in $L^2[\partial D]$. Let Ω be a disk centered at the origin and contained in D such that k^2 is not an eigenvalue of the Laplacian in Ω (i.e., choose the radius a of Ω such that ka is not a zero of the Bessel function $J_n(z)$ for $n = 0,1,2,\ldots$). Then, using the addition formula for Bessel functions, for $\mathbf{x} \in R^2\backslash\overline{\Omega}$ we can represent any finite linear combination of functions from the set (6.123) in the form of a single layer potential

$$\int_{\partial\Omega} \psi(\mathbf{y})\Phi(\mathbf{x},\mathbf{y})ds(\mathbf{y}),$$

where $\psi(\mathbf{y})$ is a trigonometric polynomial. From the completeness of the set (6.123) we see that the boundary data $u^i(\mathbf{x}) = -\exp[ik\mathbf{x}\cdot\boldsymbol{\alpha}_n]$ can be approximated in $L^2[\partial D]$ by

$$\int_{\partial\Omega} \psi_n(\mathbf{y})\Phi(\rho(\hat{\mathbf{x}})\hat{\mathbf{x}},\mathbf{y})ds(\mathbf{y}),$$

where $\hat{\mathbf{x}} = \mathbf{x}/|\mathbf{x}| = (\cos\theta, \sin\theta)$, $\psi_n(\mathbf{y})$ is a trigonometric polynomial, and ∂D is parameterized by $\mathbf{x} = \rho(\hat{\mathbf{x}})\hat{\mathbf{x}}$. Furthermore, since the far field pattern depends continuously on the boundary data (cf. Exercise 19), the far field pattern $F(\theta;k,\boldsymbol{\alpha}_n)$ can be approximated in $L^2[-\pi,\pi]$ by

$$\frac{e^{i\pi/4}}{\sqrt{8\pi k}}\int_{\partial\Omega} \psi_n(\mathbf{y})\exp[-ik\hat{\mathbf{x}}\cdot\mathbf{y}]ds(\mathbf{y}).$$

Hence, if $U(M)$ is the compact set previously defined, we can formulate the optimization problem

$$\begin{aligned}\gamma(F,M,N) = \min_{(\psi_n,\rho)\in U(M)} \Bigg\{ \sum_{n=1}^{N} \int_{-\pi}^{\pi} \Bigg| & F(\theta;k,\boldsymbol{\alpha}_n) \\ & - \frac{e^{i\pi/4}}{\sqrt{8\pi k}}\int_{\partial\Omega} \psi_n(\mathbf{y})\exp[-ik\hat{\mathbf{x}}\cdot\mathbf{y}]ds(\mathbf{y})\Bigg|^2 d\theta \\ & + \sum_{n=1}^{N}\int_{-\pi}^{\pi} \Big|\exp[ik\rho(\hat{\mathbf{x}})\hat{\mathbf{x}}\cdot\boldsymbol{\alpha}_n] \\ & + \int_{\partial\Omega} \psi_n(\mathbf{y})\Phi(\rho(\hat{\mathbf{x}})\hat{\mathbf{x}},\mathbf{y})ds(\mathbf{y})\Big|^2 d\theta \Bigg\},\end{aligned} \tag{6.124}$$

where $\hat{\mathbf{x}} = (\cos\theta, \sin\theta)$. From the compactness of the set $U(M)$, we can again establish that the solution $\rho(\hat{\mathbf{x}})$ of the optimization problem (6.124) exists, depends continuously on the far field data $F(\theta;k,\boldsymbol{\alpha}_n)$, $n = 1,2,\ldots,N$, and for exact far field data and $\rho \in C^2[-\pi,\pi] \cap U_2$,

$$\lim_{M\to\infty} \gamma(F,M,N) = 0.$$

The optimization scheme (6.124) provides an alternate method to (6.120) for constructing an approximate solution to the inverse scattering problem. It is more flexible than (6.120) in the sense that it is easily generalized to the case where the given data is some functional of the far field pattern instead of the far field pattern itself, e.g., when the measured data is the total scattering cross section defined by (6.99). Furthermore, the restriction on k that $0 < k^2 < \lambda_1$ is avoided. On the other hand, (6.124) requires the determination of $N + 1$ functions, whereas (6.120) involves only two unknown functions. Hence, if the far field pattern is known for many incident plane waves with wave number k such that $0 < k^2 < \lambda_1$ and it is desired to make use of this information, (6.120) is more efficient from a numerical point of view.

In closing, we would like to point out to the reader that the inverse scattering problem considered in this section is in many ways typical of current research in partial differential equations in the sense that the problem is rooted in physical applications, is nonlinear, and ultimately depends on numerical methods for its complete resolution. We hope that the material presented here will encourage the reader to continue studying partial differential equations and their role in understanding the world around us. In fact, having completed this book, you are already well on your way in such a journey!

Exercises

1. Define the **beta function** $B(x,y)$ by

$$B(x,y) = \int_0^1 t^{x-1}(1 - t)^{y-1}dt \qquad (x > 0,\ y > 0).$$

Show that

$$B(x,y) = \frac{\Gamma(x)\Gamma(y)}{\Gamma(x + y)}.$$

2. Use Exercise 1 to show that for $x > 0$

$$\frac{\Gamma(x)\Gamma(x)}{\Gamma(2x)} = 2^{-2x+2}\int_0^1 (1 - s^2)^{x-1}ds = 2^{-2x+1}\frac{\Gamma(x)\Gamma(\frac{1}{2})}{\Gamma(x + \frac{1}{2})}\,.$$

Deduce the **Legendre duplication formula**

$$2^{2x-1}\Gamma(x)\Gamma(x + \tfrac{1}{2}) = \sqrt{\pi}\,\Gamma(2x).$$

3. Let α_n be the volume of the ball of radius one in R^n. Show by induction and repeated integrals that

$$\alpha_n = 2\alpha_{n-1}\int_0^1 (1 - t^2)^{(n-1)/2}dt$$

and deduce that

$$\alpha_n = \frac{\pi^{n/2}}{(n/2)\Gamma(n/2)}.$$

4. The numbers

$$\alpha = \int_0^1 (1 - t^4)^{-1/2} dt, \qquad \beta = \int_0^1 t^2(1 - t^4)^{-1/2} dt$$

are known as **lemniscate constants.** Show that

$$\alpha = \frac{1}{4}\frac{1}{\sqrt{2\pi}}(\Gamma(\tfrac{1}{4}))^2, \qquad \beta = \frac{1}{\sqrt{2\pi}}(\Gamma(\tfrac{3}{4}))^2,$$

and deduce that $\alpha\beta = \pi/4$.

5. Let n be a positive integer.

(a) Show that for any positive integer k,

$$\int_{k-1/2}^{k+1/2} \log t\, dt = \int_0^{1/2} \left(\log k^2 + \log\left(1 - \frac{t^2}{k^2}\right)\right) dt = \log k + c_k,$$

where $c_k = 0(1/k^2)$. From the identity

$$\log \Gamma(n) = \sum_{k=1}^{n-1} \log k$$

deduce now that

$$\log \Gamma(n) = \int_{1/2}^{n-1/2} \log t\, dt - \sum_{k=1}^{n-1} c_k = (n - 1/2)\log n - n + c + \gamma_n,$$

where c is a constant and $\lim_{n\to\infty} \gamma_n = 0$.

(b) Use Legendre's duplication formula (see Exercise 2) to evaluate the constant c in part (a) and deduce **Stirling's formula**

$$\Gamma(n) = n^{n-1/2}e^{-n}\sqrt{2\pi}(1 + \epsilon_n),$$

where $\lim_{n\to\infty} \epsilon_n = 0$.

6. Show that

$$\int_0^z J_0(t)dt = 2\sum_{n=0}^{\infty} J_{2n+1}(z).$$

7. Use the generating function for $J_n(z)$ to show that

$$J_0^2(z) + 2\sum_{n=1}^{\infty} J_n^2(z) = 1.$$

Deduce that for real x, $|J_0(x)| \leq 1$ and $|J_n(x)| \leq 1/\sqrt{2}$ for $n \geq 1$.

8. Prove the addition formula

$$J_n(x+y) = \sum_{m=-\infty}^{\infty} J_m(x) J_{n-m}(y),$$

where n is an integer.

9. Show that

$$J_{2n}(z) = \frac{1}{\pi}\int_0^{\pi} \cos 2n\theta \cos (z \sin \theta) d\theta$$

$$J_{2n+1}(z) = \frac{1}{\pi}\int_0^{\pi} \sin (2n+1)\theta \sin (z \sin \theta) d\theta.$$

10. Use Exercise 1 to show that

$$\frac{1}{\Gamma(k+\nu+1)} = \frac{1}{\Gamma(k+\frac{1}{2})\Gamma(\nu+\frac{1}{2})}\int_{-1}^{1} t^{2k}(1-t^2)^{\nu-1/2} dt \qquad (\nu > -\tfrac{1}{2})$$

and then show that for $x > 0$, $\nu > -1/2$,

$$J_\nu(x) = \frac{\left(\frac{x}{2}\right)^\nu}{\Gamma(\frac{1}{2})\Gamma(\nu+\frac{1}{2})}\int_{-1}^{1} (1-t^2)^{\nu-1/2} \cos xt \, dt$$

$$= \frac{\left(\frac{x}{2}\right)^\nu}{\Gamma(\frac{1}{2})\Gamma(\nu+\frac{1}{2})}\int_0^{\pi} \cos (x \cos \theta) \sin^{2\nu} \theta \, d\theta.$$

11. Use Exercise 10 to show that for $n = 0,1,2,\ldots,$

$$j_n(z) = \frac{(-i)^n}{2}\int_0^{\pi} \exp[iz \cos \theta] P_n(\cos \theta) \sin \theta \, d\theta,$$

where $P_n(\cos \theta)$ is Legendre's polynomial and $j_n(z)$ is the **spherical Bessel function** defined by

$$j_n(z) = \sqrt{\frac{\pi}{2z}}\, J_{n+1/2}(z).$$

Deduce the plane wave expansion

$$\exp[ikr \cos \theta] = \sum_{n=0}^{\infty} i^n (2n+1) j_n(kr) P_n(\cos \theta).$$

12. Show that

(a) $J_{\nu-1}(z) + J_{\nu+1}(z) = \dfrac{2\nu}{z} J_\nu(z)$.

(b) $J_{\nu-1}(z) - J_{\nu+1}(z) = 2J'_\nu(z)$.

13. Show that

$$J_{n+1/2}(z) = (-1)^n\left(\frac{2}{\pi}\right)^{1/2} z^{n+1/2}\left(\frac{d}{zdz}\right)^n \frac{\sin z}{z}.$$

14. Deduce that

$$\frac{d}{dx}(x^\nu J_\nu(x)) = x^\nu J_{\nu-1}(x)$$

and

$$\frac{d}{dx}(x^{-\nu} J_\nu(x)) = -x^{-\nu} J_{\nu+1}(x).$$

Use these results and Rolle's theorem to show that the positive zeros of $J_\nu(x)$ and $J_{\nu+1}(x)$ interlace, i.e., there is a zero of $J_\nu(x)$ between each pair of positive zeros of $J_{\nu+1}(x)$ and a zero of $J_{\nu+1}(x)$ between each pair of positive zeros of $J_\nu(x)$.

15. Use the method of stationary phase to show that

$$J_\nu(\nu \sec\beta) = \sqrt{\frac{2}{\pi\nu\tan\beta}}\cos\left(\nu(\beta - \tan\beta) + \frac{1}{4}\pi\right) + 0\left(\frac{1}{\nu}\right),$$

where β is a positive acute angle.

16. Show that if the only stationary point of $\phi(t)$ on $[a,b]$ is at $t = a$ and $\phi''(a) = 0$, $\phi'''(a) > 0$, then under appropriate assumptions on $\phi(t)$ and $g(t)$ we have

$$\int_a^b e^{ix\phi(t)} g(t)dt = \Gamma(\tfrac{4}{3})\left[\frac{6}{x\phi'''(a)}\right]^{1/3} g(a)\exp\left(ix\phi(a) + \frac{1}{6}\pi i\right) + 0\left(\frac{1}{x^{2/3}}\right).$$

Use this result to show that

$$J_\nu(\nu) = \frac{\Gamma(\frac{1}{3})}{2^{2/3}3^{1/6}\pi\nu^{1/3}} + 0\left(\frac{1}{\nu^{2/3}}\right).$$

17. Show that

(a) $\displaystyle\int_0^\infty e^{-ax} J_0(bx)dx = \frac{1}{\sqrt{a^2+b^2}} \qquad (a > 0,\ b > 0).$

(b) $\displaystyle\int_0^\infty x e^{-a^2x^2} J_0(bx)dx = \frac{1}{2a^2} e^{-b^2/4a^2} \qquad (a > 0,\ b > 0).$

18. Use the method of separation of variables and Exercise 11 to formally solve the scattering problem of a plane wave moving along the z axis, $(x,y,z) \in R^3$, incident upon a sound-soft sphere of radius a. (Hint: In R^3 the Sommerfeld radiation condition takes the form $\lim_{r\to\infty} r\left(\frac{\partial u}{\partial r} - iku\right) = 0$; cf. Section 4.8.)

19. Show that in $L^2[-\pi,\pi]$, the far field pattern for a sound-soft scattering obstacle depends continuously on the boundary data in $L^2[\partial D]$. (Hint: Compute the far field pattern from (6.84) and represent ψ in the form $\psi = -2(\mathbf{I} + \mathbf{K}_1)^{-1}u^i$.)

20. Let $u \in C^2(R^2\backslash\overline{D})$ be a solution of the Helmholtz equation satisfying

$$\lim_{R\to\infty} \int_{|\mathbf{y}|=R} |u(\mathbf{y})|^2 ds(\mathbf{y}) = 0.$$

Show that $u(\mathbf{x})$ is identically zero in $R^2\backslash\overline{D}$. (Hint: For $|\mathbf{x}| = R$ expand $u(\mathbf{x})$ in a Fourier series.) This result is known as **Rellich's lemma.**

21. Let $u \in C^2(R^2\backslash D)$ be a solution of the Helmholtz equation. Show that $u(\mathbf{x})$ can be *uniquely* written as the sum of a solution of the Helmholtz equation in all of R^2 and a solution of the Helmholtz equation in $R^2\backslash D$ that satisfies the Sommerfeld radiation condition. (Hint: Let Ω be a disk containing D in its interior and apply Green's formula to $u(\mathbf{x})$ and $\Phi(\mathbf{x},\mathbf{y})$ over $\Omega\backslash D$.)

22. Let $0 < k^2 < \lambda_1$, where λ_1 is the first Dirichlet eigenvalue of the Laplacian in $D \subset R^2$. Use Green's formula and the method of single layer potentials to show that there exists a unique solution of

$$\Delta_2 u + k^2 u = 0 \quad \text{in} \quad R^2\backslash\overline{D}$$

$$\frac{\partial u}{\partial \nu} + i\lambda u = f \quad \text{on} \quad \partial D$$

$$\lim_{r\to\infty} \sqrt{r}\left(\frac{\partial u}{\partial r} - iku\right) = 0,$$

where $\lambda > 0$ and $f(\mathbf{x})$ is a prescribed continuous function defined on ∂D. How would you remove the restriction on k?

23. Let $u \in C^2(R^2\backslash\overline{D} \cap C^1(R^2\backslash D)$ be a solution of the Helmholtz equation satisfying the Sommerfeld radiation condition. Show that for $\mathbf{x} \in D$,

$$\int_{\partial D} \left(u(\mathbf{y}) \frac{\partial}{\partial \nu(\mathbf{y})} \Phi(\mathbf{x},\mathbf{y}) - \frac{\partial u}{\partial \nu}(\mathbf{y})\Phi(\mathbf{x},\mathbf{y})\right) ds(\mathbf{y}) = 0$$

and deduce that $\phi(\mathbf{x}) = \partial u(\mathbf{x})/\partial \nu$ is a solution of the moment problem

$$\int_{\partial D} \phi(\mathbf{y})v_n(\mathbf{y})ds(\mathbf{y}) = f_n \qquad (n = 0,\pm 1,\pm 2,\ldots),$$

where

$$v_n(\mathbf{y}) = v_n(\rho,\phi) = H_n^{(1)}(k\rho)e^{in\phi}$$

and

$$f_n = \int_{\partial D} u(\mathbf{y}) \frac{\partial v_n}{\partial \nu}(\mathbf{y})ds(\mathbf{y}).$$

These identities are the basis of the **null field method** for solving the scattering problem for a sound-soft obstacle. (Note that in this case $u(\mathbf{x})$ is known for $\mathbf{x} \in \partial D$, and so is f_n for $n = 0,\pm 1,\pm 2,\ldots$.) Deduce the corresponding identities for a sound-hard scattering obstacle.

24. Show that the set (6.123) is complete in $L^2[\partial D]$ for any positive value of the wave number k. Deduce that the solution of the null field equations defined in Exercise 23 is unique for any positive value of k, i.e., the problem of interior eigenvalues associated with the method of integral equations is not present in the null field method. (Hint: Modify the proof of the fact that $L^2[\partial D] = \overline{H|_{\partial D}}$ for $0 < k^2 < \lambda_1$.)

25. Show that there exists a unique solution of the boundary value problem (6.107).

References

L. V. Ahlfors

Complex Analysis. New York: McGraw-Hill, 1966.

T. S. Angell and R. E. Kleinman

A new algorithm for inverse scattering problems, in *Problems of Applied Analysis*. B. Brosowski and E. Martensen, eds. Frankfurt am Main: Verlag Peter Lang, 1987: 41–57.

T. M. Apostol

Mathematical Analysis. Reading: Addison-Wesley, 1957.

K. E. Atkinson

A Survey of Numerical Methods for the Solution of Fredholm Integral Equations of the Second Kind. Philadelphia: SIAM Publications, 1976.

G. R. Baldock and T. Bridgeman

The Mathematical Theory of Wave Motion. Chichester: Ellis Horwood, 1981.

C. Bandle

Isoperimetric Inequalities and Applications. Boston: Pitman, 1980.

S. Bergman

Integral Operators in the Theory of Linear Partial Differential Equations. Berlin: Springer-Verlag, 1969.

S. Bergman and M. Schiffer

Kernel Functions and Elliptic Equations in Mathematical Physics. New York: Academic Press, 1953.

H. Brakhage and P. Werner

Über das Dirichletsche Aussenraumproblem für die Helmholtzsche Schwingungsgleichung. *Arch. Math.* 16 (1965): 325–329.

J. R. Cannon

The One-Dimensional Heat Equation. Reading: Addison-Wesley, 1984.

D. Colton

1980. *Analytic Theory of Partial Differential Equations*. Boston: Pitman, 1980.

1984a. The strong maximum principle for the heat equation. *Proc. Edinburgh Math. Soc.* 27 (1984): 297–299.

1984b. The inverse scattering problem for time harmonic acoustic waves. *SIAM Review* 26 (1984): 323–350.

D. Colton and R. Kress

Integral Equation Methods in Scattering Theory. New York: John Wiley, 1983.

D. Colton and P. Monk

1985. A novel method for solving the inverse scattering problem for time harmonic acoustic waves in the resonance region. *SIAM J. Appl. Math.* 45 (1985): 1039–1053.

1986. A novel method for solving the inverse scattering problem for time harmonic acoustic waves in the resonance region II. *SIAM J. Appl. Math.* 46 (1986): 506–523.

1987. The numerical solution of the three dimensional inverse scattering problem for time harmonic acoustic waves. *SIAM J. Scientific Stat. Comp.* 8 (1987): 278–291.

R. Courant and K. O. Friedrichs

Supersonic Flow and Shock Waves. New York: Interscience, 1948.

R. Courant and D. Hilbert

1953. *Methods of Mathematical Physics,* Vol. 1. New York: Interscience, 1953.

1961. *Methods of Mathematical Physics,* Vol. 2. New York: Interscience, 1961.

L. M. Delves and J. Walsh

Numerical Solution of Integral Equations. Oxford: Clarendon Press, 1974.

B. Epstein

Partial Differential Equations. New York: McGraw-Hill, 1962.

A. Erdélyi

Asymptotic Expansions. New York: Dover, 1956.

G. E. Forsythe and W. R. Wasow

Finite-Difference Methods for Partial Differential Equations. New York: John Wiley, 1960.

P. R. Garabedian

1964. *Partial Differential Equations*. New York: John Wiley, 1964.

1970. An unsolvable problem. *Proc. Amer. Math. Soc.* 25 (1970): 207–208.

R. P. Gilbert

Function Theoretic Methods in Partial Differential Equations. New York: Academic Press, 1969.

C. W. Groetsch

The Theory of Tikhonov Regularization for Fredholm Equations of the First Kind. Boston: Pitman, 1984.

P. Hartman and C. Wilcox

On solutions of the Helmholtz equation in exterior domains. *Math. Zeit.* 75 (1961): 228–255.

I. Herrera

Boundary Methods: An Algebraic Theory. Boston: Pitman, 1984.

L. Hörmander

Linear Partial Differential Operators. Berlin: Springer-Verlag, 1963.

G. C. Hsiao

On the stability of integral equations of the first kind with logarithmic kernels. *Arch. Rat. Mech. Anal.* 94 (1986): 179–192.

F. John

1955. *Plane Waves and Spherical Means Applied to Partial Differential Equations*. New York: Interscience, 1955.

1982. *Partial Differential Equations*. Berlin: Springer-Verlag, 1982.

D. S. Jones

Integral equations for the exterior acoustic problem. *Quart. J. Mech. Appl. Math.* 27 (1974): 129–142.

D. S. Jones and B. D. Sleeman

Differential Equations and Mathematical Biology. London: George Allen and Unwin, 1983.

O. D. Kellogg

Foundations of Potential Theory. New York: Dover, 1953.

H. Kersten

Grenz und Sprungrelationen für Potentiale mit quadratsummierbarer Flächenbelegung. *Result. Math.* 3 (1982): 17–24.

A. Kirsch and R. Kress

An optimization method in inverse acoustic scattering, in *Boundary Elements IX*, C. A. Brebbia, W. L. Wendland, and G. Kuhn, eds. Berlin: Springer-Verlag, 1987: 3–18.

R. E. Kleinman and G. F. Roach

1974. Boundary integral equations for the three dimensional Helmholtz equation. *SIAM Review* 16 (1974): 214–236.

1982. On modified Green's functions in exterior problems for the Helmholtz equation. *Proc. Royal Soc. London* A383 (1982): 313–332.

M. Kline

Mathematical Thought from Ancient to Modern Times. New York: Oxford University Press, 1972.

P. D. Lax

Hyperbolic Systems of Conservation Laws and the Mathematical Theory of Shock Waves. Philadelphia: SIAM Publications, 1973.

N. N. Lebedev

Special Functions and Their Applications. New York: Dover, 1972.

R. Leis

Zur Dirichletschen Randwertaufgabe des Aussenraums der Schwingungsgleichung. *Math. Zeit.* 90 (1965): 205–211.

H. Lewy

An example of a smooth linear partial differential equation without solution. *Ann. Math.* 66 (1957): 155–158.

P. Linz

Theoretical Numerical Analysis. New York: John Wiley, 1979.

D. G. Luenberger

Optimization by Vector Space Methods. New York: John Wiley, 1969.

M. Z. Nashed
Approximate regularized solutions to improperly posed linear integral and operator equations. *Springer-Verlag Lecture Notes in Mathematics,* Vol. 430. Berlin: Springer-Verlag, 1974: 289–332.

R. L. Ochs, Jr.
The limited aperture problem of inverse acoustic scattering: Dirichlet boundary conditions *SIAM J. Appl. Math.*, 47 (1987): 1320–1341.

O. I. Panich
On the question of the solvability of the exterior boundary value problems for the wave equation and Maxwell's equations. *Russian Math. Surveys* 20 (1965): 221–226.

L. E. Payne
Improperly Posed Problems in Partial Differential Equations. Philadelphia: SIAM Publications, 1975.

M. H. Protter and H. F. Weinberger
Maximum Principles in Differential Equations. Englewood Cliffs: Prentice-Hall, 1967.

A. G. Ramm
Scattering by Obstacles. Dordrecht: Reidel, 1986.

R. Reemsten and A. Kirsch
A method for the numerical solution of the one dimensional inverse Stefan problem, *Numer. Math.* 45 (1984): 253–273.

B. D. Sleeman
The inverse problem of acoustic scattering. *IMA J. Appl. Math.* 29 (1982): 113–142.

J. Smoller
Shock Waves and Reaction-Diffusion Equations. Berlin: Springer-Verlag, 1983.

I. N. Sneddon
Fourier Transforms. New York: McGraw-Hill, 1951.

I. Stakgold
Boundary Value Problems of Mathematical Physics, Vol. 2. New York: Macmillan, 1968.

E. C. Titchmarsh
The Theory of Functions. Oxford: Oxford University Press, 1939.

G. P. Tolstov
Fourier Series. Englewood Cliffs: Prentice-Hall, 1962.

F. Ursell
1973. On the exterior problems of acoustics. *Proc. Cambridge Phil. Soc.* 74 (1973): 117–125.
1978. On the exterior problems of acoustics II. *Proc. Cambridge Phil. Soc.* 84 (1978): 545–548.

I. N. Vekua
New Methods for Solving Elliptic Equations. Amsterdam: North-Holland, 1967.

G. N. Watson
A Treatise on the Theory of Bessel Functions. Cambridge: Cambridge University Press, 1944.

H. F. Weinberger
Partial Differential Equations. Waltham: Blaisdell, 1965.

A. J. Weir
Lebesgue Integration and Measure. Cambridge: Cambridge University Press, 1973.

D. V. Widder
The Heat Equation. New York: Academic Press, 1975.

Index